Process Equipment Malfunctions

Process Equipment Malfunctions

Techniques to Identify and Correct Plant Problems

Norman P. Lieberman

McGraw Hill

New York Chicago San Francisco
Lisbon London Madrid Mexico City
Milan New Delhi San Juan
Seoul Singapore Sydney Toronto

The McGraw-Hill Companies

Process Equipment Malfunctions

1 2 3 4 5 6 7 8 9 0 QFR/QFR 1 7 6 5 4 3 2 1

ISBN 978-0-07-177020-0
MHID 0-07-177020-8

This book is printed on acid-free paper.

Sponsoring Editor
Michael Penn

Editing Supervisor
Stephen M. Smith

Production Supervisor
Pamela A. Pelton

Acquisitions Coordinator
Michael Mulcahy

Project Manager
Neha Rathor, Neuetype

Copy Editor
Kevin Campbell

Proofreader
Monoleena Misra

Indexer
Robert Swanson

Art Director, Cover
Jeff Weeks

Composition
Neuetype

Dedication

I was watching the Nature Channel last night. The wandering albatross spends 9 months in solitary flight over far-flung seas. Then, without fail, it returns to the Falkland Islands in the wild Antarctic Ocean. Invariably it seeks and finds the same mate it had the previous season. And so it goes on, fulfilling nature's plan for 30 or 40 years.

It reminds me of Liz and me. Wandering across the face of the earth to far-flung refineries and chemical plants. Gathering tales of process equipment malfunctions. Invariably returning to our home in New Orleans to renew our time and life together.

About the Author

Norman P. Lieberman is a chemical engineer with 46 years of experience in process plant operation, design, and field troubleshooting. An independent consultant, he troubleshoots oil refinery and chemical plant process problems and prepares revamp process designs. Mr. Lieberman teaches 20 to 25 seminars a year on "Troubleshooting Process Plant Operations," and this book is based on his long experience in field troubleshooting refinery and process plant problems.

Contents

Preface

The training provided to the process operator and to the chemical or process engineer often does not seem to apply in the plant. It's as if both formal education and training are irrelevant to actual process plant problems. The difficulty lies in an implied assumption made by instructors, professors, textbooks, and training manuals that the equipment is working correctly and within its normal operating range.

But in the real world, the process engineer and operating supervisor do not concern themselves with properly performing equipment. It's the malfunctioning pumps, control valves, pressure transmitters, compressors, fractionators, and fired heaters that occupy their attention. To identify a malfunction, the technician must first understand the normal function of that equipment. Such understanding may come from training or experience. In this book, I've assumed that you already understand the basic operating principles of steam reboilers, air coolers, distillation trays, reciprocating compressors, knock-out drums, and heat exchangers.

A reasonably intelligent person can be taught to design, monitor, or operate correctly functioning process plants. Competent maintenance personnel can efficiently execute equipment repairs. But to identify and troubleshoot equipment malfunctions requires a different and higher level of understanding and analytical reasoning. In that sense, this text presents an advanced type of training not available in universities or operator training programs.

The information and ideas I've presented are based on my own 46 years of field experience. If I have not seen it myself, I have not included it in this book. The examples are drawn from my work in oil refineries and, to a lesser extent, petrochemical plants, LNG facilities, and gas field production.

If you have an erratic bottoms level, or a flooding fractionator, or a surging steam jet, this is the text that can help you, provided that you're willing to go out into that noisy, hot, hostile, confusing, and evil-smelling world on the other side of your office door. And don't forget your wrench, infrared surface temperature gun, screwed fittings, and pressure gauge.

Disclaimer

While all my stories are true and related in a technically correct sequence of details, I have often forgotten where they occurred. Thus, references to specific companies and locations are meaningless and should be regarded as pure fiction.

I have written mainly from my personal experience. On the odd occasion where I refer to the technical literature, I have so noted. Other than references to myself and my family, all other references to individuals are also totally fictional. That is, the names have been changed to protect the guilty.

Note on Term Definition and Glossary

There are a large number of terms that are in common use in the process industry but have no particular meaning in the larger world. When I use terms that I imagine the novice process technician has not been exposed to, I have boldfaced the term at least once. Then, in the glossary, I have defined that term. Particularly when you work with older operators or maintenance personnel onsite, communication can be a big problem for the new man or woman. I have also tried to define such terms, in less detail, in the text, but not every time I use them. So, when in doubt, consult the glossary.

Other Texts by Author

To an extent, more-detailed descriptions of some of the examples cited in this book are contained in other books I have authored. I have referenced such examples throughout this text.

- *A Working Guide to Process Equipment*, 3rd ed., McGraw-Hill, 2009 (with Elizabeth Lieberman).
- *Process Engineering for a Small Planet*, Wiley, 2010.
- *Troubleshooting Process Operations*, 4th ed., PennWell, 2009.
- *Troubleshooting Natural Gas Processing*, PennWell, 1987.
- *Troubleshooting Process Plant Control*, Wiley, 2008.
- *Process Engineering for Reliable Operations*, 2nd ed., Gulf, 1995.
- *Troubleshooting Refinery Processes*, PennWell, 1980.

Acknowledgments

- Irene Hebert—my daughter, who assembled and corrected this manuscript into a publishable form.
- Roy Williams—who drafted the process figures from my scribbled penciled sketches.
- Liz Lieberman—my wife, fellow chemical engineer, and coworker, who reviewed the final draft.
- Just a few of my colleagues who have helped me over the years: Dale Wilborn, Ken Block, Mark Allen, Mike Angela, Henry Kister, Scot Golden, Gerry Carlin, Joe Gurawitz, Gerry Obluda, Cedric Charles, Terry Henderson, Nelson English, Prasnanta Kumar, Dennis Schumede, Jean Paul Mauleon, Robert Haugen, Andries Burger, Tariq Malik, Steve Hill, Jim McQuire, Archie Elam, Mike Nodier, Oscar Wyatt, Jack Stanley, Ken Rickter, Heinz Block, Telroy Morgan, Joe and Jim Deprisco, Vaidas Dirgelas, Trung Quan, Probkar Reddy, Charlie Schultz, Richard Doss, Bill Hurt, Dan Summers, Raj Malik, Ohad Rotan, Sandy Lani, Paul Schrader, Janet Wilson, Joe Petrocelli, Bobby Felts, Henry Zipperian, Greg Hevron, and Tom Varadi.

Introduction

"Norman, the toilet won't flush!"

"Use the toilet in the upstairs guest room," I replied. "I'm watching the game."

"You need to fix it! It won't flush." Liz insisted, "There's zero tolerance for failure in our home!"

Liz and I live alone in a house with seven toilets. I can't quite grasp the problem if one or two are out-of-service. But Liz sees things differently. So, during halftime, I listed the potential reasons for the toilet malfunction:

- The 3-inch connection between the toilet and the 4-inch sewer line under our slab has plugged. That's a job for my power plunger.
- The 4-inch sewer line under our slab has plugged. That's a job for the Roto-Rooter man. Cost = $250, minimum.
- The 4-inch sewer line under our slab has broken. That's a job for the Hydro-Tunnelers. Cost = $21,800 (not an estimate).
- The chain connecting the flush handle to the rubber stopper in the water closet has come loose. Cost = 2 cents for a new rubber band.
- The rubber stopper in the water closet is stuck in an open position. I should be so lucky.
- A bird has built a nest on top of the roof vent. Air trapped in the toilet drain line has vapor-locked the toilet. Likely I'll fall off the roof trying to evict the fowl.
- The sewer line between my home and the city sewer is plugged. This just happened last month; I cleared it with a "bladder" attached to my garden hose.
- The City of New Orleans has shut off the water to my house because I've again forgotten to pay my water bill. Cost = Liz will be really angry.

- The washing machine is pumping soapy water into the 4-inch sewer under our slab, and backing-out the toilet flow. I'll wait until the washer stops, and then claim to have fixed the toilet.
- I'll procrastinate and eventually Liz will fix the toilet herself.

As you can see, I'm a real expert in defining process equipment malfunctions. It comes from 46 years of home ownership. To be successful in troubleshooting, one must:

- Understand how the equipment works.
- Anticipate possible types of equipment malfunctions.
- Discriminate between these malfunctions by direct field observations.
- Devise and execute a test to prove that a particular malfunction is truly the cause of the equipment failure.

My book is written at the working level. Having a university technical degree is rather irrelevant to one's ability to understand this text. Having hands-on field experience in a petrochemical plant or petroleum refinery will certainly help the reader. But, if you're really stuck on a problem, give me a call at 1-504-887-7714, or e-mail me at norm@lieberman-eng.com. Just call during halftime.

Process Equipment Malfunctions

CHAPTER 1

Distillation Tray Malfunctions

I search for something once well known, but long since forgotten.

Don't mess with the Internal Revenue Service. I have just been audited, and it's no joke. The IRS examiner was quite unreasonable. I had written off, as a business expense, my vacation in Costa Rica. I explained to Mr. Himmel that I needed to recover from a stressful incident at the Coffeyville Refinery.

"So, Mr. Lieberman, did you sustain an injury at the refinery? Resulting medical expenses are deductible."

"Yes, but the injury was emotional, rather than physical. Kind of a cerebral stress injury. Hence the trip to Costa Rica—for therapy."

"Cerebral, stress-type injury? It rather sounds like . . ."

"No, Mr. Himmel! Let me explain. One of my fundamental beliefs was shattered!"

"Mr. Lieberman, the IRS cannot concern themselves with the beliefs of taxpayers."

"Kindly let me explain. I've always believed that as you increase the vapor flow through a distillation tower, the pressure drop across the trays increases."

"Mr. Lieberman, are we now talking about stills? Let me warn you that everything you say can and will be used against you. Did you have a federal license to operate this distillation apparatus?"

"No, I'm not talking about whiskey. I'm talking about a crude distillation tower in Coffeyville, Kansas."

Decreasing Tower Delta P

Every calculation procedure for predicting the pressure drop through trays predicts that the tray delta P increases with vapor flow because the velocity of vapor through the tray deck caps or orifices increases. Higher velocities must always result in larger pressure drops. The fact that pressure drop varies with velocity squared:

$$\text{Delta P} = (V)^2 \times (\text{Density}) \times K$$

is a fundamental belief I carry in my heart. And yet, for many years, I have actually had hidden doubts. For instance, I know for sure that often there is a negative delta P across heat exchangers even when the inlet and outlet pressures are measured at the same elevation. I'm sure because I've measured it myself, using calibrated pressure gauges in Aruba. It's related to velocity reductions through the exchanger.

But the response of certain trays in some towers is quite different. As the vapor flow is increased, delta P increases in a normal and predictable manner. But suddenly the delta P slips down, above a certain vapor rate, and then stabilizes at a lower value!

"Mr. Lieberman, I've been a tax examiner for 20 years," Himmel objected, "And I've never heard such nonsense."

"About my legitimate tax deduction in Costa Rica?"

"No! About pressure drops in trays going down as flow goes up. Don't try to bamboozle the IRS. Even the layman knows that resistance to flow goes up when more gas flows through a restriction. That's just common sense."

"So, Mr. Himmel, you will admit that an observation that contradicts 'common sense' is stressful." And to prove my point, I sketched my observations in the Coffeyville Refinery shown in Figure 1-1.

Mr. Himmel looked at the sketch, looked at his watch, and looked at me. "Mr. Lieberman, I'm going to disallow your deduction of $12,984.38 for your trip to Costa Rica. It does not qualify under the tax code as a business-related expense. And on a personal level, I find your sketch to be an affront to my intellect. Just like your tax deduction, it's just nonsense."

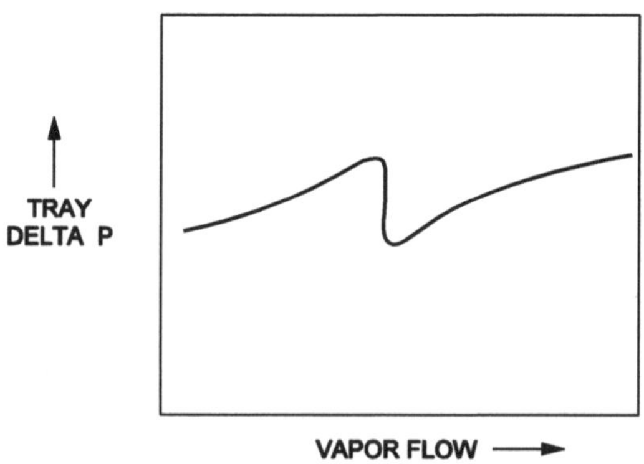

Figure 1-1 Effect of restoring downcomer seal on flooded distillation trays.

Resealing Downcomers

To understand the malfunction that leads to the nonlinear response of tray delta P to increasing vapor flow, we need to understand two terms:

- Delta P dry
- Delta P hydraulic

Delta P dry is the pressure drop of the vapor flowing through the orifices or valve caps on the tray floor. Delta P dry must always increase with the vapor flow rate, squared.

Delta P hydraulic is the height, or the depth, of the liquid sitting on the tray deck. It's mainly a function of the overflow or outlet weir height. If there are 3 inches of liquid sitting on the tray deck, then the vapor has to push those 3 inches of liquid out of its way and thus loses 3 inches' worth of pressure.

The problem is that if delta P hydraulic gets much bigger than delta P dry, then the tray decks will begin to leak. Valve trays may be a little better in retarding leakage than sieve or grid trays, but not by much.

As long as some of the liquid is overflowing the outlet weir, tray deck leakage just reduces tray efficiency. But if all the liquid is leaking through the tray deck, then the liquid level on the tray will fall below the bottom edge of the downcomer from the tray above. I have shown this problem in Figure 1-2.

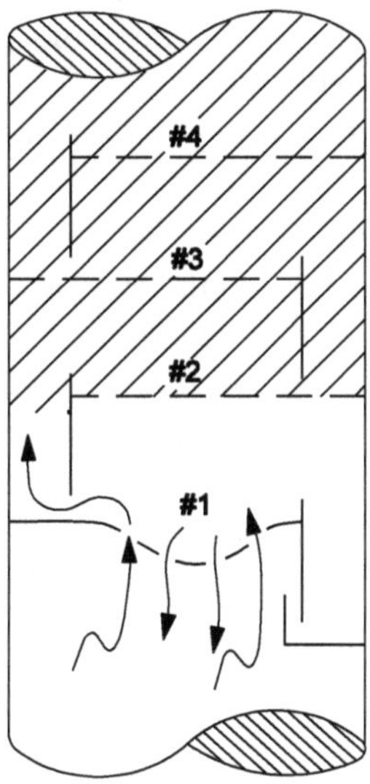

FIGURE 1-2 An unsealed downcomer causes the trays above to flood.

Tray deck #1 is sagging. The depth of liquid at the sag has caused 100% of the liquid flow to bypass tray #1's outlet weir and thus uncover the bottom edge of the downcomer from the tray above. The liquid in the downcomer from tray #2 is pushed up onto tray #2's deck, which then floods. The flooding progresses up the tower, until all the trays above tray #1 are flooded.

How could I be so smart on this subject? Because, at the Chevron Refinery in Port Arthur, Texas, they have a 4-inch diameter glass distillation tower. Unsealing any downcomer caused all the trays above to flood. Also, when we opened the tower for inspection at the Coffeyville crude distillation unit, we found tray #1 sagging.

How does this then explain the nonlinear response of pressure drop to the vapor flow shown in Figure 1-1 and through the stripping trays shown in Figure 1-2?

- Liquid drains through the tray deck #1 because of the low delta P dry and the sag in the middle of the tray deck.
- The liquid level on tray deck #1 drops below its outlet weir.
- The bottom edge of the downcomer from tray #2 becomes unsealed. Meaning, it is no longer submerged in the liquid level on tray deck #1.
- Vapor starts to flow up through the unsealed tray #2 downcomer. This vapor displaces the liquid out of the tray #2 downcomer. The liquid is pushed up onto tray deck #2.
- Tray #2 floods. As flooding progresses up the tower, trays #3 and #4 will also flood with time.
- As the vapor flow increases, the ability for liquid to flow down through the tray #2 downcomer becomes less and less. Flooding becomes progressively worse, and the tower delta P becomes progressively larger.
- However, the increased vapor flow, as shown in Figure 1-1, causes an increase in delta P dry through tray deck #1, and reduces the amount of liquid leaking through tray deck #1.
- The height of the liquid on the tray deck #1 increases until the liquid begins to overflow its weir.
- The downcomer seal from tray #2 is reestablished. Now the liquid can once again drain freely from the tray #2 downcomer, onto tray deck #1.
- Also, trays #3 and #4 drain down. The reduced weight of liquid on tray decks #2, #3, and #4 reduces tower delta P.
- As vapor flow continues to rise, the tray delta P goes up in a normal manner as shown in Figure 1-1, due to the increased delta P dry.

Effect on Fractionation

At the Coffeyville Refinery, we found that fractionation was bad and became worse as we increased the vapor rate. However at some point, fractionation efficiency would suddenly become better as we increased the vapor rate past some magic point. And this magic point would coincide with the sudden reduction in tower pressure drop, as the vapor flow increased. Further increases in the vapor flow did not have much effect on fractionation efficiency.

What I have been describing is an illustration of the malfunction called **turndown**. A fractionation tray cannot be run at too low a vapor rate before a downcomer seal is lost. This minimum vapor rate, or turndown ratio, is a function of tray levelness. So the dual morals of this story are:

- Level up your tray decks during tower turnarounds.
- Make sure your vacations coincide with business trips.

NOTE *For old-style bubble-cap trays, the preceding discussion does not apply as bubble-caps do not leak.*

Effect of Displaced Downcomer

I was working on a crude unit capacity limitation in a refinery in Lithuania. The problem was flooding. Flooding in the sense that black resid bottoms would be carried up the stripping trays by the stripping steam. To mitigate this problem, I issued instructions to the console operator to reduce the stripping steam rate from 6,000 to 5,000 kg/hr.

"Comrade Engineer," the former Soviet shift foreman complained, "Reducing the stripping steam flow will only make the flooding worse. The delta P across the bottom trays will increase and the distillate products will become darker."

"Look," I argued, "I've been working on these units before your father was born:

- More steam will increase delta P dry.
- A bigger delta P dry will increase the liquid level in the downcomer from the bottom tray.
- The tower radiation scan (i.e., the TruTec-type survey) indicated that the flooding is caused by liquid backing up from the seal pan into the bottom tray downcomer."

So we ran a test, at both 5,000 kg/hr and 7,000 kg/hr of stripping steam. Just as my Soviet foreman predicted, distillate color was better

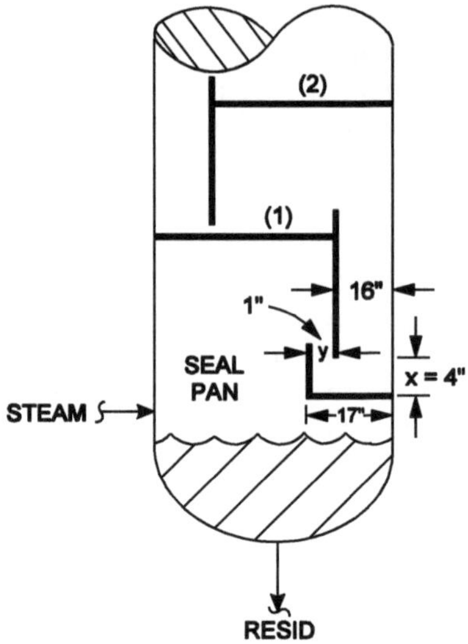

Figure 1-3 Excessive pressure loss in the seal pan causes flooding.

and the stripping tray delta P was lower at the higher steam rates. So I tried 8,000 kg/hr of steam and the distillate product went black.

Now what?

"See Comrade Engineer, American capitalists don't know everything."

I now did what I should have done in the first place. I pulled out the vendor tray drawing for the downcomer from tray #1, shown in Figure 1-3.

The downcomer detail showed a downcomer width of 16 inches, which seemed to be a reasonable dimension for the liquid flow.

But I could not find the vendor drawing for the seal pan. What I did find was a note on the downcomer detailed drawing:

"Client to Reuse Existing Seal Pan—Field Check for adequate clearances."

Who was supposed to "Field Check?" And which clearances were supposed to be checked? What criteria were supposed to be used to determine if the clearances were adequate?

After hours of searching, my wife Liz, who was working with me on the project, found the old design drawing for the seal pan. Referring to Figure 1-3, we found that:

- The seal pan was welded, not clipped, to the vessel wall, and thus could not easily be changed when the tower had been retrayed.
- The gap between the downcomer and the overflow lip (dimension y) was only 1 inch, but the downcomer clearance (dimension x) was 4 inches.

The calculated head loss under the downcomer (i.e., pressure drop of the flowing liquid) was:

$$\text{Delta P} = 0.6 \times (\text{Velocity})^2$$

where delta P = inches of liquid
velocity = feet per second
0.6 = typical coefficient for a smooth, sharp-edged orifice

Therefore, the head loss (x) under the downcomer = 1 inch

But the head loss (y) between the downcomer and the seal pan overflow lip = 16 inches

The 16-inch head loss would result in excessive downcomer backup from the seal pan and would flood the bottom downcomer. The flooding would progress up the tower and turn the distillates black.

But why did more stripping steam partially relieve this flooding? The answer was also provided by the old seal pan drawing.

Downcomer Bracing Brackets

The side edges of downcomers are rigidly supported by the downcomer bolting bars that are welded to the vessel wall. If the width of the downcomer is not more than 4 feet, this is sufficient to prevent the downcomer from flexing. If the downcomer (typically made from 2 mm or 14 gauge steel) is wider than 6 feet, then it may be quite flexible. And this usually is bad. Bad in the sense that the delta P of the flowing vapor through the tray deck above tends to push the vertical downcomer wall up against the vessel wall and reduce the open area at the bottom of the downcomer. This might result in excessive downcomer backup.

Normally this is prevented by the **downcomer bracing brackets**. (Note to reader: Terms in bold are explained in the Glossary, at the end of this text.) These are "L"-shaped brackets that bolt onto the seal pan floor or the tray floor, and to the downcomer's bottom edge. This keeps the bottom edge of the downcomer rigid. But, in my Lithuania stripper, the designer had left out the downcomer bracing brackets, even though the width of the downcomer was 10 feet.

The flexibility of the downcomer had an unexpected benefit. When the stripping steam rate was increased, the bottom edge of the downcomer was pushed away by the steam pressure from the seal pan's overflow lip. Dimension y in Figure 1-3 was slightly increased. This reduced the downcomer backup. Of course, too much steam at some point would cause normal tray flooding.

A few months later, we shut the tower down to replace the seal pan. I had the width of the seal pan extended from 17 to 20 inches. The lesson is to be careful when retraying towers. Don't try to reuse existing components (i.e., the seal pan) in conjunction with the new trays, unless you plan to inspect the final installation yourself.

Top Tray Flooding

What are the indications of a distillation tower flooding?

1. Fractionation gets worse instead of better, as the reflux and reboiler duty increase.
2. The delta T across the tower (bottom minus top temperature) gets smaller, as reflux rates are increased.
3. Increasing the reflux rate does not cause the reboiler duty to increase, even though the reboiler is in Auto.
4. Increasing the reflux rate does not cause the bottoms product flow rate to increase, even though the reboiler duty is fixed (i.e., in the manual mode of control).
5. Opening the vent at the top of the tower causes liquid, rather than vapor, to vent to the atmosphere.

And how about the delta P? That differential pressure which we learned about in class—should not the differential pressure drop across the trays also increase and indicate flooding? Maybe not.

Forty years ago, as a young process engineer, I had this problem on a 60-tray propylene-versus-propane splitter in Whiting, Indiana. How could this tower be flooding, yet the delta P (which I measured myself) be rather normal?

The tower was 150 feet (i.e., 30-inch tray spacing) across the trays. The SG of propane is about 0.5. Aerated, the normal condition of liquid on the trays, the SG of the liquid between the trays would be roughly 0.3. Thus, for 150 feet of height, the observed pressure drop across the 60 trays, if they flooded, would be:

$$(150 \text{ feet}) \times (0.3) \div (2.31) = 20 \text{ psi}$$

NOTE *There are 2.31 feet of water in each 1 psi of head pressure.*

My observed delta P was only 4 psi! How could this tower be in flood, with a normal tower pressure drop of only 4 psi?

In this calculation, I have made an assumption that the flooding is starting at the bottom tray of the tower. But suppose the flooding is starting at the top tray. Here's the source of confusion:

- Flooding progresses up a tower.
- If the top tray floods, then an increment of reflux does not go down the tower, but recirculates, in a liquid state, back to the reflux drum. The 59 trays below the top tray are *not* flooded. They simply do not fractionate efficiently because of a low **internal reflux** rate.

It's true that the reflux rate is high. But only the top tray realizes this. The other 59 trays and the reboiler think that the reflux rate is still low. In the case of my propylene-propane splitter in Indiana, the top tray flooded because the tray deck was fouled. Corrosion deposits and salts from the reflux drum accumulated on the top tray deck. This raised only the top tray delta P, promoting entrainment of the top reflux. Water washing the top of the tower corrected this malfunction.

The lesson is that delta P surveys are not a definitive method of determining if a tower is flooded. Perhaps the best method is by heat balance. That is, if the reflux can be increased without a proportional increase in the reboiler duty, then the tower is flooded. And if this observation does not coincide with an increase in the tower delta P, then the problem is flooding starting at an upper tray deck.

Loss of Liquid Level on Tray Decks

I was working on a diesel oil recovery tower in Convent, Louisiana, that had 20 trays. The design vapor flow through the trays was 100,000 lb/hr. The trays were modern grid-type MVG-type decks. The design pressure drop per tray was:

- Delta P dry—The pressure drop of the vapor flowing through the tray deck perforations = 0.1 psi per tray.
- Delta P hydraulic—The equivalent height of the liquid on the tray due to the weir = 0.1 psi.
- The total tower design delta P was then:

$$20 \text{ trays } (0.1 + 0.1) = 4 \text{ psi}$$

At an operating vapor flow through the trays of 50,000 lb/hr, what delta P do you think I observed?

Well, the vapor delta P varies with velocity squared. Since the flow had gone down by 50%, the new delta P dry should have been 0.025 psi per tray.

The weight of liquid on the tray due to the weir should not have changed with the reduced vapor rate. Therefore the observed delta P should have been about:

$$20 \text{ trays } (0.025 + 0.1) = 2\tfrac{1}{2} \text{ psi}$$

But the observed delta P, which I measured on the tower, was zero! Now what? How could the pressure drop of the trays, at half the design vapor flow, be too small to measure? The following explanation applies to:

- Sieve trays
- Valve trays

- Jet tab trays
- Grid trays
- Any modern type of proprietary perforated tray deck

However, it does not apply to old-style bubble- or tunnel-cap trays, which are immune to tray deck dumping, leaking, or weeping. I'll explain why this is so later.

Valve-type caps, contrary to vendor claims, leak almost as badly at low vapor rates as do sieve or grid trays. When vapor flow falls to 50% of design, delta P dry falls to 25% of design as explained above. But a small delta P dry causes the tray to leak. In larger-diameter towers (2 or more meters), a small amount of tray deck out-of-levelness will cause the problem to be magnified. Typically, when the vapor flow rate is 30% to 50% of design, the flow of liquid over the weir drops to zero. Why? Because all of the liquid is **dumping** through the tray deck.

Now the depth of the liquid on the tray falls below the bottom edge of the downcomer of the tray above. Vapor now begins to blow through this unsealed downcomer. The vapor is bypassing the tray decks through the downcomer. This further reduces delta P dry and promotes more tray deck dumping. The larger tray deck dumping rate further reduces the hydraulic delta P (i.e., the weight of liquid on the tray).

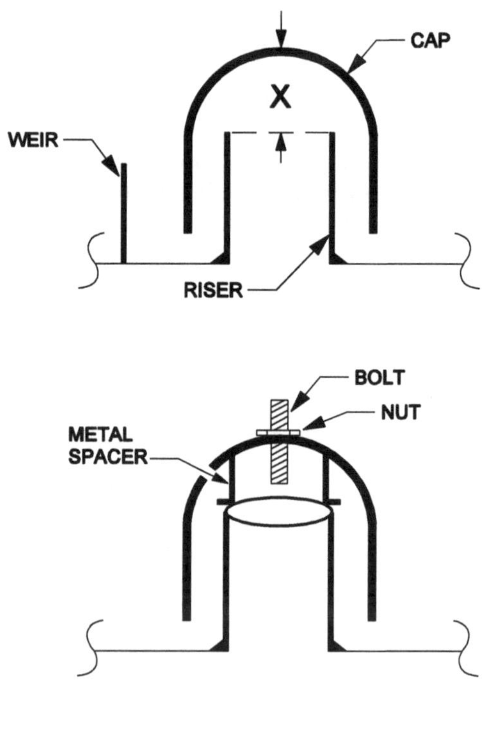

Figure 1-4 Bubble-cap trays are not subject to tray deck dumping at low vapor flows.

In summary, delta P dry is further reduced because vapor is short-circuiting the tray decks through the downcomers. Delta P hydraulic is further reduced because of increasing tray deck leakage. The overall delta P on the 20-tray tower becomes too small to measure with a single gauge, delta P survey. I'll leave it to the reader to understand how the blown downcomer seal and leaking tray deck affect fractionation efficiency.

The bubble cap shown in Figure 1-4 is not subject to tray deck leakage, as long as the top of the chimney is an inch or so above the outlet weir. For this reason, bubble-cap trays are inherently superior to valve, sieve, or grid trays, except for their lower vapor handling capacity.

Lost Bubble Cap Clearance

The bubble cap is fixed to the top of the chimney by a bolt sticking up through the chimney. A metal spacer is used to maintain dimension x, shown in Figure 1-4. This dimension determines the delta P dry of the tray. It is maintained by a metal spacer around 1 to 2 inches high.

At a visbreaker fractionator in Convent, Louisiana, I was troubleshooting a tower flooding problem. I suspected that coke had accumulated underneath the cap and restricted vapor flow. This caused a higher delta P dry, which backed the liquid up in the downcomers and caused the tower to flood at 40,000 BSD of feed.

I had all the caps removed, cleared out the accumulated coke, and with complete confidence, had the tower restreamed. The next day, Ken Starr, the operating manager, called me.

"Lieberman. Thanks to you, instead of the frac flooding at 40,000 BSD, it now floods at 30,000 BSD. Get out here and fix the problem."

I phoned the plant from Singapore where I was working, and spoke to John Henry, the maintenance supervisor. "Mr. Henry, I want you to take off every cap. And this time, clean properly underneath each cap. Also, make sure each riser is clear and free of coke."

"Look, Lieberman, we did that last time."

"Well," I said, "You must have missed some of the coke underneath some caps, because the tower is flooding." When I returned home the following week, I found this message from Ken Starr.

"Lieberman. Thanks to you, instead of the tower flooding at 30,000 BSD, it now floods at 20,000 BSD. Get out here and fix the problem."

When Liz and I crawled through the tower, I noticed something odd. The bolts that stuck up from the tops of the bubble caps protruded by 2 inches above the caps. The first time I had been in the tower, the bolts only stuck up about an inch. So I unscrewed one of the nuts with my wrench and pulled off a cap. The spacer between the top of the chimney and the inside of the cap was crushed (see Figure 1-4). The cap was jammed up against the chimney.

"Yeah, Lieberman," Mr. Henry explained. "I sure didn't want to have you complain that we didn't tighten up them caps. So I got my guys to use an air gun wrench on them nuts. Kinda looks like we overtightened a few caps."

The lesson we learn from this story is not to try to fix bubble-cap tray malfunctions long distance. You've got to get real close to the problem.

Directional Flow Tray Panels

A modern grid-type tray might use:

- MVG Caps—Sulzer (Nutter) (Good)
- Provalves—Koch-Glitsch (Better)

Such trays will have up to 10% more capacity than older-style valve or sieve trays. This benefit is largely a consequence of the use of **push**-type perforations on the tray deck. These perforations cause the gas to escape from the tray deck with horizontal components of velocity directed toward the outlet weir. This keeps the liquid level from backing up at the inlet side of the tray (i.e., near the downcomer from the tray above). Having a higher liquid level at the tray inlet side promotes entrainment and flooding at the inlet side of the tray. When I was young in the 1960s, we used to use **step-down** trays.

The grid decks accomplish the same purpose as the archaic step-down trays, but without any added mechanical complexity. With this objective in mind, a tower in Aruba was modified with directional flow grid trays, to replace the older sieve decks. Instead of an increase in capacity of 10%, a 10% decrease in capacity was observed.

Liz, my coworker and wife, crawled through the tower to determine the malfunction. There were 50 trays. All were installed correctly, except for tray #28. As Liz noted, the panels on this tray were installed *backwards!*

The installation contractor claimed that he had done 98% of the job correctly; that no one is perfect. Unfortunately, with the push valves installed backwards, the natural liquid gradient on the tray deck #28 was increased, which caused tray #28 to flood. As flooding progressed up a tower:

- The trays above tray #28 also flooded and lost fractionation efficiency.
- The trays below tray #28 began to dry out, due to low internal reflux rate, and also lost fractionation efficiency.

As fractionation got worse, the operators cranked up the reflux ratio. But this just made the flooding worse. So, to restore product purities to the required specifications, they reduced the feed rate to the tower. We reoriented the misguided tray panels, and when the tower was restreamed, all was well.

Especially on multi-pass trays, it's difficult to see if an MVG-type grid tray panel has been installed in the proper direction of liquid flow.

Missing Reboiler Return Impingement Plate

On the same tower on which Liz had found that the grid tray panels were reversed, we had also encountered a more serious flooding initiated from the bottom tray, rather than just tray #28. On startup, this new debutanizer flooded at less than half of its design rate.

Both a radiation scan and a delta P survey indicated the tower was flooding from close to the bottom tray. As the debutanizer feed was contaminated with water-insoluble iron sulfide particulates, I concluded the flooding was most likely a consequence of tray fouling.

The upstream distillation tower that provided the debutanizer feed had a carbon steel overhead condenser tube bundle. Wet H_2S reacted with the tubes to produce the water-insoluble iron sulfide particulates. Most likely, I thought, an iron sulfide sludge had accumulated on the bottom tray of the debutanizer. Even more probable was that the sludge had accumulated in the seal pan below the bottom tray (see Figure 1-5).

The seal pan tends to act as a dirt trap. Solids flushed down the column tend to accumulate in the seal pan and cause downcomer backup and flooding of the bottom tray. That's why, when I design a seal pan in fouling service, I'll provide at least a single 1-inch hole in the floor of the seal pan. This permits dirt to drain out of the pan.

After the debutanizer was shut down, I crawled through the vessel top manway. As I descended down the tray deck man-ways,

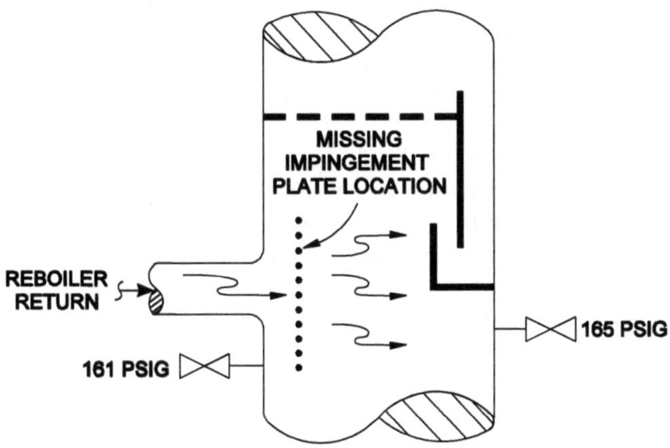

FIGURE 1-5 Missing impingement plate causes flooding.

I noted that all the trays were reasonably clean—as was the seal pan! It's true that the tower had been water washed. But iron sulfide is not soluble in water.

Now what?

Feeling bad, I slumped down in the bottom of the dark tower. Other people in Aruba were relaxing on the white sand beach or snorkeling in the crystal clear blue water. But not me. When I snapped my flashlight back on, I found myself staring at a round 16-inch hole in the opposite wall of the vessel.

"That's the reboiler return nozzle," I recall thinking. "But why is it, that I can see this nozzle? Shouldn't it be covered over by an impingement plate?" (see Figure 1-5).

But there was no impingement plate. A circular 24-inch impingement plate was shown on the vessel sketch. But it was never installed in the debutanizer when the tower was fabricated. How could the missing impingement plate account for the tower flooding?

I suddenly recalled a pressure survey that I had conducted the previous week. That survey indicated:

- The pressure of the tower just opposite the reboiler return nozzle was 165 psig.
- The pressure of the tower adjacent to the reboiler return nozzle was 161 psig.

I had ignored this 4 psig discrepancy because it didn't make any sense. But now it made lots of sense. Without the impingement plate to dissipate the momentum (mass times velocity) of the reboiler outlet flow, the returning vapor-liquid mixed phase would rush across the 10-foot-diameter tower. It would hit the opposing wall, near the bottom tray seal pan. The momentum of the fluid, in accordance with Bernoulli's equation, would be converted to pressure. Pressure, in the sense that a localized pressure 4 psig above the surrounding pressure, would be created. Localized in the sense that pressure in the region of the seal pan would be 4 psig greater than the pressure of the vapor flowing up through the bottom tray.

If the SG of the fluid in the downcomer was about 0.70, then the liquid level in the downcomer would have been pushed up an additional 13 feet. But the downcomer length was only 2½ feet. Thus, the liquid from the seal pan would have backed up onto the bottom tray. As flooding progressed up a tower, the entire tower would have flooded. To suppress the flooding, the operators would have had to reduce reboiler duty. This would force them to cut reflux. The lower reflux rate would in turn cause a reduction in feed rate to control the heavier components in the butane overhead product.

I had the 24-inch-diameter impingement plate installed 12 inches in front of the 16-inch reboiler return nozzle. (Unfortunately, I failed to inspect the rest of the tower, and missed the incorrectly oriented grid deck panel on tray #28.) As a precaution, I had several 1½-inch holes drilled in the floor of the seal pan. The number of holes was determined so that 25% of the liquid flow would drain through the holes, to keep sludge from accumulating in the seal pan.

Flow Path Length

In 1965, I began work as a process design engineer for American Oil. My first project was an absorber revamp at the El Dorado, Arkansas, refinery. The idea was to expand the lean oil circulation rate, so as to increase recovery of propylene from a catalytic cracker wet gas stream. The current lean oil circulation rate was limited by flooding and consequent lean oil carryover into the fuel gas system. The tower was rather small at 4 feet, 6 inches I.D.

My calculated percent of jet flood (flooding due to entrainment) was 90%, consistent with the observed tower operating limit. Percent jet flood is a function of:

- Liquid and vapor density
- Gas rate
- Weir loading

For higher-pressure towers with high liquid flows and small vapor volumes due to the high pressure, weir loadings are important when calculating percent of jet flood. Weir loading is GPM (hot), divided by the **weir length**, to obtain GPM per inch. For this absorber, which had one-pass trays, the weir loading was quite high. So I had a good idea. I would convert the existing one-pass tray to two-pass trays, as shown in Figure 1-6. This would greatly reduce my weir loading. I would now have not one weir, but two weirs!

My computer simulation showed that I would then be able to circulate 50% more lean oil. Propylene recovery would increase from 70% to 85%. My boss Bill Duvall approved my revamp design based on my computer simulation. New pumps, heat exchangers, and trays were ordered. But then I forgot all about the project because I was working in the Planning Division when the unit started up after the retrofit.

Another five years slipped by. I was working at the American Oil refinery in Whiting, Indiana. My officemate Jerry Edwards had been transferred from the El Dorado refinery when it shut down in 1970.

"You know, Norm, that your design didn't work," said Jerry.

"What design? You mean my absorber revamp?"

"Yeah, Norm. It didn't work worth a damn. Propylene recovery got worse rather than better. I'll tell you where you screwed up. You made the **flow path length** too short [see Figure 1-6].

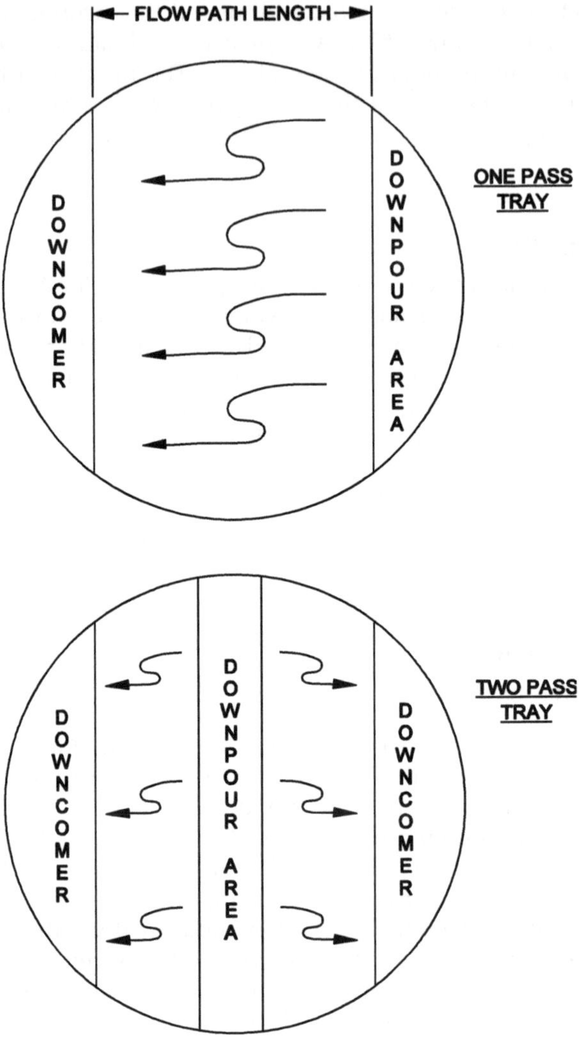

Figure 1-6 Reduction of flow path length can hurt tray efficiency.

"That old flow path length on the 4-foot, 6-inch ID tower was okay. It was 28 inches. But the new flow path length on the two-pass trays was only 12 inches wide. Real short flow path lengths for valve trays mean that there are only a few rows of caps. So, some of the liquid can bypass the caps, and then it doesn't contribute to absorption efficiency. So propylene absorption got worse. You know, Mr. Norm, the minimum tower ID for using two-pass trays is something over 5 feet ID," Jerry concluded.

But American Oil never followed up on the results of projects to see if they actually worked. My project was judged a good job by my supervisor because I had simulated the tower on my computer model with great success.

Discriminating Between Flooding and Dumping

To summarize, perforated tray decks are subject to two malfunctions:

- Dumping or weeping
- Flooding or excessive entrainment

Perforated trays means all types of modern trays, including valves, sieves, or grids. But not ancient bubble-cap trays, which cannot dump.

All perforated trays of industrial size diameter—that is, more than 1 meter—are both dumping and entraining to some degree, at the same time, and thus degrading tray fractionation efficiency. But how can I tell, when I walk into the control center, which is the controlling malfunction? I can perform two tests:

1. *Lower the tower pressure.* Does fractionation get better or worse? If I had online gas chromatographs (GCs) for the products, that would be fine. But most towers do not have such luxuries. Also, I am too impatient to wait for lab results. What I do look at is the temperature difference between the top and bottom of the fractionator. If this delta T goes up, then fractionation is improving. This indicates trays were losing fractionation efficiency due to tray deck dumping, or weeping, or leaking. This test must be carried out at a constant reflux rate.
2. *Raise reflux rate.* The presumption here is that the reboiler duty is on automatic temperature control, thus the tower bottoms temperature is constant. If raising the top reflux flow causes the fractionator top temperature to increase, then fractionation efficiency is degraded because of flooding or excessive entrainment. This test must be carried out at a constant operating pressure.

If a tower is shown to be flooding by this test, a delta P survey helps to identify the malfunctioning tray. A big delta P means flooding starting at a lower tray. A small delta P means flooding starting at an upper tray.

How about an Isoscan (TruTec or radiation scan)? Not for me. My rules for identifying tray malfunctions are:

- You have to do it in one day.
- You have to be able to do it with the tools at hand.
- You have to be able to do it yourself.

After 46 years, I have accumulated hundreds of these stories. Many of the other tray and packed tower distillation malfunction incidents are described in the books I have authored. But the most comprehensive summary is in Henry Z. Kister's book, *Distillation Troubleshooting*, Wiley, 2006.

Shed- or Baffle-Type Trays

There is another whole class of trays that does not allow the vapor to flow through their decked area. These trays are called **baffle trays**. Trays that fit this description are:

- Side-to-side baffles
- Disk-and-donut trays
- Shower decks
- Shed trays

Baffle-type trays do not work at all well in fractionation or steam stripping service. They simply do not have the ability to bring the vapor and liquid into intimate contact. I've been working this month on a 40-baffle steam stripper. Field tests conducted by varying the stripping steam rate suggest essentially no stripping efficiency. If there is a way to make baffle trays effective, I've not found it—and I've tried often.

Baffle trays, especially shower decks, do a reasonably good job in heat transfer pumparound service. They are widely used in slurry oil pumparound (i.e., bottom pumparound) in fluid catalytic crackers. A half dozen such baffles will act like one theoretical stage, as far as heat transfer rates are concerned.

Author's Observations: Concepts versus Calculations

Now that you have read the first chapter of this text, permit me to make a suggestion. My intent is not to write a reference book. There are better books available to guide one in making engineering calculations. As my coworker, Dave, observed:

"Norm, when you're out in the plant, alone, at midnight; when you're too hungry, cold, wet, and discouraged; when a tower won't fractionate, and you don't know why; when black smoke is belching out of the heater stack and the O_2 analyzer shows 6%; when hydrocarbon vapors are boiling out of your cooling tower and the plant manager has just classified you as expendable, it's not calculations that are needed. What you want is a basic understanding of process and chemical engineering concepts."

When I'm faced with field malfunctions of a pump, fractionator, or heater, I'll first try to classify the problem. How does this set of symptoms relate to other experiences I've had with similar equipment? Is this a problem with heat transfer, vapor-liquid equilibrium, mixing, hydraulics, or entrainment? What field measurements or samples should I obtain? What questions should I ask the plant operators? Maybe it's not a process equipment malfunction at all, but a process control problem.

Pretty far down on my list of concerns is how to quantify the malfunction. As Dave also said, "I can calculate anything, Norm, if only I know what it is I'm supposed to calculate."

Yes, Dave, that's the problem. If we only knew what is the question, the engineering or operational answer would follow quite easily. Understanding the nature of the question is the real challenge in correcting process equipment malfunctions. Thus, my suggestion to you, the reader, is to read the entire text. The process concepts fit together. Like any puzzle, you'll have to have all the pieces in the right spots to assemble the process solution completely and correctly.

CHAPTER 2

Packed Tower Problems

The price we pay for success is the willingness to risk failure.

—Phil Jackson, NBA basketball coach

I never really liked packed towers, mainly because they cannot be inspected in the same way as trayed towers. Once the packing is installed in a fractionator, foreign objects buried in the packed bed cannot be observed until the tower starts up. Then the buried obstruction manifests itself in an unpleasant manner, meaning, the tower floods and fails to fractionate.

Trayed towers are different. After the trays are installed, I can crawl through every tray and inspect each component for proper installation and cleanliness. Even if there is only a single vessel manway, I can, and often will, check every detail for proper assembly. The one exception is closure of the **tray deck manway**.

I imagine that packed towers have a potential for greater capacity than trayed towers. But the advantage is small. When a tower is 1 meter or less in ID, the use of trays becomes awkward and packing is preferred. Packing has a lower delta P than trays and hence may be favored for vacuum tower service. For wash oil service (i.e., de-entrainment) and especially in heat transfer pumparound service, I prefer to use structured-type packing. Some services are quite corrosive, and ceramic-type packing is required. However, in normal fractionation service, for new towers in nonvacuum service, the use of packed towers is a poor design practice.

I'm quite sure this statement may be refuted by vendor-published correlations, which compare tray capacity and efficiency to those of modern types of packing. But these correlations fail to take into account real-world installation, inspection, and fouling problems associated with packed beds.

Syn Gas Scrubber Flooding

In 1972, I was working for American Oil in Chicago. I was assigned to consult on a research project at the University of Chicago. They were developing a coal-to-gas process, and their product gas scrubber was flooding. The scrubber was a 36-inch-ID packed tower. I studied the tower's operation and design for several weeks and issued the following brilliant report:

"The scrubber was flooding due to unknown circumstance."

My report was ignored. But when the scrubber was opened later, a plastic bag, which had been used to load the loose packing, was found in the middle of the bed.

A rather similar but more complex problem occurred last year in Lithuania. A fractionation zone in a crude distillation tower was flooding. The fractionation zone consisted of 12 layers of structured-type packing. This is a type of packing that is purchased in blocks. Each block is about 10 inches high, 15 inches wide, and 8 feet long. The packing consists of thin sheets of perforated, crimped metal pressed together to form a block.

A radiation scan (Gamma scan or TruTec scan are common trade names) was performed. A source of radiation is placed on one side of a tower. The percent of absorbed radiation is measured on the opposite side of the tower. This measurement is done continuously up the length of the tower. Areas of high absorption correspond to a dense liquid phase. Areas of low absorption correspond to the vapor phase. In a packed bed, the transition from a vapor phase to an upper liquid phase determines the elevation in the packing where flooding is initiated.

The radiation scan for the crude tower showed clearly that flooding was initiated between layers seven and eight of the blocks of structured packing. Obviously, one of two malfunctions had transpired to cause the flooding:

- An obstruction had been left between the layers of packing.
- More likely, as the packing was placed in the tower, someone had stepped on a thin sheeted block and crushed its upper surface, thus reducing the open area of the packing between layers seven and eight.

I personally supervised the removal of each layer of structured packing. During my breaks, my wife, Liz, carefully observed the disassembly of the packed bed. As layer number seven was removed to expose the top of layer eight, I found . . . nothing!

Now what?

Maybe I had misinterpreted the results from the radiation scan as to where the flooding was being initiated. So, I had another, and another, and another layer removed, until I was standing on the packing grid support. Still no sign of any obstruction to vapor flow. Still no explanation as to the cause of the flooding.

It was getting dark and cold. All work had stopped. I had to make a decision.

"Discard the old packing. We'll install all new layers of structured packing," I told the foreman.

"Very well, comrade engineer. But what is wrong with this structured packing?" The foreman had been carefully stacking the blocks of used packing neatly near the tower, as they were removed.

"They are suffering from structural fatigue. Microscopic changes in their pores render them unfit for further service," I explained.

So the new packing, which was identical to the old packing, was installed. The flooding problem vanished. This incident bothers me to this day. It forms part of my bias against the use of packing in fractionation service.

Reduction in Percent Open Area

Packed beds must be supported. A distillation tray, at least in towers smaller than 10 feet in diameter, can be designed to be self-supporting. Packed beds must be supported on a grid support. This is true for structured packing as well as more conventional dumped or random-type packing.

A typical random-type packing is 1-inch pall rings. Let's assume that the percent open area of this packing is 80%. By the percent open area, I mean:

- A single layer of randomly placed rings is floating in the sky.
- Sunlight is passing through 80% of the area covered by the rings.
- Sunlight is obscured by 20% of the area covered by the rings.

Naturally, the rings can't magically float in the air. They have to be supported by a packing grid support. The openings in the grid support have to be less than 1 inch to prevent the rings from slipping through the grid. If the grid support is constructed from ⅛-inch steel rods, the open area of the grid might be around 75%. Hence, the open area of the packed bed, where the rings contact the support grid, is:

$$(75\%) \times (80\%) = 60\%$$

But the grid itself must be supported by a tray ring and a cross I-beam. Let's assume their open area is 90%. Hence, the open area of the packed bed, where the grid is supported, is:

$$(60\%) \times (90\%) = 54\%$$

In most process applications, fouling can be expected. In my long and unpleasant experience with packed towers, I have found that these deposits accumulate at the interface between the packing itself and the grid support. How do I know this?

"Mr. Lieberman, the absorber is clean," said Cathy, the unit engineer. "I washed it with clean, hot steam condensate, even though the packing was clean. I think your theory, that the absorber flooded due to fouling, is wrong. The packing was clean."

"Cathy, dear girl," I said, "The fouling was iron sulfide ($Fe[HS]_2$). Iron sulfide is not soluble in water. You need to acidify the absorber."

"But the absorber's clean anyway!" Cathy was becoming angry. "I climbed into the top of the tower and inspected it myself."

"But my dear girl, the iron sulfide solids tend to accumulate at the interface between the packing and the grid support. Toward the bottom of the packed bed, where concentrations of hydrogen sulfide in the sour feed gas are greatest."

"I am not your 'dear girl'," Cathy hissed. "How do I acidify the absorber? Circulate from the top down?"

"No. Circulate from the bottom up. If you do not fill the entire tower with acid solution, the circulating acid will promote channeling. The acid will bypass the most fouled portion of the packing. Then, Cathy, the absorber's vapor-liquid contacting efficiency will be degraded."

"What!" Cathy fairly screamed. "Do you have any idea, Lieberman, how much acid it will take to fill my absorber to overflowing? There must be an alternative."

"Look, Cathy. It was you who ignored my advice to use a trayed tower and not packing when you designed this absorber four years ago. And there is an alternative to acid washing the packing."

"Which is?" she asked.

"Take the packing out through the tower top manway in plastic buckets," I responded.

Cathy's beautiful, fair face flushed red with fury as she screamed, "Get out of my office!"

Almost the entire 20-foot packed bed proved to be reasonably clean. It wasn't until the last few feet of packing was removed in the plastic buckets that the packing was found to be mixed with large amounts of black, slippery, iron sulfide corrosion deposits. We spread the packing onto the concrete slab and washed off the iron sulfides with a fire water hose. The rather complex, corrugated grid support was also removed and cleaned.

When the cleaned packing and grid support were replaced, the absorber was returned to service. It flooded far worse than ever. Cathy had all the packing removed in plastic buckets a second time. When I inspected the tower, I saw that the corrugated grid support (see Figure 2-1) had been misassembled after it had been cleaned and replaced. After this malfunction was corrected and the packing was again reloaded, the absorber worked just fine—for a while.

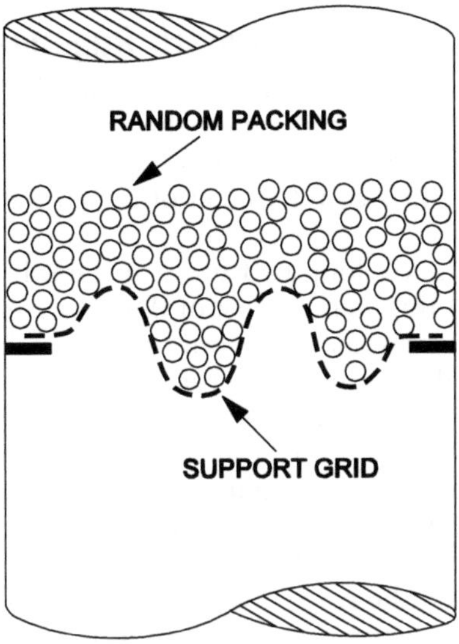

FIGURE 2-1 A corrugated grid support increases the open area at the bottom of a packed bed.

Cathy, having demonstrated determination in the face of disaster, was promoted to division manager. So all's well that ends well.

Corrugated Grid Support

Packed towers are limited not by the open area of the packing, but by the open area of the interface of the grid support and the packing itself. To reduce this limitation, a corrugated packing support is used, as in Figure 2-1. If properly designed and installed, the corrugated support can eliminate this capacity pinch-point, unless it fouls. But it's just at this point that fouling deposits tend to accumulate. Also, in larger-diameter towers, the structured support of the corrugated grid may be complicated, and its reinstallation after cleaning, problematic.

There is a reasonable, if not a complete, solution to this malfunction. Between the grid support and the regular packing, load a layer of larger-size random packing. For example, below a 20-foot bed of 1-inch pall rings, load 1 or 2 feet of 2-inch rings, which have a larger percent open area. Purchase these larger rings with the maximum thickness available. Rings crush rather easily if handled roughly. That's also the reason I avoid aluminum rings.

At the Good Hope Refinery in Louisiana, we experienced continued failure of packed beds of random packing due to the failure of their grid supports. The maintenance manager developed an excellent method to rigidly secure these beds using layers of sturdy grids laid cross-wise and vertical half-inch steel rods. I've given a detailed description of this very successful retrofit in my book, *Process Design for Reliable Operations*, 3rd edition.

Packed Bed Failure in a Catacarb Regenerator

I'm sitting on the beach in Aruba as I write this story. Six miles away is the idled Valero Refinery where this story unfolded. The packing in the Catacarb Regenerator Tower, according to the plant manager, had disappeared. "How could 15 feet of 2-inch metal rings vanish?" he asked me.

Actually, the packing had not vanished. It had been ground up into tiny metal fragments. Most of these fragments had plugged the shell side of the circulating thermosyphon reboiler (see Chapter 9, "Process Reboilers—Shell and Tube"). The remainder of the broken and ground-up rings were lying in the bottom of the regenerator. What force had ground up these metal rings into such tiny fragments?

My inspections indicated a small portion of the packing support grid had come loose. The rings had drained through this relatively small opening. The circulating catacarb (potassium carbonate solution) had carried the rings into the reboiler. There must have been a channel somewhere in the reboiler bundle large enough for the rings to pass through. The broken bits of rings spun round and round through the reboiler and through the bottom of the regenerator, until they were ground up. Not a dozen intact rings could be found. The cause of the complete loss of the regenerator's stripping efficiency was due to a minor failure to the packing grid support. A similar failure in a trayed tower would have had relatively small consequences and could not have led to a loss in thermosyphon circulation in the regenerator reboiler.

Bed Hold-Downs

Structured packing or grids are often used in the wash oil or deentrainment sections of vacuum and crude distillation towers. This is an excellent application for such packing, especially when they are constructed of sturdy layers of grid. The grids, while quite strong, do not weigh very much. A surge of vapor flow may easily dislodge them from their lower support. The liquid distributor above the grid wash oil section may then be damaged by impact with the grid. At the ARCO refinery near Houston, a delayed coker wash oil spray pipe distributor was badly bent upward by such an impact. The resulting unbalanced wash oil distribution flow coked the wash oil grid and turned the heavy coker gas oil product black.

To prevent this sort of upset, a strong hold-down grid placed atop the packing is critical. Sometimes the packing vendors will claim that the weight of the packing will, in itself, be sufficient to resist a pressure surge. This is simply not true. I say this not by calculation, but from unhappy experience. Always insist that the upward force that the packing hold-down structure must resist must be equal to at least the weight of the grid itself. Drawn from my experiences with this problem at the Good Hope Refinery from 1980 to 1983, I have summarized in *Process Design for Reliable Operations* one practical mechanical design to handle this critical problem.

Liquid Distribution to Packed Beds

Packing is employed in towers in three distinct services:

- Pumparound (heat removal) (see Chapter 6)
- Wash oil (de-entrainment)
- Fractionation (distillation)

In wash oil and pumparound services, liquid distribution is accomplished by a spray header. This is a pipe grid. For example, an 8-foot-diameter tower will have a center pipe connected to the inlet nozzle and typically six arms. Attached to the center pipe and arms are perhaps 15 to 20 spray nozzles. These are like shower heads, but with no adjustment possible. The standard spray nozzle used in the industry has the following characteristics:

- 120° spray angle.
- Full cone, meaning complete wetting within the spray cone.
- Model number corresponds to the maximum free passage of the nozzle.

The term *maximum free passage* means the maximum-size particle that can pass through the nozzle without plugging the nozzle. For example, if the model number is FMP281, the 281 number means a particle with a maximum dimension of 0.282 inches will likely get stuck in the spray nozzle. This is bad, as it will plug the nozzle. Nozzle plugging is by far the major malfunction encountered with packed beds in pumparound and wash oil service. More on this critical subject later.

The term, *full cone* is basically a lie. The lie is that the liquid is equally dispersed in the area encompassed by the spray cone. One day, while my wife was away, I removed every drinking glass from the kitchen. I set up a solid array of glasses in my driveway. I tested several reputed 120° full-cone spray nozzles from three different vendors by attaching each nozzle to my garden hose. In all cases, the vast majority of water accumulated in the outer ring of glasses. Admittedly, all my glasses had some water in them, but toward the center, there was very little water accumulation. The least guilty nozzle in this liquid maldistribution problem was the Bete nozzle. So I've always specified Bete nozzles on my designs. But because of this inherent distribution problem, spray nozzles should not be used for fractionation service. This is not just my opinion, but is generally accepted in the hydrocarbon processing industry. For fractionation service, a **gravity distributor** is required. I'll discuss this in detail later.

The term *spray angle* is just the angle at which the spray leaves the nozzle. For example, for the 120° nozzle, the liquid spray angle from vertical is 60°. A wider spray angle increases the wetted perimeter on the packing. However, a wider spray angle may also increase the amount of liquid hitting the vessel wall, which is bad.

Gravity Distributors Used in Fractionation Service

Pilot plant tests conducted by FRI **(Fractionation Research Incorporated)** have indicated that the ability of any sort of packing—rings, saddles, grids, structured packing, etc.—to fractionate is largely a function of good initial liquid distribution. Tests have shown that packed beds do not redistribute liquid, but instead promote liquid channeling. Finally, spray nozzles do not provide sufficiently dispersed liquid flow, as the liquid is concentrated around the periphery. Therefore, a gravity distributor is required, as shown in Figure 2-2. Liquid is redistributed through progressively smaller and more numerous orifices. Gravity distributors for larger-diameter towers are very complex and very costly. A properly designed distributor can cost more and take longer to install than the packed bed itself. The most common malfunctions with gravity distributors occur when their various components are not installed level. Or when they plug due to fouling deposits. Or when they are damaged due to pressure surges. Or when they are removed for cleaning and are not reinstalled properly. Or, when designed for a high reflux rate, they are run at a far lower reflux rate. Or when they are poorly designed in the first place. Or . . .!

But maybe you have read enough. The point is, it's best not to get involved with complex mechanical features that must function

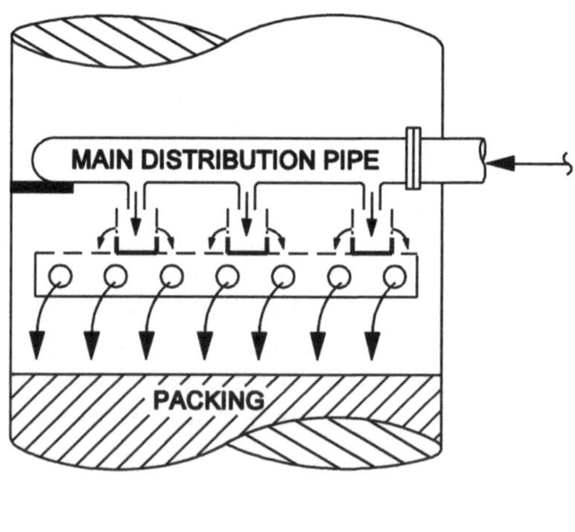

FIGURE 2-2 A three-stage gravity distributor used in fractionation service.

without adjustment inside distillation towers. This is an environment suitable only for rugged, simply designed components, components that need not be precisely aligned and that can withstand fouling, corrosion, pressure surges, and abuse during installation and inspection. Thus, my preference for trays.

Spray Nozzle Malfunctions

I had decided to properly check the spray nozzles used in the wash oil section of Coastal's refinery vacuum tower in Corpus Christi, Texas. I had each nozzle unscrewed for testing using the equipment shown in Figure 2-3. I applied the design water pressure of 15 psig to each nozzle and measured:

- The water spray angle.
- The water flow rate (i.e., the rate of accumulation in the bucket).

Two of the nozzles failed to develop any spray at all. The nozzle internals were missing. Most of the remaining nozzles were partly or totally plugged with green glass from a broken beer bottle. To prevent spray nozzles from plugging, a dual element **(duplex)** filter is needed. The elements in the filter screen should have one-third the maximum free passage of the nozzles. Smaller openings will cause the filter to plug too rapidly. Larger openings will result in the nozzle plugging. The one-third value is derived from experience. That is, the standard

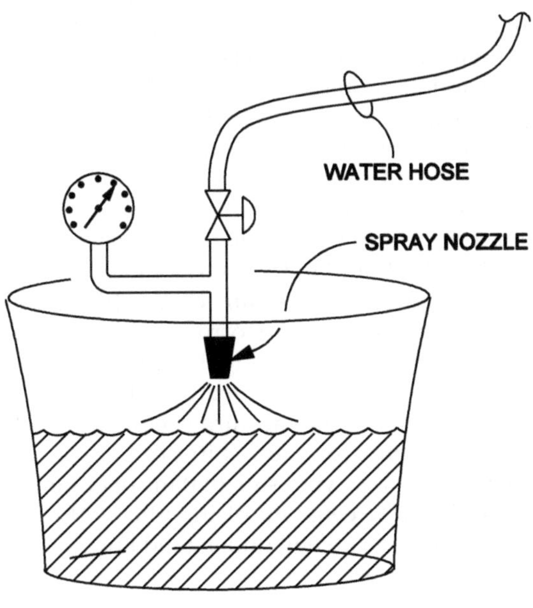

FIGURE 2-3 Testing spray nozzles for spray angle and flow rate.

one-half value is not small enough. Make sure there are no holes in the filter screen. I mean zero holes! The dual element filter must never, ever, have a bypass. A reasonable fouled delta P before the filter is cleaned is 25 to 30 psi. Filter reassembly should be verified by the unit engineer in writing. At the Coffeyville Refinery where I'm heading as I write these words, I believed the pipefitter's word that a filter was reassembled correctly, and lived to regret my ill-founded trust. Hence, my current return visit.

Spray nozzles may also plug due to a loss of flow. This happens most often during an electric power failure. A backup source of flushing oil, from an uninterruptable source, is required. This flushing oil must come on automatically on low spray header pressure. Do not use steam as a backup to the normal wash oil flow. Most likely you will wind up with a slug of water when the steam valve is tripped open, which will cause a pressure surge that disrupts the packed beds.

Finally, it's a good practice to observe the spray pattern inside the vessel using water at the intended operating pressure. However, make sure that all piping has been flushed clear before conducting such a test.

Spray Nozzle Pressure Drop

Spray nozzles do not have a very large operating range. If the delta P is less than 5 to 8 psi, a full-cone spray angle will not develop. If the delta P is high (perhaps above 50 psi), the nozzle will form a mist. As the flow through the nozzle increases, the incremental flow will not spray down onto the packed bed below, but instead will form a mist that will entrain to the packed bed above the spray header.

Another problem with a diminished spray angle occurs when using subcooled liquid. This is not a problem I have observed personally. But in theory, the vapor inside the spray cone will condense and cause an area of low pressure to develop inside the cone. This may cause the spray cone to collapse, which ruins the liquid distribution. For this reason, it's best to use saturated, bubble point, hotter liquid in the spray header.

All piping downstream of the filters must be constructed of corrosion-resistant steel. Corrosion products which form downstream of the filters are sure to plug the spray nozzles.

On startup, don't be surprised if the filters plug after an hour or less. Scale left from the turnaround has to be flushed out of the system. It's only for a few shifts, or for a few days, that this problem will persist. Make very sure that the operators do not get discouraged and pull the screens out of the filters. Nozzle plugging will surely follow. And you all will understand how I've become so smart on this particular subject.

Plastic Packing

At the Good Hope Refinery, we melted the plastic or Teflon packing in two absorbers. If you wish to duplicate our achievement, you may select one of the following two methods:

- Steam out the packing under pressure.
- Allow air to enter a packed bed contaminated with $Fe(HS)_2$. The pyrophoric iron will autoignite when dry.

The lesson is, never use a nonmetallic packing in H_2S-amine absorption service or in sour water stripper service. Of course, I did not realize that the packing had melted until we started backup and the tower flooded.

Incidentally, I had purchased a large quantity of plastic packing, which I could not decide where I could use after the melting incidents. The plastic rings were dumped onto a big pile in the equipment storage area. Exposed to sunlight, the rings turned into a fine, white powder after a year or so.

Packed Towers in Offshore Applications

There is one area where the use of packed towers in fractionation service appears to have a distinct advantage over trayed towers. That is, on offshore platforms where natural gas condensate is processed. Or, where relatively small amounts of diesel oil are recovered from crude produced on the platform, for use on the platform itself.

I have designed such packed towers using standard correlations. However, the one component that is different for offshore use is the liquid feed and/or the reflux distributor. Assuming the angle of displacement of the tower from vertical is less than 10° due to ocean swells, these specially designed liquid distributors will still produce a normal packing fractionation efficiency.

I did not design the distributor myself, but purchased it as a standard item from Koch-Glitsch. There is a surprising amount of information and published papers on the Internet relating to offshore fractionation technology using packed fractionator towers.

In summary, my negative view of packed towers in fractionation service is entirely a consequence of my many bad experiences. While the theoretical advantages for capacity, and for separation stages per unit of height for packing versus trays, cannot be denied, experience teaches caution.

CHAPTER 3

Distillation Tower Pressure and Composition Control

There's the right way, the wrong way, the Amoco way, and then there's Norm's way.

—Chief Operator Leroy Wilkes, Texas City, 1976

The objective of controlling a stable tower pressure is to provide a basis for controlling product specifications. An erratic tower pressure creates several problems in meeting a column's fractionation objectives. For example, let's say I have a tower with 50 trays operating at 100 psig. The liquid on each tray is at its boiling or bubble point. That is, saturated liquid. If the tower pressure suddenly drops to 90 psig, the liquid on each of the 50 trays would boil up. An analogy is rapidly opening a warm bottle of beer. The beer would rapidly expand and flood out of the bottle. If the liquid on each tray foams-up, then the tower will flood, and fractionation will be ruined.

I recall a packed tower that I was working on for a Texaco Chemicals plant. They were fractionating between benzene and toluene. Both products had to meet an exacting specification. The packing was structured-type material which has a high (90% plus) open area and hence a very low **liquid hold-up** time. Relatively small decreases in the tower pressure would cause a surge of heavier toluene vapors to pass up through the packing and contaminate the benzene overhead product with toluene. Relatively small pressure increases would cause a surge of liquid to drain down through the packing and contaminate the toluene bottom's product with benzene.

From these observations, I can draw several conclusions about the relative importance of pressure control in various distillation columns:

- The tighter the specs in both the distillate and bottoms, the more important is precise pressure control.
- The smaller the residence time on the trays or in the packing, the more critical is precise pressure control. For example, bubble-cap trays require less precise pressure control than beds of structured packing.
- The closer the tower runs to the **point of incipient flood** (see Chapter 1, "Distillation Tray Malfunctions") or jet flood, the more critical precise pressure control becomes.
- Towers running with materials that tend to foam (i.e., with lower surface tensions) are relatively more sensitive to tower pressure changes.

I fixed the problem at the Texaco benzene-toluene fractionator not with better pressure control, but with a trick I learned form Koch-Glitsch. We installed chimney trays between the packed beds. This increased the liquid hold-up inside the column and tended to damp out the effects of the minor pressure swings as they effected changes in composition. The chimney trays increased the inertia of the system. That is, a ping-pong ball is more easily disturbed from equilibrium than a cricket ball.

In summary, the time and expense in providing precise pressure control in some towers cannot be justified, while in other services, it is critical. And, stabilizing the tower heat input is the most important aspect of controlling a tower's pressure, as discussed in the following section.

Steam Trap Stability Malfunctions

For a steam trap to function, the supply steam pressure to an exchanger must be somewhat greater than the condensate collection system pressure. If it is not, the condensate cannot drain through the trap, even when the trap is opening and functioning properly. For an exchanger with steam on the tube side, the minimum pressure needed for drainage is set by the pressure above the lowest-pass partition baffle. For an exchanger with steam on the shell side, the delta P of the flowing steam is typically small. Thus, the minimum pressure needed for condensate drainage is a few psi above the condensate collection system pressure.

Without sufficient differential pressure between the supply steam and the condensate drain, the condensate will back up and submerge the tubes in water. This reduces the exchanger surface area exposed to condensing steam. The rate of heat transfer will drop. If the steam inlet flow is controlling the temperature of the reboiler outlet, the steam inlet control valve will open. This will blow the condensate out through the reboiler's steam trap, which is already wide open, with a resulting surge in heat input to the tower.

I was working last week with a young engineer who had this problem on a liquid O_2 vaporizer. He had observed that the steam trap was not constantly draining. Believing that the trap was malfunctioning, he opened the bypass valve around the trap. As the condensate was being drained to an open sewer, I could see that he had made the problem worse. Steam was blowing out of the drain. That is, he had blown the condensate seal, which the steam trap automatically maintains, but not when bypassed.

To correct the problem with the existing facility, we set the steam inlet control valve so that its downstream pressure would be about 5 psig minimum. Note that the pressure downstream of the steam trap in this application was atmospheric pressure. This sort of override control did stabilize the steam flow and steam trap operation. On occasion, the vaporized oxygen temperature drifts above its set point, but this does not affect the downstream operation where the O_2 is being used to regenerate catalyst.

Incidentally, the use of steam to vaporize liquid oxygen is, in itself, a process design error. My client should have used warm water from a cooling application, as the liquid oxygen at the working pressure of 150 psig was completely vaporized at less than 80°F.

The general lesson of the preceding example is that the use of a steam trap on a reboiler is almost certainly going to cause distillation tower pressure instability malfunctions, unless the pressure on the steam side is substantially above the condensate drain pressure **downstream** of the steam inlet control valve. By "substantially above," I mean 10 or 20 psi, depending on the variability of the condensate collection header pressure, and the delta P through the steam side of the reboiler.

I've dealt with this problem in greater detail in Chapter 14, "Steam Condensate Collection Systems."

Heat Makes Pressure

In my book, *Troubleshooting Process Plant Control*, Wiley, 2008, I reviewed the good, the bad, and the ugly ways to control tower pressures:

- Good: By the heat input; flooded pressure control; venting from the reflux drum.
- Bad: Hot vapor bypass; holding backpressure with a control valve in the vapor line.
- Ugly: Introducing fuel gas or N_2 into the reflux drum; throttling on the cooling water to the overhead condenser.

You can consult my Process Control book if you wish to read about the bad and ugly methods. Here, I'll only detail malfunctions with the good methods. Venting from the reflux drum always works fine, if there is noncondensable gas to vent. The more difficult cases involve total condensation of the overhead product. For this common application, both flooded condenser control and heat input pressure control work fine and should be used in preference to the bad or ugly methods tabulated above. Both good methods work on the principle that adding heat increases pressure and subtracting heat reduces pressure. I call this the Heat Makes Pressure theory of control.

Leaking Vent Valve

Figure 3-1 shows a distillation tower served by flooded condenser pressure control. The pressure control valve on the overhead product flow is partly closed to raise the level in the condenser shell. This covers some tubes in the shell. The surface area exposed to the condensing vapor is reduced. The rate of condensation falls, and the tower pressure rises. The level in the condenser shell is not known, nor is it directly controlled. There are no level taps on the shell.

I have had decades of experience with flooded condenser pressure control and have found it to work satisfactorily. This applies both to water-cooled shell-and-tube exchangers (with the condensing fluid on the shell side), as well as air-cooled fin-fan exchangers (with the condensing fluid on the tube side). There is a theory in our industry that flooded condenser control does not produce stable tower pressure control for air coolers. Thus the proposed need for hot vapor bypass pressure control. This supposed theory has not been proven by my field experience in light hydrocarbon distillation service.

FIGURE 3-1 A leaking vent valve is a serious malfunction for "Flooded Condenser Pressure Control."

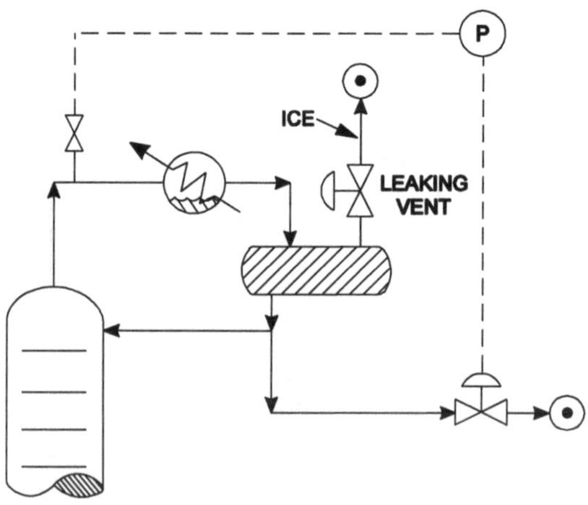

There is a potential for serious malfunction in using flooded condenser control, and it has nothing to do with a stable tower pressure. This malfunction is a leaking noncondensable vent on the top of the reflux drum (see Figure 3-1). I first encountered this problem on my alkylation unit debutanizer tower in Texas City in 1974. The overhead product from this tower was typically only butane. But, sometimes noncondensables would slip into the tower's feed. Or, on occasion, the overhead condenser cooling water became too warm to fully condense the debutanizer overhead product. Thus, a vent valve from the reflux drum was required.

Regardless of the sort of pressure control selected, a vent valve from the top of the reflux drum is needed. But at Texas City, this vent valve leaked. Because the reflux drum was liquid full, it leaked liquid butane. When liquid ethane, propane, or butane leaks past a vent valve connected to a low-pressure system (flare or fuel), the downstream line becomes coated with ice. That is, moisture is extracted from the atmosphere and freezes to the exterior of the pipe. A little bit of liquid hydrocarbon leakage will cause a large accumulation of ice downstream on the vent valve as shown in Figure 3-1. This extremely obvious loss of product is justifiably objectionable to management.

If the reflux drum is not liquid full, as it must be for flooded condenser control to function, the vent valve will leak vapor, which is not subject to the auto-refrigeration just described. So there are two methods to stop the ice formation downstream of the reflux drum vent valve:

- Install a leakproof vent valve that really seals tightly when shut.
- Convert to one of the Bad or Ugly sorts of pressure control that I have listed, all of which do maintain a liquid level in the reflux drum.

At Texas City, our flooded condenser distillation tower pressure control always worked fine, except when we had a level appear in the reflux drum. Then, my operators had to rush outside to line-up (i.e., to open the isolating gate valve—the valve to the flare), to dispose of noncondensable vapors that had accumulated in the reflux drum.

Excess Condensing Capacity

For any pressure control scheme to work, there must be an excess of condenser surface area. For example, you will have exhausted condenser capacity when:

- Flooded condenser control: the liquid level is not in the condenser shell but appears in the reflux drum.
- Hot vapor bypass: the hot vapor bypass valve is shut.

- Throttling on cooling water: the water supply valve is wide open.
- Gas blanketing: the gas make-up valve is closed.
- Backpressure control on vapor line: the backpressure valve is wide open.

For flooded condenser pressure control, once the level appears in the reflux drum, the distillation tower pressure is going to increase uncontrollably. This is not a malfunction of the pressure control instrumentation. It's just that you're out of condenser capacity. To restore the condenser capacity, the gas vent on top of the reflux drum must be opened. Venting from the reflux drum will then cause the reflux drum liquid level to rise. The trick is to close the vent when the liquid level reaches a maximum elevation in the drum, to avoid blowing liquid hydrocarbons out of the vent.

As I explained in my book, *Troubleshooting Process Plant Control*, this may all be automated using **split range control**. Once the noncondensable gas has been vented from the reflux drum, with the liquid level having risen back so as to almost refill the condenser shell, flooded condenser pressure control may be resumed for the distillation tower.

Oversized Control Valves

Placing a backpressure control valve in the overhead vapor line of a distillation tower is a poor method of pressure control. It introduces an unnecessary delta P between the tower and the reflux drum. Then, for a given tower pressure, the reflux drum will run at a lower pressure. This makes it more difficult to fully condense the overhead product, especially during hotter ambient conditions.

I recall working as a subcontractor for Glitsch, who had supplied the internals for a packed naphtha fractionator in Montreal. Their client had complained of poor separation efficiency in the tower between normal pentane and iso-hexane. The lab data supported their complaint. What I found, however, from a field inspection of the tower, was that the tower pressure was varying rapidly by several psi (the column itself operated at about 20 psig). I checked the backpressure control valve on the overhead vapor line. It was barely open. Very small movements in the valve position resulted in relatively large changes in the upstream tower's pressure. This was a consequence of the designer oversizing the control valve. I suppose the intent was to minimize the pressure loss in the overhead system between the reflux drum and the tower. The designer should have used **flooded condenser pressure control**, which has no valve in the overhead vapor line.

The correct way to resolve this malfunction was to change the control scheme. But I didn't have the time or patience to explain this concept to the plant people in Montreal. So I slowly closed the isolation gate valve upstream of the backpressure control valve. When this control valve was about 50% open, I stopped.

A control valve does not produce a linear response throughout its range of operation. Large movements when it is mostly open produce little change in flow or in backpressure. Small movements when the control valve is mostly closed produce large changes in flow or in backpressure. It is best to force the typical control valve to work somewhere between 25% and 75% of its range of adjustment.

With the control valve now operating around the half-open position, the tower stability was greatly improved and fractionation between the naphtha products was satisfactory. The pressure loss introduced by partly closing the gate valve was offset by the lower control valve delta P.

Effect of Light Ends on Pressure Control

Process effects are often mistaken for instrument malfunctions. For example, let's say the feed to a debutanizer has the following mole percent composition:

Hydrogen	0.1%
Methane	0.8%
Ethane	2.5%
Propane	5.8%
Butane	14.3%
Pentane	15.9%
Hexane plus	60.6%

The preceding analysis is typical for the composition of the overhead product from a refinery crude distillation unit, if the lab sample has been taken carefully to avoid the loss of lighter (H_2, C_1, C_2) components. Referring to Figure 3-2, let's say I now increase the reboiler duty, with the tower reflux rate fixed. Obviously, the tower top temperature will go up. Clearly, the amount of the overhead product will increase. But what will happen to the tower pressure?

Well, the tower pressure will drop. I recall my friend Steve, who worked for Texaco, interpreted the falling tower pressure, as the reboiler duty decreased, as some sort of instrument or control malfunction.

"Norm! Don't you always teach that heat makes pressure?" Steve asked. "The instrumentation's screwed up on my debutanizer."

"No, Steve," I replied. "Here's what happens:

- Raising the reboiler duty raises the tower top temperature, because the reflux rate is kept constant.
- More pentane-type material is distilled over into the reflux drum.

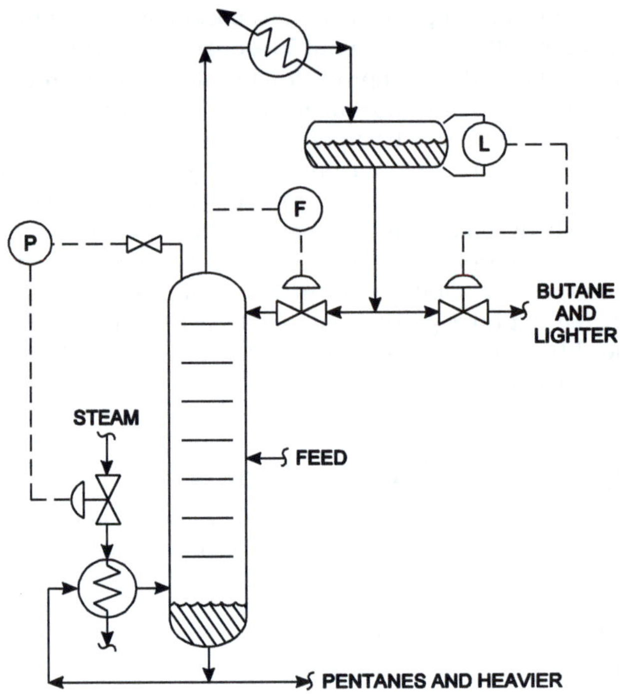

Figure 3-2 Increasing the reboiler duty will lower the tower pressure, if the reflux rate is kept fixed.

- The higher molecular weight components help absorb the hydrogen, methane, and ethane into the liquid phase. Or, the increased pentane content of the distillate reduces the bubble point pressure for a particular reflux drum temperature.
- As the reflux drum pressure falls, so does the tower pressure, which floats on this reflux drum pressure."

On the other hand, if the operator had increased the reflux rate, as she increased the reboiler duty, to hold a constant top temperature, the tower pressure would have increased. The higher reboiler duty would have increased the condenser duty, but the overhead distillate product would not have become heavier. The bigger condenser duty would have increased the condenser outlet temperature and thus the pressure in the reflux drum. Again, the tower pressure floats on the reflux drum pressure.

As I explained in *Troubleshooting Process Operations*, 4th edition, Steve's error was obtaining the feed sample for his design of this debutanizer in an open glass bottle, which permitted much of the methane and ethane to flash-off prior to the lab analysis. He should have used a sample bomb.

Effect of Lighter Components on Overhead Composition Control

I noticed on my depropanizer in Texas City that the butane content of my propane product was erratic, even though the tower top pressure and temperature were constant. However, when I studied the lab data closely, I noticed an interesting trend. When the ethane content of the propane product was high, so was the butane. As the ethane content of the depropanizer feed became erratic, so did the butane content of the propane. But why?

The vapor leaving the top of a tower is at its dew-point temperature and pressure. As the ethane content of a propane vapor stream flowing at its dew point goes up, the dew-point temperature will decrease. The ethane makes the propane more volatile. If the reflux rate is on tower-top temperature control, the lower dew-point temperature will reduce the reflux to restore the temperature set point. The reduced reflux flow permits more butane to be distilled overhead.

The resulting variable butane content of the overhead product reflects not so much a malfunction of the temperature control, but a natural variation in the overhead ethane content. What I eventually did on my depropanizer was to integrate the online gas chromatograph (GC) to reset the tower-top temperature set point for ethane and butane composition. Now, with advances in computer control, this technique has become common.

Use of Balance Lines

When using high-pressure steam to a reboiler, the correct method to control the steam flow for tower stability is by throttling on the steam inlet. When using lower-pressure steam, it's best to vary the condensate level in the condensate collection drum to control tower pressure or heat input. By "lower-pressure," I mean the supply steam pressure is 20 to 50 psi above the condensate collection header pressure. The use of a condensate collection drum requires the use of a balance line between the steam-side channel head and the condensate drum to achieve tower stability. However, the balance line must be connected below that channel head pass partition baffle for this to work. After all, the pressure we want in the condensate drum is not the supply steam pressure, but the pressure of the channel head where the steam condensate accumulates prior to drainage. Placing the balance line on the vent on top of the channel head connection (where it too often is connected) will promote tower pressure instability due to the resulting variable heat input, because of the variation of the condensate level in the channel head.

I use the concept of tower pressure and heat input control as if they were the same parameter. And in my mind, they are that way.

Another common use of a balance line is in the distillation tower overhead system. Typically, this is an open pipe, half the diameter of the tower's overhead vapor line, directly connecting the tower top and the reflux drum. Sometimes this balance line is somewhat smaller and connects the top of the condenser shell with the reflux drum.

Neither balance line, as far I can observe in the field, serves any useful function. I have never understood their intended purpose. When possible, I have blocked these lines in. Typically this will cause a small increase in condenser capacity or no noticeable effect at all.

Ambient Conditions and Air-Cooled Condenser

My experience with air-cooled exchangers during sudden rainstorms is quite limited. Most of the plants I have operated had water-cooled overhead condensers, which are not greatly affected by sudden ambient changes. My understanding of the problem with air coolers is that tower pressure instability during a sudden rain can be best avoided by the use of:

- Variable-speed fans, or at least two-speed fans.
- Louvers placed on top of the air-cooled bundles. The louvers have to be automated. When they close, airflow through the finned tubes will be restricted. My experience with louvers is that they are quite effective when in working order. However, I have also found them too often to be mechanically unreliable and difficult to maintain.

Sticking Control Valve

On some towers, pressure is controlled by venting excess, noncondensable vapors from the reflux drum. This is the simplest method to control tower pressure if noncondensable vapors are always produced from the reflux drum.

One tower had a long history of instability, and thus poor fractionation efficiency. At my recommendation, the plant manager contracted two experienced process control experts to resolve the problem. When I read their report, it dealt with, "Gains," and "Off-set," and "Re-sets," and "Proportional Bands," and "Multi-Variables," and a bunch of other stuff I didn't understand. So I approved their recommendations. After all, they were both experts. And when their recommendations were all implemented, it didn't make any difference.

I then did what I should have done in the first place. I went out to look at the vent valve on top of the reflux drum. I asked the panel operator to increase the reflux and reboiler duty. The drum pressure slowly rose above its set point. As the vent valve was an **air to open** valve (i.e., air pressure was supplied below the diaphragm), the air pressure to the valve diaphragm very slowly increased. It's much easier to feel a very small control valve movement than to see it. So I rested my finger on the valve stem position indicator to sense any small control valve response to the increasing instrument air pressure signal. But I felt none, and the reflux drum pressure continued to creep up past its set point. Suddenly, I felt the valve stem jump up a quarter of an inch. The drum pressure then rapidly dropped below its set point.

"It looks like the valve is sticking," I thought.

So I went back to my truck and got a can of WD-40. As the panel operator stroked the valve (i.e., opened and closed the valve on manual control), I sprayed it with the lubricant, and tower pressure control stability was restored.

Instability in Cyclic Operations

Refinery delayed coking is a semicontinuous operation. The coke drums work in pairs. While one drum is being filled, the other drum is being cooled, decoked, steam purged, and finally reheated. The empty drum is reheated from the hot vapors diverted from the drum that is filling. This diverts part of hot vapors away from the fractionator. With a sudden partial loss in hot vapor feed, the fractionator pressure and temperature slump. Product specifications are often not met during these periods.

Experienced plant operators learn to cope with this problem by not taking samples during coke drum warm-up. But I'd like to propose a more productive method.

When I watch how different console operators react to the cyclic disturbance of a delayed coker fractionator, I observe two classes of reactions. Some operators will make adjustments to try to stabilize the periodic upsets by reducing the heavy coker gas oil pumparound flow and throttling back on the wet gas compressor suction pressure control valve.

The better or more experienced console operators will begin making these same corrective moves as soon as the warm-up vapors are partly diverted into the cold and empty coke drum. The objective of the process control engineer is to automate the response of the experienced operators. The proper way for the control engineer to proceed is to observe, on several occasions, the parameter adjustments made by these operators as a basis for their automated computer control sequence.

I have seen this sort of approach used on delayed cokers at several Amoco refineries with good results for both the fractionator and the coke drum water quenching cycle. This is a somewhat more complex application of standard feed-forward process control design.

Instability Due to Flooding

Often, when I try to run a pressure drop survey on a column, I find the tower pressure is quite erratic at certain locations, making it impossible to make meaningful pressure measurements. Reducing the reboiler duty (i.e., vapor rate through the tray decks) restores the stability of the pressure readings. This is an indication of flooding in that liquid is filling up the space between the tray decks. When a sufficient liquid head has accumulated, the weight of the liquid dumps down through the tray. Thus the observed pressure instability. This is an indication of quite a serious process malfunction. Further increases in vapor flow may restore pressure stability, but they will make fractionation worse as the flooding becomes more entrenched.

Thermosyphon Circulation Instability

Most of our towers are served by thermosyphon or natural circulation reboilers. Should the bottoms liquid level rise above the reboiler return nozzle, the rate of the thermosyphon circulation will become erratically low. Heat input will also become erratically low as a consequence of the reduced rate of liquid flow through the reboiler. As the heat balance on a distillation tower ultimately sets the operating pressure, the tower pressure will also trend toward becoming erratically low due to the high bottoms liquid level. If the reboiler is served by a forced circulation pump, these observations will not apply.

A second source of thermosyphon reboiler instability is a consequence of oversizing the reboiler return line. I call this line the "emulsion outlet." If the reboiler outlet line operates above 15 feet per second (25 is even better), then there will be a mixed phase, or emulsion, flowing back to the tower. If the return line velocities are much lower, there will be phase separation. Slugs of vapor and liquid will flow back to the tower. The erratic vapor flow causes erratic heat flow and consequently tower pressure instability. It is probably best to size this line a little too small rather than a lot too big.

Feed Instability

I had a similar problem at the Texaco refinery in Convent, Louisiana. The feed to a tower had to flow, as a vapor–liquid mixture, from a feed preheater at grade to a substantially greater height to reach the tower feed tray. I placed a pressure gauge on the tower feed preheater effluent at grade. The pressure of the feed slowly rose by 3 or 4 psi. After a few minutes it would rapidly drop. The cycle continued hour after hour. I interpreted this as liquid slowly accumulating in the feed riser line. At some point the liquid would blow clear and the cycle would be repeated. The uneven flows of vapor and liquid into the tower promoted pressure instability. This is an example of phase separation in the rise line.

To correct this malfunction, I had the operators increase the duty of the feed preheat exchanger. At the resulting higher feed vaporization rate and line velocity, the observed pressure at the base of the feed riser stabilized, as did the distillation tower operating pressure.

In summary, erratic tower pressure control is always going to diminish fractionation efficiency, and thus it should be minimized as far as practical. The tighter the final product specifications, the more critical stable tower pressure becomes.

Author's Observations: Chemical Engineering Education—Reality versus Theory

It seems to me that chemical engineer graduates are poorly equipped to perform their technical function in refineries and petrochemical plants. The way chemical engineering is currently taught in universities is a waste of time and money. Too often, the wrong subjects are taught from the wrong books by the wrong people.

Partly, it's the use of computer modeling that is at fault. Engineering calculations should never be done with computer programs, unless the student has written the program. How can the novice engineer grasp the underlying principles if he or she is only performing the clerical function of computer-aided calculations?

The larger issue is that instructors concentrate on the theoretical and mathematical aspects of the process phenomenon without explaining the practical application or the conceptual basis of the principle involved. A student should first be instructed in the visual concept of the process application, followed by an explanation of what the technology under study is used for. For example, thermodynamics is used to study how work can be extracted from steam in turbines and steam ejectors. The second law of thermodynamics is used to explain how gas compressors work. Vapor–liquid equilibrium is needed to distill brandy from wine.

Only after the concepts and applications are explained should calculation procedures be presented. The calculation procedures should be simplified by making engineering assumptions, which also must be stated by the instructor. There is no place in undergraduate chemical engineering for differential equations. We do not use higher math in industry.

Instructors themselves should have practical process plant experience. A significant portion of their courses ought to be based on first-hand plant experience. Textbooks should be written by professors who have actually worked with the technology that they are describing. Too many university texts are simply reworks of older university texts.

I've had over 10,000 chemical engineers pass through my seminars. In 46 years, I've worked with 5,000 more on field process problems or revamp designs. My experience is that chemical engineering is taught incorrectly in universities.

CHAPTER 4

Salt and Rust Formation on Tray Decks

A politically acceptable theory presented to management as the cause of a process malfunction must have these characteristics:

1. *It's no one's fault.*
2. *Nothing can be done to correct the problem.*
3. *There's no way to test the theory.*

My first project as a process design consultant was with the NutraSweet Company in Augusta, Georgia. Aspartame or NutraSweet is now mostly marketed as Equal (in little blue packages). It's really just 3% NutraSweet. The other 97% is inert material. In 1983 when I first became involved with the plant in Augusta, pure NutraSweet sold for $90,000 per ton.

The NutraSweet product, at a rate of 5 tons a day, was extracted from a reactor by circulating methanol. By-products from the NutraSweet production reaction were water and acetic acid.

The water was removed in a trayed dehydration tower, where water was rejected from the bottom of the column. The acetic acid was neutralized by the addition of caustic (10% NaOH) in the tower's feed. The methanol overhead product spec was not less than 99.6% methanol. The tower was flooding at feed rates needed to sustain 5 tons a day of NutraSweet production. Flooding in the sense that increasing the tower's feed rate increased the water content of the overhead methanol. Increasing reflux just made the water content problem worse.

When the tower was commissioned in the previous year, it supported the production of 7 tons a day of NutraSweet, but now it supported only 5. The J.D. Searle Corporation, owner of the plant, had hired me to expand or replace the tower to relieve this bottleneck.

I have presented this story in greater detail in my book, *Process Design for Reliable Operations*, 3rd ed. If a tower's efficiency and capacity have declined gradually over a year, the malfunction is fouling. If caustic is injected into the feed on a continuous basis, the obvious culprit is caustic salting on the ¼-inch sieve tray decks.

The objective of the caustic injection was to neutralize the acetic acid in the tower's feed. The instructions given to the operators were to check the pH of the aqueous bottom's product once a shift and adjust NaOH injection into tower feed to hold a 6½ to 7½ pH. However, the operators were not using litmus paper, but multicolored pH indicator strips that they did not understand how to interpret. So when I checked the water bottom's product, its pH was 14, not 7. The operators were using 10 times the amount of NaOH in the feed than they should have.

Thinking that the trays were salted out with NaOH, I proceeded as follows:

Step 1—Measure the tower delta P.

Step 2—Raise the reboiler duty.

Step 3—Lower the top reflux rate.

Step 4—Monitor the falling tower delta P.

Step 5—Continue until the tower delta P lined out.

The purpose of steps 2 and 3 was to increase the water content in the normally dry, upper trays, where the NaOH deposits would accumulate. After 75 minutes, the rectification section delta P had dropped from 1.5 to 0.6 psi (across the top 11 sieve trays). Normal operations were then resumed at a 40% higher feed rate. NutraSweet production was then restored to 7 tons a day.

This incident, at the start of my career as a consultant, made a big impression on me for two reasons:

1. My clients, I observed, often missed obvious symptoms of a problem.
2. Money could be made with a little effort, if you thought about a problem logically.

I realize this story makes it seem that the malfunction, salting-out with caustic on the tray deck panels, was obvious from the start. Many malfunctions sound that way, after they are recognized and corrected. Incidentally, I estimated that the operators' confusion over the pH paper cost the J.D. Searle Co. $30,000,000 in 1982 and 1983. Plus my $5,673 consulting fee.

Naphtha Reformer Stabilizer

Aromatics (xylenes) are used to make polystyrene. The aromatics are mainly produced in refinery naphtha reformer units. The catalyst employed is impregnated with chlorine. The chlorine converts the straight-chain paraffins to propane and butane. Unfortunately, if there is nitrogen in the reformer feed, the nitrogen is converted to NH_3, which is a polar compound. Ammonia will strip off the chlorides from the catalyst to produce NH_4Cl. This salt will **sublime out** (vapor changing to a solid) on the upper tray decks. Fouling or plugging of the trays will cause flooding.

There are two methods used to remove sublimed salts from the trays. One method, which I prefer, is to increase the tower top temperature by 30°F to 50°F for several hours. This method takes a while, maybe a shift. The overhead product will be quite heavy during this period. Monitor the NH_4Cl content of the reflux drum water draw-off boot to decide when to stop.

The second, more common method is onstream water washing. For the naphtha reformer stabilizer, the reactor effluent is dry. Therefore it is necessary to establish a water level in the reflux drum and wash off the salts. This will only move the NH_4Cl deposits from the upper trays down to the middle trays. The tower bottoms temperature must be reduced so that a free water phase can leave the bottom of the column. The tower bottoms temperature has to be less than the boiling point of water at the tower's operating pressure. For a 150 psig, the tower's bottom will have to be less than 350°F. At one refinery, the stabilizer functioned as a depropanizer. Washing these trays just transferred the NH_4Cl salts to the downstream debutanizer and plugged the trays in that column.

At a Citgo refinery, the tower bottoms were reboiled by a fired heater. The circulation pumps had a tendency to cavitate when free water entered their suction during onstream water washing. This cavitation blew out the pump's mechanical seal, which resulted in dangerous hot hydrocarbon leaks.

I convinced them to stop the potentially dangerous water washing. But then they complained about the pentanes and the NH_4Cl contamination of their overhead stabilizer product. So I suggested that they operate their upstream feed hydrotreater at a high enough hydrogen partial pressure to remove nitrogen from the naphtha reformer feed, which was the root cause of their NH_4Cl salting problem in the stabilizer.

The Tenneco Refinery near my home had a continuous water wash on their stabilizer. This prevented tray salting with NH_4Cl. However, this practice also caused aqueous phase corrosion in their stabilizer tower.

Continuous Water Wash

On some columns, the concept of continuous water washing of the trays to remove salts is practical. My experience on this subject is limited to **fluid catalytic unit (FCU)** fractionators. The culprit is ammonia salts. It's a serious problem when running high-nitrogen

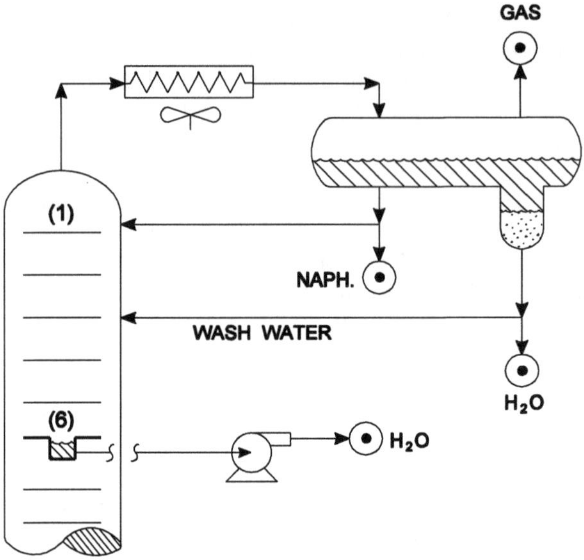

Figure 4-1 Continuous water wash circulation to remove salts.

FCU feed. Salting-out and flooding occur toward the top of the fractionator. Water is circulated from the reflux drum, back to the fractionator, as shown in Figure 4-1. The trays above the wash water inlet are protected from salt deposits because the salt has been removed in the lower trays. All this works fine, except for two possible malfunctions:

- If too much flow is drawn off the water draw tray, then hydrocarbons will be drawn off in excessive quantities. Some hydrocarbon entrainment in the water is unavoidable. Ten percent is a typical target.
- The bigger problem is allowing water to overflow the water draw-off tray. Then salts are flushed down the tower, where they are certain to foul the lower tray decks. As tray deck flooding progresses up a tower, this will make a bad situation much worse.

I've designed but never operated the system shown in Figure 4-1. At the Hess Refinery in St. Croix (U.S. Virgin Islands), they have found it very difficult to run the water draw-off on automatic interface level control (i.e., tray six draw-off sump).

KOH Carry-Over

In 1972, Amoco purchased a used propylene–propane splitter from Union Carbide in Whiting, Indiana (see Figure 4-2). The tower was equipped with Linde high-capacity trays. These are sieve trays with

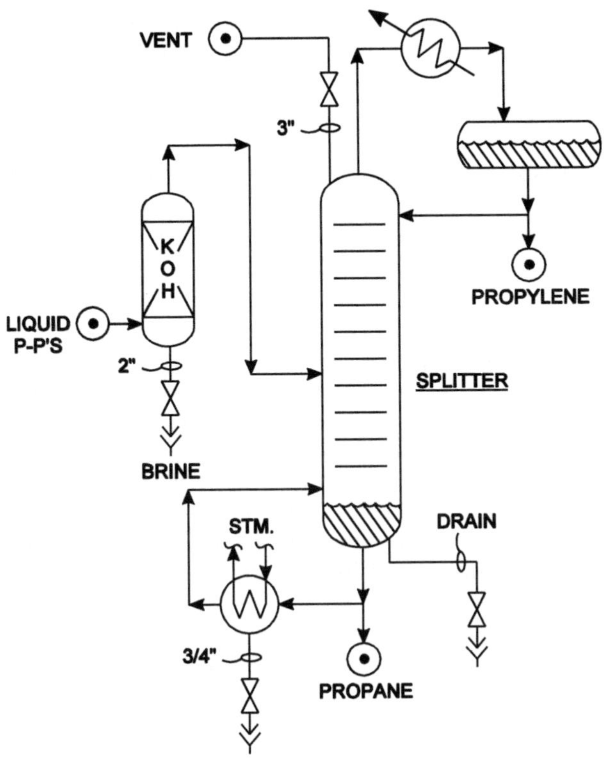

Figure 4-2 Carryover of KOH brine salts up lower trays in splitter.

¼-inch holes and many small rectangular downcomers arranged across the tray decks.

Union Carbide had operated the tower for 30 years with good fractionation. Yet, when we at Amoco tried to operate the tower, it flooded at 60% of the feed and reflux rate at which Carbide had operated. We knew the cause of the problem immediately. The tale was told in the snow.

Upstream of the 60-tray splitter, there were dual KOH dehydrator vessels. The purpose of the KOH was to extract all traces of moisture from the splitter's feed. The splitter overhead product was polymer-grade propylene that was shipped to the Amoco chemicals plant in Houston. The splitter bottoms product was LPG propane, sold for residential purposes. Both products had to be essentially free of water before they were shipped out of the Whiting refinery in Illinois Central rail cars.

The KOH dryers were large, vertical vessels full of walnut-sized pellets of potassium hydroxide, which is a very powerful desiccant. The wet, light liquid hydrocarbon flows upwards through the dryer, where both entrained and dissolved moisture are extracted. An aqueous KOH brine is formed, which accumulates in the bottom of the dryer. Periodically this brine phase is supposed to be manually drained to the sewer by an operator. If this doesn't happen, then the brine will back up into the KOH bed and be re-entrained by the upflowing liquid lighter phase.

"Gil," I asked, "I wonder if they have been draining down the KOH brine?"

"If they have, they have been real neat about it," observed Gil. "Didn't leave any footprints in the snow."

The outside plant operator walked past and eyed us in an unfriendly manner.

"You got a permit to be here?"

"We're not on the unit. We're on the road. We don't need a permit," Gil said. "You been draining down that dryer?"

"Say, who the hell are you guys? You need a permit to be on this unit! You got any questions, take them up with the chief operator. We drain down the brine regular. You got safety shoes on?"

"I'm the tech manager of this plant, Gil Gerlach. I don't need any permit to be on this unit. When was the last time this dryer was drained?"

"We drain it down regular," the operator mumbled, as all three of us looked at the untrodden snow around the dryer.

"Norm, go crack open the ¾-inch drain on the shell side of the reboiler. See if propane or brine drains out," Gil instructed (see Figure 4-2).

"You need a permit to open a valve on my unit."

"We're getting a nonroutine sample for technical evaluation. Read your union contract. If you people spent more time working and less time complaining, I wouldn't be out here freezing. Engineers can get any nonroutine sample. It's in the OCAW-Management Agreement of 1964. Go ahead, Lieberman. Get a sample off that reboiler shell."

The sample was water, not LPG. It was slippery and removed the dead skin from my fingers. This was caustic—KOH brine. A drip from the valve turned pH paper purple-blue.

"Yeah, Mr. Gerlach," said the operator. "I was just gonna check that dryer drain. Just making my first round. I'll get it drained right away. I been off last week. Back spasms and such. Still hurtin'. Damn cold weather."

The operator looked sadly at the purple paper lying on the dirty gray snow, as he opened the dryer's 2-inch drain valve to the sewer.

"Man. Look at all that brine! Lucky we came by," observed the operator. "Could of carried all that stuff into the splitter. Would've salted up them trays for damn sure. I bet all that KOH would of dried out on them trays. Could of flooded the P–P splitter. I seen that happen when I worked for Union Carbide in the 1950s. But they done laid me off. Didn't have a plant union over there," the operator pointed across slushy Indianapolis Blvd. as the brine gushed out of the dryer.

"So you all had a flooding problem with the splitter when you worked for Union Carbide?" I asked. "What'd you do about it?"

"Oh! We just steamed it out. Happened maybe once a year."

"Steam?" asked Gil.

"Yes, sir. The steam would wash off that KOH. Had to cut out all the feed. Opened the top 3-inch vent on the splitter. Actually, most of the steam would condense in the splitter. Drain out the bottom (as shown in Figure 4-2). Would steam it out until the steam condensate didn't turn the litmus paper blue any more. You guys reminded me about that with that pH paper you used. Same color I remember from my Carbide days. My wife worked for Carbide, too. She was a secretary in the front office. We got to talking one day and then I . . ."

"How long did you steam for?" Gil interrupted.

"Not too long. About a shift. That wasn't the problem, Mr. Gerlach. Problem was getting the propylene back on spec. The LPG bottoms would get dry right off. But that propylene overhead would be wet forever. Would take us days to get it dried up to spec before we could line back up to on-spec propylene product. But then I got laid off from Carbide. They said I had excessive absences. Back spasms and such. But my wife, she was . . ."

Dry Amine Salt Sublimation on Trays

When crude oil is distilled, to some extent, HCl will be evolved:

$$MgCl_2 + H_2O = Mg(OH)_2 + 2HCl$$

The $MgCl_2$ comes from residual seawater salts in crude. The reaction proceeds at 650°F. But it can occur at 500°F if catalyzed by metals (nickel and vanadium) in the crude. In the crude preflash tower below 400°F, I do not find any signs of HCl. So I know this hydrolysis reaction occurs well above 400°F. The other seawater salts (NaCl and $CaCl_2$) are more thermally stable. The HCl is distilled overhead in the crude tower. We try to keep the tower overhead temperature above the water dew-point (240°F to 300°F) to prevent water condensation and absorption of HCl inside the tower. But if the crude feed or tower reflux contains amines, then temperatures sufficient to prevent water condensation inside the crude tower will not be hot enough to prevent dry amine salt precipitation on the trays.

I have experienced three sources of amine in the crude towers of refineries:

- At the Texaco plant in Convent, Louisiana, crude oil received from Aramco in Saudi Arabia would contain MEA (mono-ethanol-amine) added for corrosion control in the tankers.

- The wash water used at my crude unit at the Good Hope refinery desalter was sometimes contaminated with amines. The amine was absorbed from the desalter wash water (which was sour water stripper bottoms) into the crude.
- At a Coastal refinery, neutralizing amine (actually morpholine) was injected directly into the top of the crude tower. This proved to be a terrible practice.

The effect of the direct or indirect introduction of amines into the crude tower was the same. The amines reacted with the chlorides to form dry amine chloride salts. These sublimed out on the tray decks above the kerosene (i.e., jet fuel) draw-off and also several trays below the top of the tower. The results of these salt deposits promoted two types of malfunctions:

- The trays flooded due to the salts plugging the tray deck valve caps.
- The salts extracted moisture from the upflowing vapors at temperatures well above the calculated water dew-point temperature. This caused a localized aqueous corrosive phase to form on the trays. H_2S, NH_3, and HCl were extracted from the vapors, with the resultant product of corrosion of $Fe(HS)_2$ (iron sulfides) contributing to further fouling of the tray decks.

The dry amine chloride salts are water soluble and can be washed off, without shutting down the crude tower, by refluxing water down the tower and extracting the water in the jet fuel draw-off. Not so the $Fe(HS)_2$. Iron sulfide is quite insoluble in water. The majority of flooding problems we have in crude towers that are not corrected by onstream water washing are caused by iron sulfide scale fouling the tray decks.

At the Exxon refinery in Benicia, California, they knew all about dry amine chloride salts, and they tried to keep amines out of their crude tower. But they failed. They were using neutralizing amines in the crude tower overhead condensers. The amines would be extracted from the top naphtha reflux with the overhead wash water. But the overhead reflux drum water–hydrocarbon separation time was only 2 minutes, which is too short for proper separation of the two liquid phases. So they refluxed entrained water back down the tower, which had the same effect as adding amine to the tower. The upper trays flooded due to sublimation of dry amine chloride salts. Exxon progressively **hot tapped** new reflux nozzles down the tower to bypass their plugged trays. By the time I became involved with the problem, they were down to tray #6 with their third reflux nozzle.

Draw-Off Nozzle Plugging with NH_4Cl Salts

On some cracking units with high N_2 and Cl content feeds, the partial pressure of NH_4Cl can get quite high. Then, if the tower is kept quite dry, a white deposit of NH_4Cl will accumulate in the tower. The operating temperature range for this to occur is between 340°F and 380°F. The salt can become so thick that it can partly plug a product draw-off nozzle and restrict the side-product draw-off rate. Water washing done onstream corrects the problem and will restore the side draw-off product rate.

I have seen a similar problem on a gas oil hydrodesulferizer reactor effluent fractionator. The side fractionator draw-off was light diesel oil. I corrected this malfunction by increasing the reactor effluent water wash makeup rate. That reduced the NH_3 content of the fractionator feed. Before I increased the wash water makeup rate, the purge water from the reactor effluent cold separator had an overwhelming NH_3 odor. A reasonable target for this wash water makeup rate, so as to retard fractionator salting up, is to hold the salts in the water purge to around 40,000 ppm (4 wt%). At Amoco, our target was 3 wt% . When I taught a seminar at Tesoro last year, they claimed to run above 100,000 ppm (10%), which for me is a bit hard to believe.

Carbon Steel Tray Decks

Regardless of the service, I always warn my clients never to use carbon steel components inside distillation towers. For example, at the Coastal refinery in Corpus Christi, they had a crude fractionator with bad fractionation between diesel oil and jet fuel. The jet fuel suffered from a high end point. The tower was opened for inspection and I observed two tray malfunctions:

- Most of the trays between the jet fuel and diesel oil draw-off had several dislodged tray deck panels, which permitted vapor flow to partly bypass the tray active area.
- The trays were equipped with movable valve cap assemblies (flutter caps). While the caps themselves were 316 stainless steel, the tray deck panels were carbon steel. The carbon steel had corroded. When steel corrodes, its thickness increases by a factor of about 10. Thus, the space between the caps and the tray decks was greatly reduced. The normal clearance or lift between the open cap and the tray deck had been reduced from the design of ¼ inch to an average of ⅛ inch.

The reduced clearance between the cap and the deck would increase delta P of the vapor flowing through the trays by a large amount. This would cause the liquid to back up in the downcomers and promote flooding. I therefore asked Mr. Coker, the plant manager, to fix the loose tray panels and clean all of the corrosion products from the carbon steel tray decks. But Mr. Coker refused.

"Lieberman! We don't have time to clean all those two-pass trays. It'll extend the turnaround by a full day. We've already been down for five days. That's 300,000 barrels of crude that we haven't run."

So we only resecured the tray deck panels between the jet and diesel draw-offs and started back up. And the tower flooded. The vapor delta P through the restored trays was too high because of the reduced lift of the caps above the fouled tray decks. Then we shut back down and cleaned the trays. It required five more days to bring the tower back up. The lesson is, don't use any carbon steel in distillation tower tray components.

Movable Valve Caps

Minor amounts of salt accumulation on tray decks may cause serious flooding problems if movable or flutter-type valve caps are used in distillation tower service. My wife Liz and I were once inspecting an 8-foot-diameter, 40-tray debutanizer in a refinery in Aruba. The tower had been in service less than a year and had begun to flood at progressively reduced feed rates.

I had promised Liz that we would vacation at a luxury beach hotel at the other end of the island as soon as we completed the inspection of the tower.

"Well," Liz asked, from the other side of the two-pass tray downcomer, "when is this paradise vacation at the Royal Sonesta starting?"

"Real soon, sweetheart. Let's just finish prying the stuck valve caps off of the top of the tray decks," I answered. "There's only another few thousand more caps to go. Look, we're almost down to the feed distributor. How are you coming on your side of the tower?"

A minor amount of salts and corrosion products had caused most of the caps to stick to the tops of the tray deck panels. This increased tray delta P and downcomer backup. Flutter and movable valve caps are not widely used in the process industry any longer. They were purported to (but never did) improve tray turndown by retarding tray deck leakage. I have come to view their development as a fraud. The best way to prevent trays from leaking prematurely is to insure tray deck levelness. Now, I will only use fixed or grid-type valve caps. For severe salting services, I use a tray deck with a large cap lift. Again, lift is defined as the vertical space between the underside of the cap and the tray deck. A small lift is ¼ inch. For most trays, a maximum lift is ½ inch. For the Hess FCU fractionator that I revamped in 2009, we used a Koch-Glitsch Pro-Valve in the area most prone to NH_4Cl sublimation. This sort of cap has an extremely big lift (roughly 0.6 inch) compared to ordinary grid trays. I believe the Pro-Valve design to be mechanically weaker than grid-type trays. But the bigger the cap lift, the more fouling the tray can tolerate before being subject to downcomer backup and flooding. (Note: The Hess fractionator is now onstream and working great.)

Water in Reflux Absorbs Salts from Vapors

In many services, we calculate that volatile salts should be distilled overhead from a tower. Yet even when the operating tower top temperature is well above design, salt formation of the tray decks may still be encountered. Figure 4-3 illustrates the malfunction. One possibility is a faulty interface level control in the boot. Especially if the naphtha product is transparent, it's very difficult to locate the water–hydrocarbon interface in the gauge glass, which makes calibration of the boot's automatic level control difficult. Also, corrosion deposits settle out in the boot and often plug the level taps. This is easy to troubleshoot and correct by blowing out the level taps and watching the interface movement in the gauge glass.

A more intractable malfunction is the accumulation of corrosion deposits from the overhead condenser in the reflux drum. This was a terrible problem in the main crude fractionator at Aruba, which suffered from severe overhead condenser corrosion rates. The deposits had accumulated to the top of the 12-inch riser pipe shown in Figure 4-3. With the top of the riser pipe essentially flush with the bottom level of the reflux drum, water easily entered the reflux draw-off nozzle. The water being refluxed down the tower absorbed the

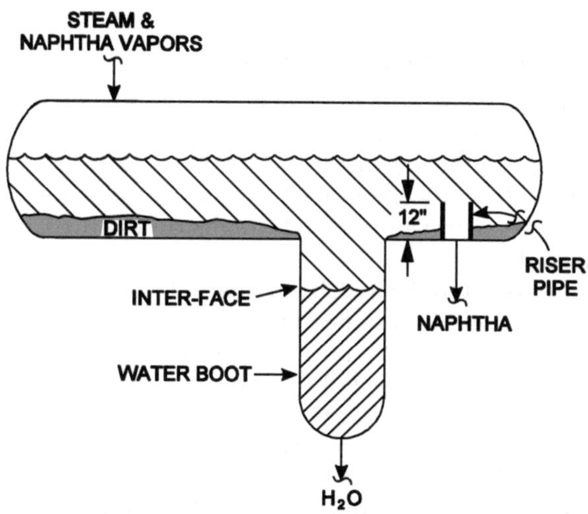

Figure 4-3 Dirt buildup above the riser causes water to be refluxed back to the tower.

volatile salts from the vapor phase. When the water was eventually revaporized lower down in the hotter part of the tower, the salts were left behind.

Why, one may ask, didn't the salts just revaporize? Well, they did. But then the salts were reabsorbed by the water in the reflux. Eventually, equilibrium was established. That is, the partial pressure of the salts increased, so the volatile salts begin to sublime or precipitate out on the tray deck panels, at tower elevations that calculations had predicted would be too hot for such salt formation.

I recall that the height of the corrosion deposits in the crude tower reflux drum in Aruba was measured by an external radiation density measurement. Refinery management decided not to clean, but to replace the reflux drum, which also had some severe corrosion-related issues. When I issued the design for the new reflux drum, I extended the height of the riser to accommodate any future corrosion products.

Unfortunately, a most common cause of refluxing water down a tower is not that the riser is too short, but that often it does not exist. Sometimes I have seen vessels with no riser and with the naphtha draw-off nozzle located on the upstream side of the water boot, that is, the opposite of the nozzle configuration shown in Figure 4-3. This is certain to leave water in the reflux.

Checking for water in the reflux is easy. Leave the suction valve open to the spare reflux pump. Wait a few hours and check the spare pump case drain for water. Obtaining a sample of reflux in a quart bottle and allowing it to settle out is a good alternative method. If you get the sample of the pump discharge and it looks milky, that's emulsified water, which may take hours to settle out.

Combustible Salts in Packed Towers

Some sulfur, nickel, or vanadium salts are combustible and will autoignite upon exposure to air when they are dry. At a chemical plant across the river from my home in New Orleans, a large tower full of stainless steel structured-type packing was opened for inspection. Apparently the combustible metallic salts autoignited and all the packing melted.

At the Amoco Oil Texas City refinery, I was running a sulfur recovery plant. The IFP-designed tail gas treater was taken offline to remove a year's worth of accumulated sulfur salts.

It was during the long 1980 strike, and I was working as the outside operator on the unit. I was told to vent and drain this giant packed tower. But the drain line was plugged with sulfur salts.

So I removed the bottom tower manway, and nothing came out. But air was drawn into the manway. I looked up at the tower's top vent, 150 feet above my head, and saw white smoke blowing from the 4-inch vent. Meanwhile the draft at the manway was getting stronger.

Apparently, the sulfur salts in the packing had autoignited. The white plume was mainly SO_2 and SO_3.

So I walked back into the control room for my lunch.

"Hey Norm," Joe Petracelli, the panel operator called, "Your tower's getting real hot all of a sudden."

"I'm eating lunch, Joe."

"Yeah Norm, but those TIs are all taking off. You need to do something pronto."

"I'm eating lunch, Joe."

"No, Norm! Look at that sulfur plume coming out of that tower's top vent. Just look at those damn TIs. The mid-TI is over 200°F. Do something!"

I hated that tower and I hated that process (IFP sulfur recovery unit tail gas treater). However, with great reluctance, I walked slowly outside and slid a piece of plywood across the bottom manway. The plywood was sucked up tight against the manway flanges by the draft that had developed in the tall tower. Slowly the tower TIs slipped back down.

And to this day, I've regretted not finishing my lunch.

CHAPTER 5

Fractionator Bottoms Product Stripping

The measure of success is functionality.

—Liz Lieberman after viewing my repair to our front gate

My first assignment in 1965 as an assistant tech service engineer was at No. 12 Pipe Still at the American Oil Company Refinery in Whiting, Indiana. This plant was built in the 1890s to process local crude production. At the time, it was the largest refinery in the world. American Oil was owned by Jacob Blaustein, a former Jewish peddler from Baltimore, who sold kerosene from a barrel in a pushcart. My grandfather was also a Jewish peddler with a barrel in a pushcart. He sold pickled herrings—unfortunately not kerosene.

In 1965, No. 12 Pipe Still was the largest crude unit in the world—180,000 BSD capacity.

Gil Gerlach, my boss, instructed me as follows:

"Norm, process engineering is a craft, not a profession. You've got to learn your trade by practice. So get out of your office and run a test."

"Gil, what should I test? How do I get started? Isn't everything on No. 12 P.S. already optimized? Don't the operators routinely adjust all operating parameters to maximize efficiency?"

"Norman, this is the real world. You're not at the university. The operators are there to handle unit upsets, startups, and shut-downs, and to coordinate with the maintenance division. We at the technical department are responsible for optimizing No. 12 Pipe Still. So, what I want you to do is to optimize the stripping steam rate to the bottom of the crude unit atmospheric tower bottom's section steam stripper."

"Gil, it's running currently at the design rate. 17,000 lb/hr. Isn't the design optimum?"

"Well, look," said Gil, "If we were running at the design crude rate, with the design crude type, at the design flash zone temperature and pressure, you could be right. But we are not. Also, you always have to suspect equipment malfunctions. Like maybe, Norm, the trays are damaged or installed wrong."

"Yeah, Gil, that could happen."

"So here's what I want you to do: Write up your plan and show it to the stillman on shift. I suggest you vary the steam rate in 5,000 lb/hr increments, from 10 to 25. Then, Norm, follow changes in these seven parameters:

1. The 660°F point temperature on the ASTM D-86 (lab test) of the tower bottoms product. Material boiling below 650°F is good diesel oil and doesn't belong in the tower bottoms.
2. Check the delta P across the four stripping trays. A typical delta P per tray is around 0.15 psi, or 6 inches of liquid. A delta P of about 2 inches or less indicates the trays are weeping or damaged. A delta P of about 10 inches or more indicates the trays are flooded or fouled.
3. The delta T (see Figure 5-1) between the flash zone and the tower bottoms should increase as the stripping rate increases. A delta T of about 25°F to 30°F is about right.

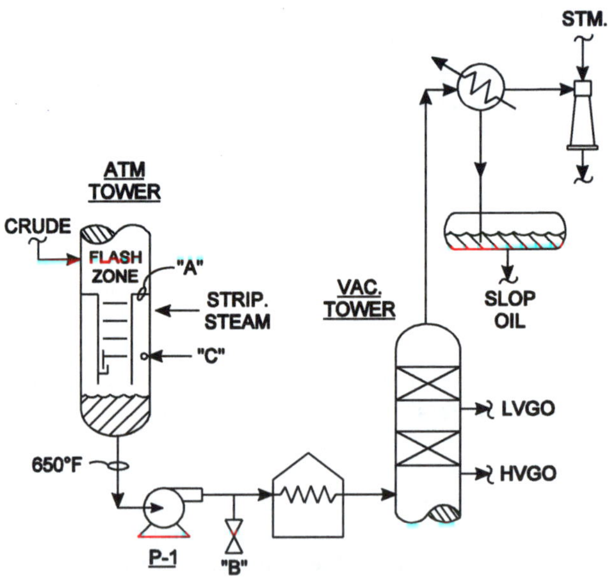

Figure 5-1 A hole in the internal head at point A ruined stripping efficiency.

4. Check the diesel content of the light vacuum gas oil (LVGO). Use the same technique as in Step 1. The diesel content of LVGO should decline with increasing stripping steam. A good diesel content of LVGO is about 15%. A bad diesel content of LVGO is about 50%.
5. The performance of the vacuum tower should get better. Meaning, the vacuum tower pressure expressed in mm Hg ought to get lower.
6. Production of slop oil from the vacuum system seal drum should drop off. Norm, that stuff's mostly jet fuel or kerosene. It's worth 30% more if we recover it as jet fuel in the crude tower than if it's rerun in the refinery slop oil system.
7. Production of diesel product should increase by a few percent, if the stripping steam is truly effective."

"Finally," Gil concluded, "Too much stripping steam will increase the flash zone pressure in the crude atmospheric tower, due to overhead condenser limitations. Too much pressure will reduce diesel vaporization in the flash zone."

"About how long should I wait between making moves on the steam flow?" I asked.

"Maybe one hour. Maybe longer if the panel board operator thinks he's not lined-out the unit. Also, get squared away with the lab before the test. You'll want them to run your D-86 **check samples** as they are received, and not on the lab's night shift."

Sampling Hot, Heavy Residual Hydrocarbons

"Gil, I'm going to need to get a sample of residual heavy hydrocarbons from the atmospheric tower bottoms (valve B in Figure 5-1). You said I need to check its diesel content in the lab. That stream is like 650°F. I know there's a sample cooler at valve B. But the sample cooler will plug unless it's cleared with the flushing oil. And the flushing oil is diesel. Won't this contaminate my sample?"

"Right, Norm," Gil answered. "Here's what to do. Forget about the sample cooler. Have the operators switch out the tower bottoms (P-1A) to the spare (P-1B). Then:

- Immediately shut off the external seal flush oil if there is any.
- Wait until the pump cools down. That is, until you can spit on the pump case without your spit sizzling.
- Then, get a sample off the pump case drain in a metal can. If the sample smokes, your sample was too hot, so wait a while longer."

"But Gil, why use spit instead of just water?"

"Cause, Lieberman, you always got spit with you," Mr. Gerlach concluded.

Bypassed Stripping Trays

So I provided the operators a box of donuts and ran the test. And nothing happened. None of the operating parameters improved. Actually, the diesel content of the vacuum tower feed went up a bit, not down, as I increased the stripping steam rate from 10,000 to 25,000 lb/hr. This was really depressing! Now what?

My calculations showed that the tray section delta P with four trays should be about 1 psi with 25,000 lb/hr of steam. Yet the measured delta P was zero. "Obviously," I thought, "The stripping section trays are blown out."

But Stan Kowalski, the old instrument tech, said, "You know, Lieberman, that delta P has always been zero. It's been zero ever since No. 12 Pipe Still started up in 1950. I also thought the tray decks were damaged. But I've been inside the tower six times, during every **turnaround**. And every time the trays are just in perfect condition."

"Gee, Stanley, could the stripping steam somehow be bypassing the trays?"

"Could be," Stan answered. "Like maybe through the hole I always see here (see point A in Figure 5-1)," Stan explained with a quick sketch. "I've seen that hole where the internal head supporting the stripping trays is ripped away from the vessel support ring."

"But Stan, don't they fix that hole during the turnaround?"

"Sure, Norm. But the hole always comes back. Maybe at a slightly different spot every turnaround. It's a big hole, too, big as my hand. Guess that stripping steam is blowing through that hole and bypassing the bottom's stripping trays."

Restoring Stripping Efficiency

In troubleshooting any process malfunction, there are always two questions:

1. What caused the malfunction?
2. What's the fix?

My analysis as to the cause of the hole at point A was:

- During startups, an erratic bottoms liquid level is unavoidable.
- If the bottoms level rose above the bottom edge of the stripping section (see Figure 5-1), the steam would be trapped in the area I've marked C on the figure.

- The steam would pressure-up area C.
- The internal head would then bulge up and create the hole observed by Stanley during his inspections.

To fix the problem, during the next turnaround, I had the steam inlet connected directly to the internal stripping tray cylinder below the seal pan. Thereafter, the stripping tray efficiency behaved in a normal and efficient manner. Diesel production increased by almost 20%. Incidentally, the use of internal piping as I employed in this project is poor design practice. The problem is differential rates of thermal expansion between the piping and the vessel, causing flanges to open. Proper expansion loops are required to avoid such leaks.

When I worked in the 1960s for American Oil, I was immersed in the knowledge and experience of people like Gil Gerlach and Stanley Kowalski. What the novice refinery process engineer is supposed to do now, I can't begin to imagine.

Faulty Stripping Tray Installation

"Dad, you must have gotten a really big promotion," Lisa said when I told my eldest daughter this story. "American Oil must have given you a giant raise."

"No, Lisa, it doesn't work that way. Actually, other than Stanley Kowalski, nobody really liked me. So in 1983, I decided to become a consultant. My first job was at the Exxon refinery in Baton Rouge."

"Dad, is this going to be another of your tedious technical tales? If so, write it down. I'm watching TV."

The problem at the Exxon plant was black gas oil (see Figure 5-2). The black gas oil could only result from black tar (asphaltines or resid) being entrained upward from the flash zone. I suspected the malfunction was flooding from the lower portion of the tower. First, I observed in the gauge glass that the bottoms level was well below the steam inlet nozzle.

Next, I measured in the field the delta P between P_1 and P_2. Note that P_1 was located 16 feet above the top tray. The delta P was 4 psi, which I then converted to feet:

$$(4 \text{ psi}) \times (2.31) \div (0.80) = 12 \text{ feet}$$

where 2.31 converts from psi to feet of water
0.80 is the SG of hot tar

As the measured delta P (12 feet) was much greater than the 6 feet of height between the tray decks (see Figure 5-2), I concluded that the flooding from the trays had backed up into the flash zone

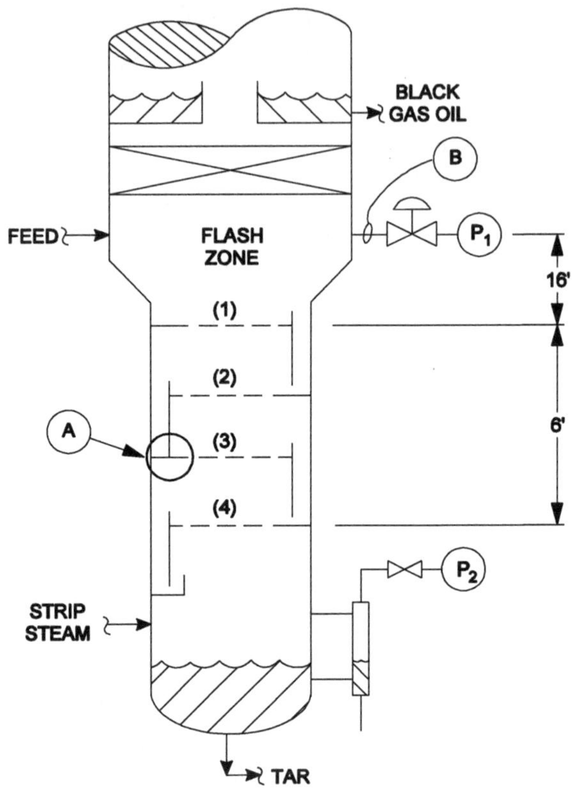

FIGURE 5-2 Loss of downcomer clearance causes black gas oil.

and thus caused the observed entrainment of black tar into the gas oil product.

Next, I shut off the stripping steam completely. While the pressure at P_1 didn't change, the pressure at P_2 at first jumped up by several psi. This really surprised me. But then, I reasoned as follows:

- The stripping steam was aerating the tar and reducing its density on the tray decks.
- When I shut off the steam, the aerated tar settled down and its density jumped up.

However, after a few minutes, the pressure at P_2 slipped down. Finally after half an hour, the pressures at P_1 and P_2 were identical. I now restarted the steam flow, but not at the previous 3,000 lb/hr. I increased the steam until the delta P between P_1 and P_2 was about 0.6 psi. (A delta P of about 0.15 psi per tray in this service is close to optimum.) But now I was only using 1,200 lb/hr of steam. This left quite a bit of diesel in the tar. However, the gas oil product was clean.

My Exxon representative, Rich Cotton, was both pleased and disappointed. "Mr. Lieberman, it's great to have clean gas oil. But the trays are designed for 4,000 lb/hr of steam. Why can't we use the design steam rate? The 1,200 lb/hr is leaving too much diesel in the tar."

"Mr. Cotton," I responded, "Something is wrong with the tray installation or the tray's mechanical condition. Maybe fouling? I do agree the trays should be able to handle 4,000 lb/hr of steam were it not for some unknown malfunction. Please call me the next time you open the tower for inspection."

"Very well, Lieberman. That I will do. But is your inspection included in your current consulting fee?"

"Sadly, Mr. Cotton, no. My socialist principles preclude that practice."

"What socialist principles?"

"You see, Mr. Cotton," I explained, "Exxon has a lot of money. And I have a lot less. To promote social justice, I'm trying to even out the income levels."

Six months later, I did have an opportunity to inspect the column internals. From Figure 5-2, you can see the malfunction. The downcomer from tray deck #2 was resting on tray deck #3 (Point A). How then could the tar be flowing down past tray #2? In one of two ways:

1. The liquid could back up into the flash zone and push the tar down through the sieve holes on tray #2, even though 3,000 lb/hr of steam was flowing through the holes.
2. The liquid could back up a few inches and push down through the sieve holes on tray #2, even though 1,200 lb/hr of steam was flowing through the holes.

To fix this problem, I had the bottom 2 inches of the downcomer from tray #2 cut off. When the tower started back up, 4,000 lb/hr of bottoms stripping steam was employed without any color degradation of the gas oil product. Not only did diesel oil recovery increase, but the pressure in the downstream vacuum tower decreased. This resulted in more heavy gas oil recovery in the vacuum tower.

One interesting aspect of this problem was that my client had been suffering from this malfunction for 8 years. They had three turnarounds before my inspection. The overly extended downcomer was due to an installation error 8 years earlier. Why was this malfunction, which was easily seen, not corrected before? There's an answer to this question, but I don't know it.

Failure of Vacuum Tower Stripping Trays

The most common problem with stripping residue from the bottom of a vacuum tower is that the trays are blownout as a consequence of two malfunctions:

- High bottoms liquid level
- Water in steam

It's easy to identify this problem:

- The vacuum resid contains a lot of gas oil. By a lot, I mean more than 15%, 1,050°F and lighter on a laboratory D-1160 **distillation test**. Anything less than 10% I consider good operation.
- The delta P across the stripping trays is small. By small, I mean less than 2 or 3 mm Hg per tray.
- Increasing the stripping steam rate does not influence either of the preceding parameters.

If the tower bottoms level gets too high, it will cover the steam inlet nozzle. Then liquid is blown by the steam through the tray deck perforations. This increases the upward force against the tray decks. This upward force will bend the trays where they are clipped to the tray ring. The trays are not bolted, but clipped to the **tray ring**, as there are no holes in the tray ring.

The other mechanism for tray failure is water in the steam. The density of hot water is 60 lb/ft^3. The density of steam, under vacuum tower conditions, is about 0.001 lb/ft^3. Thus, 1 cubic foot of water will expand to 60,000 cubic feet of steam under vacuum conditions. It's akin to setting off a bomb inside the tower.

There are two approaches to protecting the stripping tray integrity in view of these dual, and essentially unavoidable, malfunctions. The first method is to improve the mechanical integrity of the trays. I've discussed this subject in detail, and have even provided the mechanical specs in the appendix to my book, *Process Design for Reliable Operation*, 3rd edition. However, the tray vendors will now provide you with mechanically robust trays if you specify:

- Back-to-back tray panels
- Shear clips
- Bolted-in trays, with double nutting
- No friction fit washers on tray panels
- Explosion-relief doors
- 10 gauge tray thickness

- No carbon steel components (typically 316 s.s. or 410 okay)
- Fixed valve cap assembly (no sieve or movable caps)
- Trays to be bolted to cross I-beam supports or fixed with shear clips

These sorts of tray specifications will more than double the installed cost of trays.

The other part of preserving tray integrity is to trip off the stripping steam immediately and automatically when there is water in the steam or a high liquid level. The basis for this design feature relies on the **restrictive steam sparger** and the **ram's horn** level sensing. That's all I'm going to say on this critical subject. This is not a design text, and I charge for this particular bit of engineering know-how. The main idea, though, is to prevent slugs of water from dropping into the pool of 700°F oil in the bottom of the tower and to immediately cut off the steam flow before the level rises to the stripping steam nozzle inlet elevation.

Seal Pan Problems

For the tower bottom's flow to drain out of the lower tray, the seal pan must not restrict the liquid flow. I have found that dirt tends to accumulate in this seal pan and restrict the downcomer clearance. To preclude this malfunction, I design seal pans with a large drain hole in the floor of the seal pan. I size this hole for 20% to 25% of the total liquid flow.

A less common problem is lack of adequate downcomer clearance. There are two clearances to check:

- The clearance between the bottom edge of the downcomer and the floor of the seal pan. I like to make this 4 inches.
- The horizontal clearance between the downcomer and the overflow lip, or the seal pan weir, should also be about 4 inches.

Once I was working at a refinery in Lithuania. To make sure the stripping trays were assembled correctly, an inspector cross-checked the actual seal pan assembly against the tray drawings. He found that the gap between the downcomer and the seal pan overflow weir or lip was 3 inches. However, the tray design drawing showed it to be only 1½ inches. So the inspector had the clearance reduced to the 1½-inch specification. This greatly increased the delta P (by 400%) of the liquid flow through this clearance. The liquid backed up in the downcomer, and the stripping trays flooded, eventually turning the gas oil side draw product black. The operators reduced the stripping steam flow so that the bottoms product could drain through the tray deck perforations, thus bypassing the restrictive seal pan. Naturally, this also resulted in $100,000 per day of diesel oil product slipping out of the bottom of the fractionator.

Stripping Tray Efficiency

As discussed in Chapter 7, "Reboiled and Steam Side Strippers," there is a large difference in vapor rates between the top and bottom trays in a steam stripper. The usual industry practice is size all four or five trays in the bottoms stripper based on the vapor flow to the top tray. If the top tray is designed at 80% of jet flood, then the bottom tray will run at only 30% of jet flood. For **bubble-cap** trays, tray fractionation efficiency would not be particularly affected. However, for ordinary perforated tray decks (valve, sieve, grid, jet tab, pro-valves, super-frac, SVG, MVG, etc.), depending on tray deck out-of-levelness, a great deal of loss in tray fractionation efficiency may be expected, due to reduced vapor loads. And this reasoning applies, to a lesser extent, to the other intervening trays.

Let's assume, for purposes of comparison, two cases. Case 1, is that all four stripping trays have the same number and size of perforations. The average tray efficiency is 50%. That is, 75% for the top tray, and 25% for the bottom tray. Case 2 has all four trays with an optimum number of perforations. Average tray efficiency is 75%.

Using a HYSIM computer simulation model, the calculated effect of the extra tray efficiency is to recover 0.9 additional volume percent on feed from the bottoms vacuum resid into the heavy vacuum gas oil. If the vacuum tower feed is 100,000 BSD, we've recovered an extra 900 BSD from coker feed or **industrial fuel oil**. At $33 per barrel, delta product value, that's $10,000,000 a year.

Admittedly, this is all based on computer simulation results that I've extracted from the literature,[1] but still, directionally, optimizing tray hole area in stripping service is important.

Commissioning Steam to Strippers

One of the bigger problems my clients have is damaging trays in heavy hydrocarbon steam strippers. The problems that cause the tray damage most often occur on startup due to a combination of two factors:

- Water in the steam that has accumulated in the steam supply line, when no steam is flowing.
- Stripper bottoms liquid level rising above the bottom trays of the tower.

The water in the steam line expands with explosive force when it encounters the pool of hot oil in the bottom of the stripper. The higher the temperature and the lower the pressure in the stripper, the more violently the water will expand to steam. The sudden generation of vapor in the submerged trays will dislodge the tray decks from their tray support rings.

To prevent such damage, I will proceed as follows:

- Step 1—Place a pressure gauge on the top tap of the gauge glass on the bottom of the tower. Or even better, on the tower itself, at the same elevation.
- Step 2—Make sure this pressure is not more than a psi or so above the actual tower pressure. I'll call this pressure P-1.
- Step 3—Reduce P-1 to the required pressure, as per Step 2, by lowering the bottoms level in the tower. If the level shown on the console is lower, then you are likely "tapped out," as described in Chapter 16, "Level Control Problems."
- Step 4—Blow out the steam supply line to atmosphere. Hopefully, the steam will come out invisible (i.e., dry) after some reasonable period (20 minutes?). If not, proceed anyway. White steam is wet steam.
- Step 5—You can anticipate that slugs of water are lying in the steam supply line, even though the steam is blowing out dry. This water will only be entrained into the steam flow when the steam flow increases as you increase the stripping steam rate.
- Step 6—Locate the local steam isolation (i.e., gate) valve nearest the tower. Either you or a colleague must monitor the pressure at P-1 as this valve is opened.

  CAUTION: Open this valve yourself. Do not use the control valve, unless it can be operated locally. Do not be guided by the steam flow meter. It won't measure slugs of water.
- Step 7—Watching P-1, slowly open the steam supply valve. The slower the pressure at P-1 increases, the better. You cannot go too slow.
- Step 8—If P-1 suddenly jumps, immediately throttle back on the steam flow. This will happen. It's a slug of water blowing into the hot oil.
- Step 9—Have the console control operator put the steam flow on FRC. Slowly open your gate valve in the field 100% as the FRC valve assumes control of the flow. It is best to monitor P-1 as this happens.

Unfortunately, every time your stripping steam flow stops during normal operations, this procedure will have to be repeated. How one can motivate the operators to always follow this field-tested method is beyond the scope of this text and your author's experience. Having a restrictive steam sparger below the stripping trays will vastly improve chances for a successful stripping steam startup.

Screwed Connections on Vessel Walls—A Safety Note

When performing my pressure survey at the Exxon facility in Baton Rouge, Louisiana, I encountered a common but potentially lethal piping safety malfunction. In Figure 5-2, connection B, the ¾-inch nipple serving the pressure gauge P_1 was a **screwed connection**. Having a screwed nipple or fitting on a vessel in hydrocarbon service is contrary to Exxon's and all my clients' safety practices. The reasons for prohibiting such screwed connections at vessels are:

- When a piece of pipe is threaded, half the pipe thickness and strength are lost as the threads are cut.
- Should the pipe fail where it is the thinnest, at the threaded connections, there will be no way to stop the hydrocarbon leak. This happened to me on an isobutane-filled vessel in Texas City in 1975. I recall standing there in a fog of cold butane vapors with my shift foreman, Bobby Felts, screaming, "Norm! I told you to have those damned screwed connections back-welded. Now what are we going to do?"
- It's pretty easy to accidentally unscrew the threaded nipple in the vessel wall, especially if the pressure gauge won't turn out easily. Which is exactly what I almost did that morning in Baton Rouge, at the Exxon Crude Tower.

Reference

Unfortunately, most of the technical articles that now appear in the literature are advertisements for products or services. This reference is somewhat better than average for technical content.

1. Pilling, M. et al., Entrainment Issues in Vacuum Column Flash Zone, *Sulzer Chemtech. Petroleum Technology Quarterly*, Q1, 2010, pages 57–65. Published February 2010.

This article also contains an excellent description and drawing of a modern vacuum tower vapor horn distributor, of the sort that I also design.

CHAPTER 6

Heat Extraction with Pumparounds

What, me worry?

—Alfred E. Newman, *MAD* Magazine, 1954

We have two methods available to remove heat from distillation towers. These are:

- Top reflux
- Pumparounds

Many operators refer to pumparounds as circulating reflux. Liquid is drawn from the side of a tower, subcooled by heat exchange, and then returned to the tower, several trays above the draw tray. For a fixed reboiler duty or flash zone condition, the sum of the pumparound duty and reflux heat extraction is constant. Meaning, increasing a circulating reflux flow or a pumparound flow automatically causes the top reflux rate to drop to preserve the tower's heat balance.

The reason for pumparounds is threefold:

- Vapor flows above the pumparound return tray are reduced. This suppresses flooding above the pumparound return tray.
- The overhead condenser is unloaded. This permits the tower pressure to be reduced, or perhaps the feed rate to be increased.
- Heat recovery and energy economy are enhanced. Heat recovered in a pumparound is typically used to preheat unit feed. Heat rejected to the tower's overhead condenser is lost to the air cooler or cooling water.

Increased feed preheat means a lower reboiler duty or a reduced fuel consumption to the fired feed preheat furnace.

Pump-Downs versus Pumparounds

What is the difference between a pumparound (P/A) and a pump-down (P/D)? Both circuits are involved in heat extraction from a flowing vapor. In both operations hot liquid is drawn from a tray, cooled, and returned to the tower. Figure 6-1 illustrates the difference between these two design options.

In Figure 6-1A, the pump-down liquid is cooled and returned to the tower, on the tray immediately below the draw-off tray. In Figure 6-1B, the pumparound is cooled and returned to the tower, five trays above the draw-off tray.

In a pumparound, these five trays provide approximately one theoretical stage of fractionation. That is, the trays are there primarily to transfer heat between the subcooled pumparound return liquid and the rising saturated vapor. The liquid circulation rate can be selected at will. The greater the liquid circulation rate, the more heat that can be removed from the tower at a given pumparound draw-off and pumparound return temperature.

In a pump-down, all the trays, both above and below the draw-off tray, are primarily in fractionation service, rather than in heat transfer service. This is good. But, the amount of heat that can be extracted from the pump-down liquid is constrained by the liquid flow. Meaning, the maximum amount of liquid that may be extracted from the pump-down draw-off tray is limited to the amount of internal reflux appearing on the draw-off tray. Hence, if we want to increase the heat extracted from a pump-down flow, we may only reduce the liquid outlet temperature from the pump-down heat exchanger. If the primary objective is to maximize heat extraction from a side draw-off, then pumparounds are preferred. If the primary

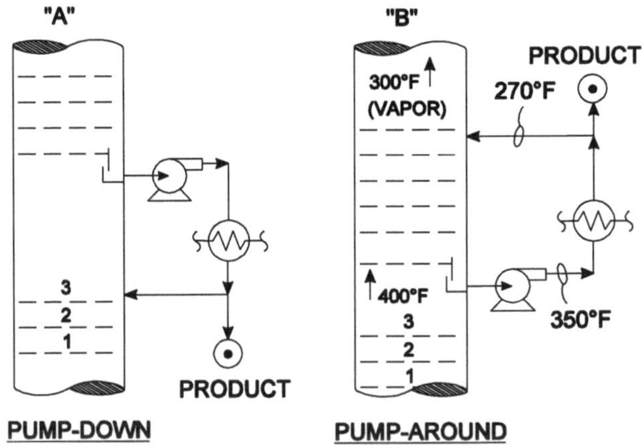

Figure 6-1 Pump-downs provide better fractionation. Pump-arounds provide better heat extraction.

objective is to maximize fractionation efficiency, then pump-downs are preferred.

I once had a project to revamp the No. 2 crude unit at the Muskiz refinery in Spain. My client told me to maximize fractionation efficiency, which I did by using pump-downs.

The Foster-Wheeler Corporation was given the contract to revamp the No. 1 crude unit in Muskiz, with the same objective. But Foster-Wheeler ignored this instruction and designed their revamp based on using pumparounds.

My Spanish client was quite distressed with the result of my revamp of their No. 2 crude unit.

"Señor Lieberman, why No. 2 crude preheat is 28°C lower than No. 1 crude?"

"Because," I explained, "Foster-Wheeler used pumparounds and I used pump-downs."

"Si, si, Señor Lieberman. Comprendo! Foster-Wheeler is very smart. They have excellente computer models."

"No! That's not it at all! You told me to maximize fractionation efficiency. Which I did. Foster-Wheeler ignored your objectives. They maximized heat recovery to crude preheat. You never told me your real objective was to save energy. I was just following your instructions. It was Foster-Wheeler that did not..!" This was beginning to sound like an argument I had with an ex-girlfriend when I failed to send her flowers, after she had told me not to send her flowers.

Effect of Tray Loading

Normally, the pumparound draw tray (B in Figure 6-1) experiences the maximum vapor and liquid loads. The liquid flow from the draw tray consists of:

- The pumparound flow
- The net liquid product
- The internal reflux

As no heat has been extracted from the vapor flowing to the draw tray, the vapor flow to the draw tray is also at its maximum.

The pump-down draw tray (A in Figure 6-1) does not experience the highest vapor rate in the tower.

The liquid flow from the pump-down draw tray consists of:

- The net liquid product.
- The internal reflux or pump-down flow.

However, heat has been extracted from the vapor flowing to the draw tray because of the subcooled pump-down flowing onto tray #3.

(Note that I'm using the term pump-down interchangeably with "internal reflux.") Thus, the vapor flow, temperature, and vapor density are all reduced between tray #3 and the pump-down draw tray. While the liquid flow from the draw tray may well be higher than the liquid flow from tray #3, the overall tray loading between both trays is very much the same.

This is definitely very far from the case with a pumparound. Here, the draw tray may easily have an overall loading 30% to 40% greater than the tray below (tray #3). Thus, the pumparound draw-off tray will likely limit the tower's capacity due to flooding.

So what's my point? Well, if you have a tower limited by flooding on the pumparound draw tray (which is typical), proceed as follows:

- Divert a portion of the pumparound exchanger effluent to return to the tower just below the draw tray, but above tray deck #3.
- The maximum amount of flow that can be diverted in this manner is limited to the internal reflux, which was overflowing the draw tray, onto tray #3, before the preceding modification.

Caution! This sounds like a neat trick to stop flooding on the draw tray and expand tower capacity. I've used this trick on many happy occasions. But if the draw-off tray is leaking badly, then the amount of liquid that can be pumped down will approach zero. It's like sending your girlfriend flowers after she has latched onto another guy. It doesn't help.

One of the reasons my retrofit design at the No. 2 crude unit at the Muskiz refinery in Spain was a disappointment was indeed that the pump-down draw trays all leaked rather badly. Therefore, the volume of liquid available for heat exchange with the crude oil feed was quite small. If I had used a pumparound like Foster-Wheeler did on the smaller No. 1 crude unit, the pumparound circulation rate could have been simply increased to exchange more heat to crude. I suppose, in retrospect, that if I had used a total trap-out chimney tray for my pump-downs, I could have had a happier ending to my Muskiz experience. I suppose if I had selected a nicer girlfriend, flowers would also have been more effective. But wisdom comes with experience in life.

It follows that if the draw-off tray leakage rate exceeds the sum of the internal reflux rate, plus the net product rate, then the draw tray will be emptied. I will discuss this unfortunate but common malfunction in the following section. I include in this discussion the other factors that cause loss of flow to the pumparound or pump-down pump, in addition to the draw-off tray leakage malfunction.

Circulation Pump Cavitation

A familiar problem that I have encountered on many occasions is cavitation of a pumparound circulation pump. There are a number of common causes:

1. Draw-off nozzle too small.
2. Required NPSH (net positive suction head) exceeds available NPSH.
3. Suction line partly plugged.
4. Draw-off tray and/or draw-off sump leaking.
5. Flooding of trays above the pumparound draw-off tray.

If there is any general principle to solving process malfunctions, this is it: Training, education, and experience aren't worth much. Field observations and experiments are golden. In this case I might first reduce the pumparound rate. If the cavitation gets worse, then the problem is item 5 above, "Flooding of trays above the pumparound draw-off tray." That is, the increased vapor flow leaving the pumparound draw tray is making the flooding worse.

Let's assume I have a level indication in the draw-off sump of the bottom pumparound tray. This is rather standard. Increasing the pumparound rate should raise this level due to increased vapor condensation. But if the draw tray is leaking, a higher circulation rate will likely reduce the level, due to higher leakage rates caused by more liquid traffic on the tray (i.e., item 4 above).

But let's say you observe that the level on the pumparound draw tray goes up as cavitation becomes worse at higher circulation rates. Then you have to check the pump suction pressure.

If the suction pressure stays constant, but the pump begins to cavitate as you increase flow, then item 2, "Required NPSH exceeds available NPSH," is the problem. However, if the suction pressure drops radically, and the level on the draw-off tray rises as the circulation rate is increased, then pehaps the draw-off nozzle is too small or partly plugged. Or, perhaps the suction line itself is undersized or badly fouled. Here's how to discriminate between these two common malfunctions:

Reduce the flow by about 5%. This will reduce the delta P or head loss in the suction piping by about 10%. If the pump's suction pressure increases by 10%, then the suction line itself is the problem.

But if the suction pressure increases by 60% to 80% as a percentage of the height of the suction line converted to psi (i.e., the elevation between the pump and the draw nozzle), then it's the draw-off nozzle itself that is restricting the flow. I'll explain.

The liquid is being pumped away from the suction line faster than it will drain into the suction line. That is, the liquid level in the suction line is being pumped down until the pump cavitates due to lack of NPSH. This can only happen due to a large pressure drop through the draw-off nozzle.

Undersized Draw-Off Nozzle

You will note that the preceding analysis as presented was rather complex. After 46 years, I've seen it all. But sometimes, the mistake is obvious and only simple observations are required.

For example:

- The place—New Orleans.
- The year—1984.
- The setting—No. 1 Crude Unit.
- The location—Tenneco Oil Refinery.
- The contractor—Walk-Kydell.
- The engineering observer—Me.
- The refinery rep—Gerry.

The contractor had designed a new and very large LVGO P/A pump. They had sized the new suction line correctly, at 8 inches. Also, the size of the new pump was appropriate for the application.

Scene I

"Lieberman, what's wrong with the vacuum tower? The top temperature's too hot," asked Gerry, the refinery representative.

"Gerry, that's it. The LVGO P/A pump is slipping," I answered.

"Slipping! Cavitation? But it's a new pump!"

"Gerry, the suction line is 8 inches. It replaced the old, undersized 4-inch line," I explained.

"But 8 inches sounds big enough to me, Lieberman."

"But Gerry, they left the old 4-inch draw-off nozzle," I explained softly.

"What!" Gerry screamed. "A tower 4-inch nozzle tied into an 8-inch suction line to a P/A pump? Impossible! Look Lieberman, this is a mess. Give me a quotation to revise the design of this tower to correct these problems. I need your quotation ASAP!"

Scene II

So I quoted $20,000. And Walk-Kydell quoted $1,000,000. And which company do you think was awarded the contract? (Hint: This story unfolded in southern Louisiana.)

"Norm, sorry you didn't get the job."

"Gee, Gerry. Why? I spent a lot of time preparing my bid."

"Yeah, Norm. It was your bid that was the problem. Your $20,000 quotation. Management felt it was out of line. It was just way too low to be taken seriously."

And this, ladies and gentlemen, is exactly what happened.

Effect of Pumparounds on Fractionation Efficiency

Let's refer to Figure 6-2. I'll make three assumptions:

1. The heat input to tray #13 is constant.
2. The heat extracted (i.e., the sum of the overhead condenser and pumparound exchanger duties) is constant.
3. The tower top temperature is constant.

My question is, what happens to the separation or fractionation efficiency between toluene and benzene as I increase the heat extracted in the pumparound or circulating reflux exchanger?

Ordinarily, removing more heat in the pumparound makes fractionation worse. As less heat flows up the tower, the tower top temperature will cool. The reflux control valve, which is on automatic temperature control, will begin to close. The reflux rate will diminish. The internal reflux rate on trays 1 through 6 (Figure 6-2) will fall. Lower reflux rates will normally reduce tray or fractionation efficiency. If the tower top temperature is constant because it is on temperature control, then the tray #11 toluene draw tray temperature will decline. The reduced delta T between tray 1 and tray 11 is a common indication of reduced fractionation efficiency.

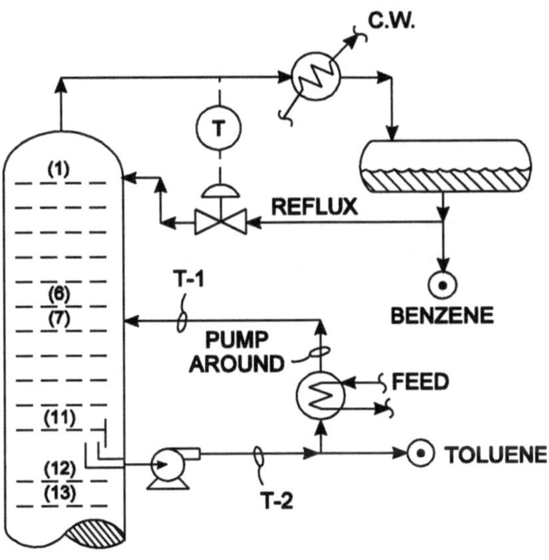

Figure 6-2 Increasing pumparound duty reduces fractionation between benzene and toluene products if the total heat input to the tower is constant.

However, in process operations, nothing is ever that simple. For instance, let's now assume trays 1 through 6 are suffering from entrainment and a consequent reduction in tray efficiency. Now, increasing the heat extracted in the pumparound exchanger will have just the opposite effect. In other words, the lower vapor velocities rising through the top six trays will suppress entrainment, and separation efficiency between benzene and toluene will improve.

To optimize the pumparound duty, one should maximize the delta T between two points in the tower, top temperature and side draw-off temperature.

In summary, if not limited by flooding, reducing pumparound heat extraction will improve fractionation, at the expense of reduced heat recovery and a higher tower operating pressure. That is, pumparounds make fractionation worse if tray efficiency is constant.

Pumparound Heat Transfer Capacity

The feed versus pumparound heat exchanger shown in Figure 6-2 is extracting heat from the tower by cooling and partially condensing the upflowing vapor. As the pumparound circulation rate is increased, the heat extracted from the vapor also increases up to a point. Then a further increase in pumparound flow will reduce the amount of heat extracted from the vapor. This is not a malfunction, but a consequence of the limited heat transfer capacity of both the pumparound trays (trays 7 through 11) and the surface area of the pumparound heat exchanger.

As the pumparound rate is increased, the pumparound return temperature (T-1) will also increase. That's because the ability of the heat exchanger to transfer heat to the feed is limited by the feed flow rate and the area of the exchanger. Thus, the larger pumparound flow cannot be cooled as efficiently as a smaller flow.

As the pumparound rate is increased, the pumparound draw temperature (T-2) will drop. That's because the capacity of the pumparound heat transfer trays (trays 7 through 11) is limited. Also, the pumparound draw-off temperature drops because more of the lighter component (benzene) is being condensed in the toluene. To calculate the pumparound duty, we would calculate:

$$\text{Duty} = (DT) \times (F)\,(SH) \tag{6-1}$$

where $DT = (\text{T-2}) - (\text{T-1})$

$F = \text{Flow, lb/hr}$

$SH = \text{Specific heat of toluene, Btu/lb/°F}$

$\text{Duty} = \text{Btu/hr}$

What actually happens on a process unit is that at some point, *DT* will go down faster than *F* is going up. Cleaning the pumparound exchanger would mitigate this limitation. Or, from the designer's perspective, increasing the number of pumparound trays would help. For the console operator, the only realistic option to increase pumparound heat extraction when limited in this way is to increase the feed flow through the pumparound exchanger.

Often, engineers observe a console operator reducing a pumparound flow to increase heat extraction from a tower. The engineer concludes that the operator is confused, or that there is an equipment malfunction. In reality, it's just a reflection of the inherent limitations of the pumparound heat transfer trays and the pumparound heat exchanger surface area.

Incidentally, for those readers who are working on computer control for heat balancing fractionators, the preceding section illuminates the complexity of your task. To some extent, in the real world, closed-loop, multivariable computer control of towers with multiple pumparounds is often an exercise in self-deception.

Structured Packing in Pumparounds

In Chapter 2, "Packed Tower Problems," I recommended the use of layers of structured packing in pumparound heat extraction service. The depth of packing required is determined by:

$$\text{Duty} = (U) \cdot (A) \cdot (\text{Delta T}) \qquad (6\text{-}2)$$

where Duty = As calculated in Equation (6-1)

U = Heat transfer coefficient. Depends on the type of the packing selected. This coefficient has to be obtained from the packing vendor correlations.

Delta T = Assume the packed bed is like a heat exchanger. The hot end delta T is the difference between the vapor to the pumparound and the draw temperature. The cold end delta T is the difference between the vapor from the pumparound and the pumparound return temperature. Just to be very clear, refer to Figure 6-1B:

Hot end delta T = 400°F – 350°F = 50°F

Cold end delta T = 300°F – 270°F = 30°F

Pumparound section delta T = (50°F + 30°F) ÷ 2 = 40°F (Note—I should really use the LMTD (log mean temperature difference), not the average, but I'm trying to keep this simple.)

A = "Area" of the pumparound. If we had trays, the area would be the tray deck active or bubble area, times the number of trays. For packed beds, this "area" is the volume of packing in the pumparound. To compensate for this strange area, which is really a volumetric heat transfer coefficient, U is also expressed in cubic feet and not the usual square feet.

To obtain the required height of the packed bed, divide A (in cubic feet) by the tower cross-sectional area (in square feet).

Let's say you have an existing bed of structured packing in pumparound service. The data from the existing operation indicates a calculated height of 2 feet. But the actual height of packing is 5 feet. We would then say that the structured packing is only working at 40% of its published efficiency.

Malfunctions that might account for the low heat transfer efficiency are:

- Poor initial liquid distribution.
- Poor initial vapor distribution.
- Improper installation of packing.
- Fouling and/or coking of the packing.
- Packing damage.

At the Good Hope refinery, a pressure surge in an FCU fractionator blew a 3-foot indentation into a 7-foot bed of a grid-type structured packing. Heat transfer efficiency was halved. I described this incident in detail in my book, *Troubleshooting Process Operations*, 4th edition, PennWell.

Fouling will lead to vapor–liquid channeling. I can't recall any firsthand experience with this problem with structured packing in pumparound service, but I have observed this malfunction with cascade mini-rings in crude tower pumparounds.

At the Texaco Plant in Convent, Louisiana, I retrofitted a vacuum tower by design to use a combination of grid (with a large percent open area) below a layer of structured packing (with a high degree of vapor–liquid contacting, but less open area). The structured packing arrived onsite first, and it was installed before (i.e., below) the grid, which arrived a few days later. Amazingly, Texaco did not blame me for the resulting poor performance of the vacuum tower.

Liquid distribution is typically achieved in this service by the use of spray nozzles. By far the most common cause of poor heat transfer efficiency of packed beds is partial plugging of the spray nozzles due to improper maintenance of the spray nozzle external duplex filters. Contrary to vendor claims, the packing does not redistribute poorly distributed liquid to any real extent. To troubleshoot liquid distribution problems, one should proceed as follows:

- At five or six points, 6 to 12 inches above the spray header, cut a 4-inch-wide hole in the external insulation around the vessel.
- Check the vessel skin temperature with your infrared gun.
- Any external area of higher-than-average temperature may indicate a portion of the nozzles that are plugged.
- Calculate the expected nozzle delta P, and compare it to the observed spray header pressure drop to confirm the nozzle pluggage.

Almost always, plugged nozzles are a consequence of filter assembly malfunctions or holes in the filter screen. Or, sometimes operators bypass the filters as they become badly plugged, resulting in nozzle plugging.

I recall many cases of liquid maldistribution reducing pumparound heat transfer efficiency, but only a few cases of the loss of heat transfer due to poor vapor distribution.

Duplex Filter Malfunctions

Things to worry about, that I've seen more than once, are:

- Small breaks in filter screen.
- Screen mesh size too big. The mesh size should be one-third the size of the maximum free passage of the spray nozzles.
- Screen mesh size too small. Causes operators to bypass or remove filters to sustain flow.
- Screen basket not placed in filter housing properly and firmly.
- Carbon steel piping downstream from filter. Causes corrosion products to be flushed into the spray nozzles.
- Filters equipped with a bypass. Sooner or later this will be used. Duplex filters should never have a bypass valve!

At a refinery in Lithuania, I became so discouraged by these problems that I had a single filter installed downstream from the duplex filters. The openings in this secondary filter were one-third the diameter of the openings in the duplex filters. To clean the secondary filter, the unit had to be shut down, as I had deliberately omitted a bypass around this secondary filter. But at least I finally stopped the pumparound spray nozzles from repeatedly plugging. I could exercise this degree of authority because I was staying at the home of the general director.

Pumparounds Do Fractionate

The standard industry assumption is that pumparounds provide a single equilibrium separation stage. This common assumption is often, if not usually, wrong. Again consider Figure 6-1B. Note that:

- The vapor leaving the pumparound top tray is 300°F.
- The liquid leaving the pumparound bottom tray is 350°F.

If the pumparound truly represented a single equilibrium separation stage, these two temperatures would be identical. But if a pumparound has five intact and correctly designed trays, the vapor leaving the pumparound is going to be much cooler than the liquid leaving the pumparound. This indicates that the pumparound is providing more than one theoretical tray of fractionation. Under certain circumstances, this leads to a large increase in the volume of vapor inside the pumparound section. Towers have flooded because of what I call the "bubble effect." I have described this problem in great detail in my book, *Process Engineering for a Small Planet* (2010, Wiley). The specific example described occurred at the Good Hope Refinery FCU Main Fractionator in 1981, in the slurry pumparound section. This is a complex subject. I'll just provide a word of caution. When modeling a pumparound with several trays, it is best to assume two theoretical stages to most correctly represent vapor and liquid loads.

CHAPTER 7

Reboiled and Steam Side Strippers

The lies and the truth are mixing together to form a potent new reality.

A sketch of an installation of an atmospheric gas oil (AGO) stripper is shown in Figure 7-1. A summary of the many malfunctions that I have seen with such strippers are:

- Vapor line delta P excessive
- Feed control valve stuck
- High stripper bottoms level
- Inadequate elevation difference between stripper and fractionator draw-off nozzle
- Lack of a loop seal
- Top tray flooding
- Moisture in stripping steam
- Excessive ambient heat loss
- Extraction of heat from stripper gas oil feed
- Low stripping efficiency due to excessive tray hole area
- Improper clearance between seal pan and bottom downcomer
- Flooding due to salt sublimation
- Blown-out tray decks

Excessive Vapor Line Delta P

Let's begin by looking at Figure 7-1. Note that the stripper top pressure is 24 psig, and the fractionator pressure is only 21 psig. How can the liquid gas oil feed from the fractionator flow from a lower

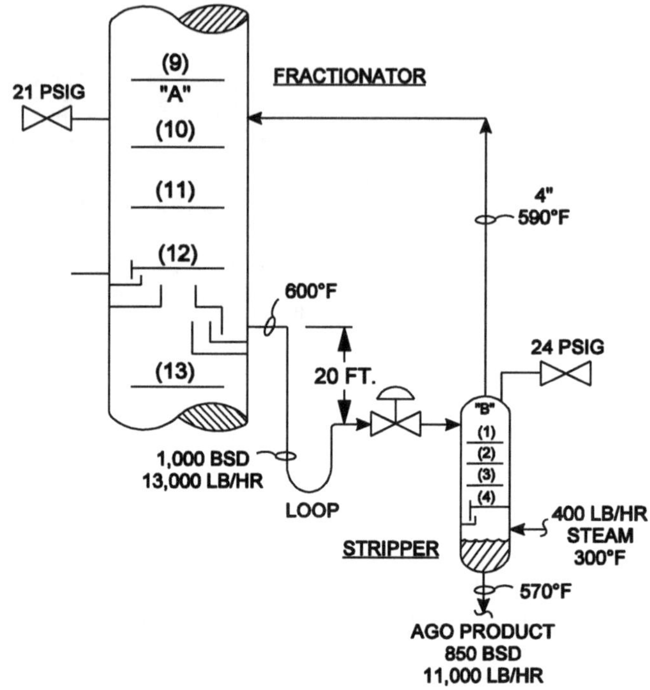

FIGURE 7-1 An atmospheric gas oil side stream stripper. The small delta P on trays 10–12 has been neglected.

pressure, to the 24 psig higher pressure at the top of the stripper, without the assistance of a pump?

It's that the draw-off nozzle on the fractionator is elevated by 20 feet above the stripper inlet:

$$(24 \text{ psig} - 21 \text{ psig}) \times (2.31 \div 0.70) = 10 \text{ feet}$$

where 2.31 is the conversion between psi and feet of water

0.70 is the specific gravity of the 600°F gas oil. The SG of water is 1.00 at 60°F.

The delta P in the vapor line of 3 psi is line and nozzle loss, which is a function of the stripping steam flow rate. For the moment, I'll neglect the pressure drop in the gas oil feed line to the stripper. Therefore, since we have 20 feet of elevation driving force and we only need 10 feet, the gas oil will flow out of the fractionator and into the stripper due to the force of gravity.

Let's now increase the stripping steam rate from 400 lb/hr to 800 lb/hr. The pressure in the fractionator is still 21 psig (Note: To simplify the discussion, I've assumed the pressure loss through trays 10 through 12 is zero. In reality, it would be about 0.3 psi.). The pressure drop through the overhead 4-inch vapor line increases from 3 to 7 psi. Thus, the pressure at the top of the stripper will rise to 28 psig. Now, the required elevation difference between the fractionator gas oil draw-off nozzle and the stripper inlet above tray #1 will be:

$$(7\text{ psi}) \times (2.31 \div 0.70) = 23\text{ feet}$$

But the actual elevation difference is still 20 feet. So as I raise the steam flow to the stripper, at some point I will observe that the stripper goes empty. The flow of gas oil to the stripper has not slowed, but stopped.

I recall that Daisy Wong, the unit engineer at the Tenneco Oil Coker in Chalmette, Louisiana, had this problem. But she calculated the pressure drop through the stripper 4-inch overhead vapor line not as 7 psi, but only 2 psi. Daisy did the calculation correctly. She allowed for:

- The frictional loss in the 4-inch piping.
- The acceleration loss in the 4-inch nozzle (i.e., exit loss) at the top of the stripper.

(I describe this method of calculation with reference to kettle reboilers in Chapter 9.) But the measured delta P between point A and point B in Figure 7-1 was 7 psi, not 2 psi.

So Daisy, being very young, asked the tray vendor what to do. They told this innocent young person, "Daisy, the stripper is flooded. You need to replace the four trays in the stripper with 8 feet of pall rings" (dumped packing). And she did it!

I tell this story because I met Daisy 20 years later. She had turned into a very smart operating manager for a refinery in California. The problem at the Tenneco Oil refinery in 1984 was that the 4-inch line was welded over a 2-inch opening in the side of the fractionator vessel. A serious vessel fabrication error! I recall Daisy's sad face when she told me, "But Norm, the packing didn't work! I did what the vendor told me. My first project, and it didn't work."

Lack of Elevation Driving Force

At the Coastal refinery in Corpus Christi, I had a similar problem: loss of stripper level with increasing stripping steam rate. The AGO product was actually 35% diesel. Here, the vapor line pressure drop was too small to measure. The calculated delta P was less than 1 psi. But the malfunction was obvious. The elevation between the gas oil draw-off nozzle and the inlet to the stripper was not the 20 feet shown in Figure 7-1, but less than 20 inches. To overcome this problem, I decided to pump a slipstream from the AGO pumparound into the stripper, instead of the futile gravity feed. We could now aggressively steam strip the AGO. The product AGO diesel content dropped from 35% to 20%. I recall this project for several reasons:

- The plant manager was so pleased with the enhanced diesel recovery that he approved an engineering efficiency study by my company for a lump sum charge of $135,000.
- The idea of pumping the gas oil into the stripper was suggested by an operator named Lester.
- I was so pleased with Lester's contribution to my project that I rewarded him with a baseball cap emblazoned with my company insignia.

Subcooled Liquid Feed

Consider Figure 7-1. Observe that the stripping steam is only 300°F. The AGO feed is hotter. About 15 weight percent (2,000 lb/hr) of the stripper feed is vaporized. The latent heat of vaporization of the lighter components in the gas oil feed is 100 Btu/lb. Therefore:

$$(2{,}000 \text{ lb/hr}) \times (100 \text{ Btu/lb}) = 200{,}000 \text{ Btu/hr}$$

This large amount of heat cannot come from the stripping steam, as the steam is colder than the hydrocarbon feed. The latent heat required comes from the feed itself:

$$(200{,}000 \text{ Btu/hr}) \div [(13{,}000 \text{ lb/hr} \times 0.62)] = 25°\text{F}$$

where 0.62 is the specific heat of the feed in Btu/lb/°F.

13,000 lb/hr is the feed to the stripper.

25°F is the temperature drop of the oil, due to the conversion of the oil's sensible heat to latent heat.

This calculation suggests that the effluent temperature of stripper bottoms is:

$$600°\text{F} - 25°\text{F} = 575°\text{F}$$

This is not quite true, as I've neglected a few factors:

The steam heating up from 300°F (inlet) to 590°F (vapor outlet):

$$(400 \text{ lb/hr}) \times (590°\text{F} - 300°\text{F}) \times (0.55) = 64{,}000 \text{ Btu/hr}$$

where the 0.55 factor is the specific heat of the steam in Btu/lb/°F.

Also let us assume that this steam is of a low quality, as described in Chapter 15, "Steam Quality Problems." That is, the steam has 5% moisture:

$$(400 \text{ lb/hr}) \times (5\%) \times (1{,}000 \text{ Btu/lb}) = 20{,}000 \text{ Btu/hr}$$

where the 1,000 Btu/lb is the latent heat of evaporation of water.

Also, the feed line and the stripper itself are not perfectly insulated and are losing another 44,000 Btu/hr to ambient heat losses. So:

$$64{,}000 + 20{,}000 + 44{,}000 = 128{,}000 \text{ Btu/hr}$$

These heat requirements have to be added onto the 200,000 Btu/hr required for the latent heat of evaporation of the feed. This additional temperature drop will be 16°F. Therefore, the stripper bottoms temperature will be:

$$600°\text{F} - 25°\text{F} - 16°\text{F} = 559°\text{F}$$

The Function of Stripping Steam

If the steam is not providing any heat to the stripper, but extracting heat from the stripper, what is the function of the steam?

- It reduces the hydrocarbon partial pressure.
- Or, I can say, the steam dilutes the moles of hydrocarbons in the vapor phase and thus makes it easier for more moles of the lighter hydrocarbons to escape out of the liquid into the vapor phase. This means that the steam reduces the hydrocarbon partial pressure, even though it actually increases the physical pressure in the stripper, due to the extra delta P in the overhead vapor line.

For the hydrocarbon to partially vaporize in the stripper, it must enter the stripper close to its saturated liquid temperature or close to its boiling or bubble point temperature. Anything that extracts heat from the feed is going to reduce the percent vaporization of the feed and the efficiency of the stripper.

Moisture in Stripping Steam

At the Coastal refinery in Eagle Point, New Jersey, the operators found one evening that they could not meet their diesel oil **flash specification**. That is, there was too much residual naphtha in the diesel stripper bottoms product. That afternoon, a freezing rain had descended over southern New Jersey. The refinery was covered with an inch of slippery ice. The steam lines in the ancient plant (part of the old Getty Oil empire) were badly insulated. The steam flowing to the diesel oil stripper looked more like warm water than steam when I blew down a ¾-inch connection. So I had an operator install a steam trap on this same ¾-inch connection. The next sample showed that the diesel oil flash spec was back to normal.

Ambient heat loss is always a problem. The more heat that is lost from the stripper and/or the feed line, the more steam is required to strip out a given quantity of hydrocarbons. But this is true only up to a point. For example, I once determined that if the bottoms temperature from a stripper was 40°F or more below the fractionator draw temperature, with no stripping steam flow, then adding any amount of stripping steam would be completely ineffective.

I recall at another plant that there was a crude preheat versus AGO product heat exchanger upstream of the stripper. The AGO was **subcooled** by 80°F before it entered the stripper. The diesel content of the AGO product was 40% with or without stripping steam. As I reported to the V.P. of Refining, "Sir, you can't strip subcooled liquid. I can't imagine what fool placed the AGO product versus crude preheat exchanger upstream, rather than downstream of the stripper!"

Unfortunately, I later found out that the fool was the same V.P., who designed it that way, 20 years prior to my career-ending presentation.

Even if a high pressure in the stripper is not interfering with the feed flow into the stripper, it's still a problem. Each extra psi of stripper pressure in diesel or AGO stripping service has the same effect as subcooling the liquid by about 2°F.

To gauge how bad your subcooling problem is due to ambient heat loss:

- Shut off the stripping steam and wait half an hour.
- Using your skin temperature infrared gun, measure the draw temperature at the tower and the stripper bottoms product temperature at the stripper itself.
- If the delta T is 5°F or 10°F, forget it.
- If the delta T is 10°F or more, then check your insulation integrity.

I'm not talking about energy conservation. I'm concerned about fractionation efficiency.

Lack of the Loop Seal

At one time or another, I had provided the process design to revamp most of the crude units in the former domestic Texaco refineries. One of the outstanding features of the vanished Texaco Corporation was its inability to learn from previous mistakes. As an example, let's look at Figure 7-2. Note that the feed control valve is in the vertical portion of the feed line. Also, unlike Figure 7-1, there is no loop in the feed line. Why the loop? Let's consider the following possibility:

- The flow of gas oil to the stripper drops below the set point for 1 microsecond.
- The control valve starts to open.

- The liquid in the feed line drains down. In Figure 7-2, the liquid can drain out of the line because there's no loop seal in the line.
- The steam at the top of the tower can now start to flow up the feed line. Once this begins, the liquid head pressure needed to restore the flow is gone.
- The only way the operator has to restore the feed flow is to shut off the stripping steam for a while.
- The more stripping steam that is used, the more likely it becomes that all feed flow to the stripper will be lost. Thus, the operators at Texaco plants tended to use far less than the optimum stripping steam rates to avoid the loss of their stripper bottom level.

So I revamped the Texaco steam strippers with loop seals in their feed lines to permit higher stripping steam rates. It all worked fine until I was hired by the Texaco plant in El Dorado, Kansas. I was working with a young female engineer on the crude unit. I asked her a question and she gave me a wrong answer. Ordinarily, if I was working with a male engineer, I would have tapped the guy on the head with my wrench (of course only if he was wearing his hard hat) for such an answer. But then I thought, "It's really sexist to treat coworkers differently based on gender." So I tapped the young lady

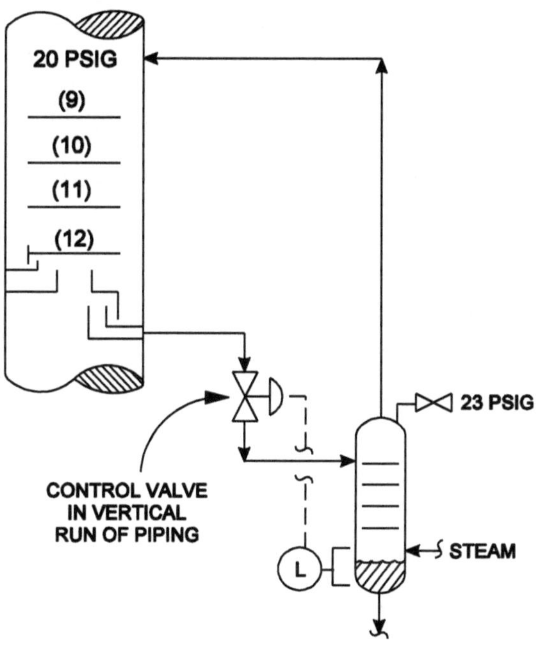

FIGURE 7-2 Lack of the loop seal reduces the ability to use stripping steam.

on the head with my wrench, and Texaco canceled my contract because of "harassment in the workplace."

Jammed Control Valve

In 1966, I designed the coker fractionator tower for the new B coker for American Oil in Texas City. It was my first major process design. When the unit started up in 1967, the production of gas oil from the side stream stripper was only 1,000 BSD, far less than my design flow of 2,000 BSD. I asked Kenny, the panel board operator, to increase the flow.

"Can't do that, Lieberman. The FRC (feed line valve) valve to the stripper's 100% open."

Ten years passed. It was 1977, and in the interim, I had become smarter. While in Texas City for a meeting, I went to the coker control room. I saw that the gas oil flow was still running at 1,000 BSD from the bottom of the steam stripper. I asked Kenny, the same panel board operator, to increase the flow.

"Can't do that, Lieberman. The FRC valve to the stripper's 100% open."

But the valve position shown on the panel or control console is not the control valve position. That's an air signal coming into the control room that represents what the control valve position is supposed to be. Next, I went out and climbed up into the fractionator structure to look at the control valve, which was stuck in a mostly closed position. So I set the **hand jack** and manually forced the valve open.

When I came back into the control room, Kenny was staring at the stripper bottoms level indicator. It had jumped up to 100% full. "Lieberman! What'd you do out there?" screamed Kenny.

"Who, me? I didn't touch anything. I always keep my hands in my pockets."

All these stories are true. And it's also true that

- Texaco
- Amoco (American Oil)
- Coastal
- Getty
- Good Hope (GHR)
- Gulf
- Unocal
- Arco
- Tenneco

are no longer in the oil refining business. Maybe you can begin to see why.

High Liquid Level

Once the liquid level gets above the stripping steam inlet, the steam forces the liquid up against the bottom tray and the seal pan. This causes all the trays to progressively flood. Tray efficiency and hence stripping efficiency is greatly reduced. At an older refinery in Spain, and at an even more ancient plant in Port Arthur, Texas, I had an identical experience on a diesel stripper. Both strippers had an archaic circular chart that indicated bottoms level. On both charts, a perfect round circle had been inscribed by the recording pen for months, or years, or decades. Both strippers' bottoms levels were "tapped out" (see Chapter 16, "Level Control Problems"). Basically, the malfunction is that the bottoms level is above the top tap of the level-sensing device. But due to a calibration problem, the indicated level is reading less than 100%, even though the actual level in the vessel is above the top level tap by some unknown amount.

In both cases, I identified the malfunction by placing a pressure gauge on the top of the gauge glass. If the pressure at this point is 3 or 4 psi above the pressure at the top of the stripper, then the stripper is flooded. If shutting off the stripping steam does not reduce the bottoms pressure closer to the stripper top pressure, then the malfunction is a high liquid level.

In both cases, I lowered the level in the diesel stripper until the recorder pen on the circular level indicator chart started to move up and down, below its former perfect circle. And here are the responses I received both in Spain and in Texas from the panel board operators:

- "Señor Norman, muy bueno. Mira! The 5% point of the diesel has increased by 30°C. Mira! We're making 300 BSD more jet fuel. Muy, muy bueno."
- "Lieberman, look at my chart. It's drawn that perfect round circle at 70% level for 10 years. Who the hell hired you, anyway? You've been on my unit for 10 minutes. Look what a mess you made of my chart. Just look at all the purple ink I'm wasting."

And don't think I made that part of the story up about the ink. I even remember the purple color.

Top Tray Flooding

Let's assume that all four trays in the AGO steam stripper are designed with the same hole area or number of caps. Typically, this is the case, even though it represents poor design practice. The vapor flow through the bottom stripping tray shown in Figure 7-1 is 400 lb/hr of steam. The vapor flow from the top tray is:

400 lb/hr stm + 2,000 lb/hr hydrocarbons = 2,400 lb/hr of vapor

The 2,000 lb/hr of hydrocarbons is the difference between the stripper feed and the stripper bottoms. The vapor flow at the top of the stripper is then seven times higher than the vapor flow through the bottom tray. Taking into account molecular weight, pressure, and temperature effects, I have calculated that the hole area at the top tray should be about three or four times greater than the hole area on the bottom tray for good vapor–liquid contacting. But if the hole area is the same for all trays, as I've assumed above, then the top tray will flood first, or the bottom tray will dump.

This is not the only reason for top tray flooding. If there are corrosion products (typically iron sulfide) in the stripper feed, they will accumulate on the top tray. If the process stream contains salts (typically NH_4Cl), these salts tend to sublime out on the top tray (sublime = change from a vapor to a solid). The problem of salt accumulation on the top tray of a side stream stripper is pretty universal on hydrodesulferizers and FCUs that are running on a high content nitrogen feed. If neutralizing or filming amines are used in the fractionator, then dry amine chloride salts (a white substance with a strong ammonia odor) will accumulate on the side stripper top tray.

As I've discussed in Chapter 1, "Distillation Tray Malfunctions," flooding progresses up a tower, but not down a tower. Therefore, if only the top tray floods, then the bottom three trays will not flood. And the overall stripper pressure drop will not become unusually large.

But the consequences of top tray flooding are particularly severe. I first noted this problem on a gas oil HDS unit (hydrodesulferizer) at the Coastal refinery in Corpus Christi, Texas. The unit had a fractionator designed to produce a small diesel product as a side draw, as is shown in Figure 7-3. Fractionation between diesel and naphtha was bad in the sense that the naphtha was contaminated with diesel and had a high end point. When the stripping steam was stopped, the naphtha end point was fine. But the diesel oil product no longer met its flash specification of 150°F (Note: A rough approximation of flash point temperature can be obtained by taking the ASTM D-86 Distillation 5% point temperature and subtracting 200°F). Without stripping steam, the diesel oil product was contaminated with naphtha.

With the stripping steam in service, I checked the delta P across the four stripping trays. The delta P was very low, which indicated that the stripping trays were not flooding. Yet, when I opened valve A on the top of the stripper, liquid diesel squirted out. Venting liquid from the top of a tower is proof that the tower is flooded. But how can a tower flood with a low pressure drop across its trays?

The answer is top tray flooding. In a steam stripper, the top tray will almost always flood first because of:

- High vapor rates.
- Salt sublimation (chloride or bisulfide salts).
- Fouling deposits from the feed.

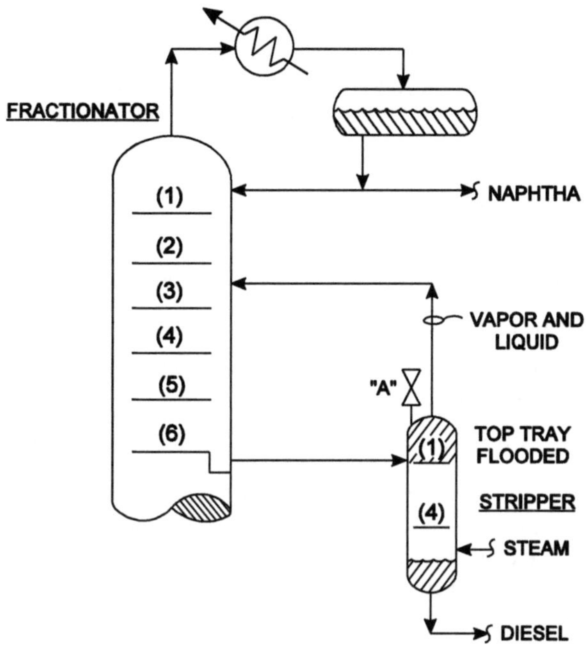

Figure 7-3 Top tray flooding ruins fractionation in fractionator between trays 3 and 6.

If the top tray floods, the feed to the stripper will be lifted up the vapor outlet line above the fractionator tray #3 (Figure 7-3). The steam acts as **lift gas**, in the same way that vapor generated in a reboiler promotes natural or thermosyphon circulation. The steam acts as lift gas in the same way that air is used in a lift gas pump. The steam actually pumps the liquid, which is flooding on the top tray of the stripper, up onto tray #3 of the fractionator. Large volumes of liquid now circulate from tray #6 in the fractionator back onto tray #3 in the fractionator.

At best, trays 3 through 6 may act like one tray due to the recirculation of the diesel. With diesel being pumped back up to tray #3, the naphtha product becomes contaminated with diesel oil.

At the Coastal refinery in Corpus Christi, we corrected this problem by adding water to the stripper feed for several minutes. This dissolved the accumulated salts from the top tray of the stripper.

The root cause of this problem was upstream of the fractionator. The HDS reactor effluent water wash flow injected into the reactor heat exchanger train was far too small to remove the NH_4Cl salts from the fractionator feed.

At the Chevron refinery in El Segundo, they had essentially the same problem on their coker kerosene stripper. There the problem was solved by eliminating the amine corrosion control chemical injection into the coker fractionator. On FCU LCO strippers, periodic on-steam water washing is my preferred solution. LCO blowing out of the stripper top vent indicates when such water washing is needed.

Hole Distribution

A typical side stream steam stripper is designed as follows:

- Six trays.
- The vapor rate to the top tray is used to determine the hole area of the top tray.
- All six trays are fabricated from the same set of drawings.

The result of this negligent design is that the hole velocity through the bottom few trays is too low to develop reasonable tray efficiency. The use of extra steam may be constrained because of top tray flooding. To prevent this sort of malfunction, I suggest the following rule of thumb:

- Trays 1 and 2 (top two trays): 100 holes or caps
- Trays 3 and 4 (mid two trays): 65 holes
- Trays 5 and 6 (bottom two trays): 35 holes

What I'm suggesting is fundamental to the work of the process engineer. When ordering new trays from the tray vendors, you have to specify the total hole area yourself for each tray deck. Of course, that involves a lot of extra work, especially if you don't know how to calculate the required hole area. But the vendors have the required correlations on their tray rating programs. Or, if I'm still alive when you read this, you can phone me in New Orleans, Louisiana, and I'll run out the answer on my slide rule.

Reboiled Side Stream Strippers

There are very few reboiled side stream strippers still in existence. We always (before the 1960s) used to reboil kerosene and virgin diesel strippers because we wanted to produce a dry product (free of water) directly from the crude unit. But when refineries started to hydrotreat kerosene and diesel, producing a dry product from the crude unit fractionator was no longer helpful. So we converted most of the reboiled strippers to steam strippers. The high-level heat used to reboil the strippers was then made available for crude preheat, which unloaded the crude unit fired heater. And that's one way a refinery expanded crude capacity.

The only difference between troubleshooting reboiled side stream strippers and steam strippers is that:

- In a steam stripper, the maximum vapor and liquid loads are on the top tray.
- In a reboiled stripper, the maximum vapor and liquid loads are on the bottom tray.

Whenever I have a problem with a reboiled side stream stripper, I hook up a small steam injection point to the reboiler outline line. Meaning, I combine reboiled and steam stripping. This is a nice trick of the trade, which works every time. Injecting the steam onto the reboiler itself is also okay.

False Steam Flow Indication

The People's Republic of China (i.e., the communists) owned a controlling interest in the now-vanished Pacific refinery, just north of San Francisco. I guess they were interested in learning more about American refinery technology. So here's one bit of technology I helped them acquire.

This 40,000 BSD plant had a small AGO stripper on the crude unit. The AGO's diesel oil content was excessive at 35% to 40%. Why? Because the operators reported that every time they introduced the steam to the stripper, they would get an erratic and uncontrollably high bottoms level indication. I later found that because of the location of the top-level tap in relationship to the steam inlet nozzle, moisture was getting into the level-sensing apparatus. I discuss this type of problem in Chapter 16, "Level Control Problems." The stripping steam was not causing an erratic bottoms level as the operators thought, but just an erratic level indication. To steady-out the bottoms liquid level indication, they would then stop the steam flow to the stripper.

But the Pacific refinery plant management, who were not proletariats like me, refused to discuss the malfunction with the workers. They issued this order:

"Henceforth, 1,200 lb/hr of steam shall be used in the AGO stripper, regardless of level problems. No deviations will be tolerated."

The operators obeyed this directive. And lo and behold, the level was no longer erratic. Yet the diesel content of the AGO stayed at 35% to 40%! Now what?

Plant management consulted with the tray vendor. They recommended retraying the stripper with a bed of structured packing to get more theoretical separation stages.

Plant management consulted with a famous and expensive engineering company. They recommended a computer-monitored, multivariable control scheme.

Plant management consulted with me. I went out to look at the stripper. The operators were using 1,200 lb/hr of steam, as follows:

- The local 2-inch steam globe valve controlling the flow was partly open.
- The 1,200 lb/hr of steam flowed through the orifice meter.
- The 2-inch isolation gate valve at the tower was shut.
- The 1-inch atmospheric vent on the steam line, downstream of the steam orifice meter, and upstream of the 2-inch gate valve, was wide open.
- The 1,200 lb/hr of steam was vented to the atmosphere!

Yes, the People's Republic of China learned a valuable lesson about American manufacturing technology at the Pacific refinery. If you don't believe me, visit your neighborhood Walmart. Just try to find something made in the U.S.A.

CHAPTER 8

Inspecting Tower Internals—Checklist

Expect what you inspect.

—U.S. Navy slogan

This chapter presents my detailed procedure for inspecting distillation trayed tower internals for mechanical integrity, proper assembly, and cleanliness. I've been doing this work on and off for over four decades, so I ought to be able to provide some reasonably good advice on the subject.

Begin with a few hours of basic preparation. Cover the following items:

- Do you have an entry permit for the tower from Operations?
- Has Operations provided a "**hole watch**?" This is the person who monitors your location inside the tower from the nearest manway. (Note to Reader: Terms in bold are explained in the Glossary at the end of this text.)
- Always conduct your inspection with a second person to aid in observations and for safety.
- Review internal drawings for details and make notes of dimensions you plan to check out in the vessel. Sometimes I'll make a cardboard gauge to match dimensions.
- Don't wear loose-fitting coveralls or anything that is likely to catch on tray parts.
- Gloves, flashlight, and small spare flashlight are required.
- Use the toilet before you start.

- Has all hot work in the tower been suspended during your inspection?
- I like to start at the top, as it's easier to climb up an external ladder than to climb up through the tray deck manways.
- How do you plan to record your observations? I used a small voice recorder once. Pen and paper are really awkward. I've never really solved this problem.

Use of a Self-Leveling Laser Light

You will want to check tray deck levelness and the outlet weir for levelness. This is done with a laser light that automatically levels itself with a small internal pendulum. I bought mine in a Lowe's hardware store for $135. The laser produces a light beam for checking weir levelness, or a plane of horizontal light for checking tray deck levelness. It's quite small, lightweight, and can fit in your coverall pocket with ease. An ordinary carpenter's laser level which only produces a single beam of light is far less useful.

What to Look for When Inspecting Trays

- Weir levelness above the tray deck
- Weir height versus the downcomer clearance
- Downcomer clearance above the tray deck for levelness

Criteria for an acceptable tray and weir levelness were discussed in Chapter 1, "Distillation Tray Malfunctions." Other critical items are:

- Is the bottom edge of each downcomer secured to the tray below with downcomer bracing brackets? Bracket spacing should be 3 or 4 feet.
- For center downcomers—are the spacers to maintain the downcomer width installed? These are usually tubes bolted between the two downcomer sides.
- Are the downcomers bulged out or pushed in?
- If there are inlet weirs, then the horizontal space between the downcomer and the inlet weir should be at least 2½ to 3 inches. Inlet weirs are not common. But if this spacing is not maintained, flooding will result.
- Are the deck plates beneath the downcomers holed through by corrosion?
- Are there holes in the downcomers?
- Are the vertical edges to the downcomers secured tightly to the downcomer bolting bars?

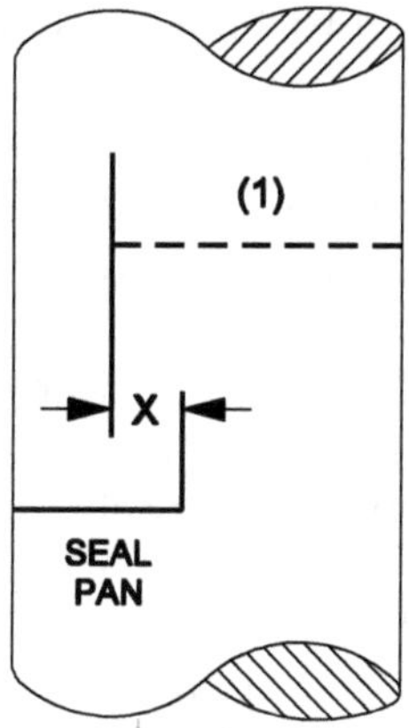

FIGURE 8-1 A restrictive clearance between the bottom edge of the downcomer and the seal pan lip caused tower flooding.

- Are the seal pans clean?
- Do the seal pans all have at least a single, ½-inch drain hole?
- Check the horizontal clearance between the bottom downcomer and the overflow lip of the seal pan. It must be larger than the downcomer clearance (dimension *x* in Figure 8-1).
- Check the vertical clearance between the bottom downcomer and the seal pan. It must be equal to or greater than the downcomer clearance of the other trays.
- Make very sure the bottom edge of the downcomer in the seal pan is rigid (see Figure 8-1).
- The top edge of the seal pan must be at least ½ inch above the bottom of the downcomer.

NOTE *If you have concluded that I've had lots of bad experiences with flooding due to seal pan malfunctions, you have drawn the correct conclusion.*

- Are all tray clips at least finger tight?
- Are all nuts securing the downcomers to the downcomer bolting bars at least finger tight?
- Are any impingement plates in front of the inlet nozzles installed as per the design?
- Distributor pipes must be self-draining. Drill a half-inch drain hole if required.
- Distributor pipes should not be rigidly fixed to the support shelf; this allows for thermal expansion of the distributor pipe.
- All portions of sumps and chimney trays must be self-draining. The exception is sumps with draw-off nozzles flush with the floor of the sump.

- Are hats on chimney trays tight and aligned directly over the chimneys?
- Chimney trays are normally sloped in the direction of flow. If they are not, drain holes are required.
- Water-test sumps and chimney trays for leakage. Weld up all the observed leaks.
- Inspect the draw-off nozzles. This is often difficult, as the seal pan will obscure your view of the nozzle. I use a cell phone camera for this purpose. Make really sure rags and gloves are not caught on draw-off nozzle vortex breakers. You can imagine how I became so smart on this subject.
- Inspect distributors and spray nozzles for pluggage with water circulation. This may be hazardous. I've described an important incident in my book, *Process Engineering for a Small Planet* (2010, Wiley), relating to an adventure that almost killed me.
- If shear clips are used, they should not be welded to the tray support ring.
- Are bolts securing the internal distributor, draw-off piping, and level indication flanges and piping at least finger tight?
- For valve trays, check the lift of caps to make sure they correspond to design. There are two standard valve cap leg lengths in common use.
- For grid trays, check that the tray decks are oriented in the correct direction, that is, with the wider part of the grid cap facing the flow. Vapor should preferentially flow toward the outlet weir.
- For trays with **explosion doors**, check that the doors, when opened, can fall back without getting jammed in the open position. You'll have to push each door open with your feet, as they weigh about 90 pounds each.
- Tray decks must be clean. However, a few missing valve caps are of no great consequence. Anything less than 5%, I would likely ignore, but not if they are all missing in the same tray deck area.
- For old-style bubble caps or older tunnel caps, remove the cap and check the clearance between the riser and the cap. Make sure the inside of the cap is clear and free of accumulated coke. Replace the caps, but do not overtighten the nut that secures the cap in place on the threaded stud protruding from the bubble cap. The outlet weir height should be no higher than the top of the cap risers, but above the bottom edge of the bubble caps.

Tower Closure

Unfortunately, experience has shown that we cannot always trust our coworkers. Having generated a list of tower internal malfunctions, you may find that it will be ignored. Thus, you have to insist that the mechanical folks give you an opportunity to re-inspect the tower prior to final closure. However, to preclude your follow-up inspection, the mechanical people may close all tray deck manways on the graveyard shift. That's when the young engineer finds out what process engineering is all about. Time and determination. Take your wrench and start unbolting the tray deck manways yourself.

Finally, closure of each individual tray deck manway must be witnessed and approved, one manway at a time. Otherwise, only the tray deck manways adjacent to the tower manways will be closed. Again, see my book, *Process Engineering for a Small Planet*, for an example of when tray manways were left off. For smaller-diameter towers (5 feet), observing this closure is quite difficult, as only one person can be working at the level of a tray at one time. The engineer has to watch the manways being secured from above. An awkward but necessary procedure.

Inspecting distillation tower internals is the most important function of a process engineer in a chemical plant or refinery. It should not be left in the hands of a junior engineer or a nontechnical person. At the Good Hope refinery, Jack Stanley, the refinery owner, insisted that I, the tech service manager, inspect each tower prior to the final closure. At Amoco Oil, the manager of the Process Design Division, Joe Gurawitz, insisted that no design was considered complete until the process design engineer inspected and approved the completed tower internal installation.

I'll admit that it's an exhausting, dirty, bruising job, which I've come to dislike. But the alternative is to start up with tray malfunctions that will certainly degrade fractionation efficiency and reduce tower capacity.

Seal Pan Design Error

In 2005, I was working for the Mazeikiu NAFTA refinery in Juodeikiai Lithuania. The problem was flooding in the bottom of the crude distillation column. The sequence of events was:

1. The tower, built in the 1980s, was able to handle 100,000 BSD of charge prior to 2005.
2. My friend, Nelson English, became the plant's general director.
3. Mr. English, being a thorough sort of engineer, hired an inspector from Koch-Glitsch out of the Czech Republic. The

crude unit had just been shut down for a turnaround. The inspector's instructions were to check the tower internals for proper assembly, to insure maximum tray fractionation efficiency.

4. The Koch-Glitsch inspector found that the lower edge of the bottom tray downcomer was loose, meaning that it was not secured to the existing downcomer bracing brackets. All the nuts and bolts that attached the bottom of the downcomer to the floor of the seal pan, as the inspector reported, were "missing."
5. The four sets of nuts and bolts needed to properly secure the downcomer to the floor of the seal pan were restored, as per the Koch-Glitsch tray design drawings.
6. The crude unit was restreamed. The tower now flooded at only 85,000 BSD of charge. What had happened?

Liz, Vaidus, and author in Lithuania.

On many towers, the seal pan is part of the vessel itself, like the tray rings or impingement plates. This vessel was built in Romania; the trays were from Czechoslovakia; the engineering had been done in Russia. The horizontal clearance (Figure 8-1, dimension x), could not be determined without the help of an expert in three different Eastern European languages.

Mr. English directed me and my wife, Liz, to enter the tower and find out what had happened. Liz measured dimension x at about 1 inch. My calculations indicated it should have been 3 inches. As delta P varies with velocity squared, the pressure loss of the liquid escaping from the seal pan would then be nine times too high. This would result in bottom downcomer backup and flooding on tray #1.

I imagine that what had happened was that many years ago, during the early 1980s, a Lithuanian engineer was troubleshooting the same problem. He made the same observations as Liz and performed the same calculations that I've just related. He then reached into his coveralls for his wrench and removed the four sets of nuts and bolts securing the downcomer bottom edge to the floor of the seal pan. Maybe he then used a few 3-inch sections of angle iron to increase the clearance (x) shown in Figure 8-1, to force the clearance to increase from 1 inch to 3 inches. After all, my paternal grandmother came from Lithuania. Who knows what genes I've inherited.

I've learned some important lessons from this incident:

- First, as we say in Texas, "If it ain't broke, don't fix it."
- Second, it takes Liz two hours to remove crude tower tar from her hair.
- Finally, you should really understand how the tower internals work when you inspect trays. I could detail many examples where a distillation tower malfunctioned due to a design error. While I had been unable to identify the error from the drawings, I quickly recognized the problem during the tower internal inspection. It's often very hard for me to envision a problem that exists in three dimensions from a two-dimensional tray drawing. And, to be totally honest, I dislike reviewing vendor tray drawings. Thus, I will carelessly sign off in the client drawing approval box without really studying all the tray dimensions. I assume that I can always correct any problems later on in the field when I inspect the tower internals.

[illegible]

CHAPTER 9

Process Reboilers—Shell and Tube

To learn, read; to understand, write; to master, teach.

The term *reboiler* is often misleading. It implies that all or most of the heat transfer is devoted to latent heat transfer and that only a small amount is devoted to sensible heat transfer. If the distillation tower bottoms product is a pure component like propylene or benzene, then 99% of the reboiler duty will be latent heat transfer. But say the bottoms product is a mixture of hydrocarbons like natural gasoline or crude naphtha. Let's assume:

- 40% is vaporized in the reboiler.
- The reboiler temperature rise is 100°F.
- The latent heat of vaporization of the hydrocarbon is 100 Btu/lb.
- The specific heat is 0.60 Btu/lb/°F.

The latent heat component of the reboiler duty is then calculated as the following:

$$(40\%) \times (100) = 40 \text{ Btu/lb}$$

The sensible heat component of reboiler duty is calculated as:

$$(100°\text{F}) \times (0.6) = 60 \text{ Btu/lb}$$

Thus, 60% of the reboiler duty is devoted to sensible heat transfer and only 40% to latent heat transfer.

Effect of Velocity on Heat Transfer Coefficient

In 1965, I was sitting at my desk at the Amoco Oil Engineering Center in Whiting, Indiana, reading a report from the R&D department. The report was about a xylene isomer splitter. The ability to fractionate between the two xylene isomers (needed to make Styrofoam coffee cups) was limited by a low reflux rate. The reflux rate was constrained by a low reboiler duty. The reboiler duty was low because of a low reboiler heat transfer coefficient:

$$U = \text{Btu/hr/°F/ft}^2$$

The research engineer assigned to rectify this malfunction had placed a restriction orifice at the reboiler inlet (xylene on the shell side; steam on the tube side) to reduce the shell-side circulation rate. This was contrary to my training as a new chemical engineer. I had learned in school that high velocities will increase heat transfer coefficients:

$$U \propto (M)^{0.7}$$

where M is the mass flow rate in lb/hr. This was also stated in our bible on heat transfer, *Heat Transfer Fundamentals*, by Donald Kern.

However in this case, the reduced liquid xylene circulation rate resulted in a higher heat transfer coefficient. And my experience over the years has also confirmed that in clean services, when reboiling a pure component, slowing the liquid circulation rate does increase heat transfer. Of course, if the flow gets too low, some of the tubes may dry out, which will impede heat transfer. This is true for condensation as well. High velocities impede condensing heat transfer rates. The common experience of blowing a steam condensate seal illustrates the principle.

How or why this happens, I cannot explain. It only applies when almost all the heat transfer is in the form of latent heat (which it will be for a pure component such as xylene) and almost none of the heat transfer is in the form of sensible heat.

Flux-Limited Situations

I never actually read Donald Kern's book. It's too depressing. I just look stuff up in it. In 1966, I unfortunately read that reboilers were flux limited to a rate of 12,000 Btu/hr/ft^2. The ft^2 term refers to the outside area of the tubes. I say that my reading this was both unfortunate and depressing because I had on my desk a performance test from the Amoco refinery in El Dorado, Arkansas. In particular, I was evaluating the performance of an old propane–butane splitter. The actual flux rate I had calculated from the performance data on this old reboiler was about 18,000 Btu/hr/ft^2.

So the one and only fact that I had ever checked in Mr. Kern's text on heat transfer was wrong! What a way to begin my chemical engineering career. Hence my depression.

But Donald Kern was not wrong. His 12,000 Btu/hr/ft^2 flux limitation was correct. It's just that I had to wait another 40 years to see the light. Kind of like the Jews wandering in the Sinai Desert for 40 years.

Brent Hawkins was the tech manager for the Valero plant in Houston. He was reboiling a tower with a butane bottoms content of about 95%. Thus, almost all the reboiler duty was in the form of latent heat transfer rather than sensible heat transfer. During a unit turnaround, the 20-year-old carbon steel bundle was replaced with a new bundle. This was done because about 10% of the old tubes were plugged, due to corrosion failures and pitting on the process (i.e., the shell) side. 30 psig steam was the tube-side heating medium. When the tower was returned to service, the maximum reboiler duty capacity had dropped by half, even though the 10% loss in tube surface area had been eliminated. Brent repiped the steam supply to use 100 psig steam instead of the 30 psig. This doubled the LMTD (log mean temperature driving force) of the reboiler. But this reduced the reboiler duty even more.

Brent then retained my services. I recall the incident rather precisely because I was never paid for my advice. First, I calculated the current flux rate. It was almost exactly 12,000 Btu/hr/ft^2. This was the flux-limited situation, correctly described in Mr. Kern's old text, *Heat Transfer Fundamentals*. Also, just as noted in this venerable text, when limited by flux rate in latent heat transfer service, making the heating surface area hotter retards the heat transfer rate—just as Brent observed.

The problem is called a **nucleate boiling limitation**. You can see this problem if you try to boil clean water in a new pot without any scratches. If you want to promote higher rates of boiling in your pot, you have two choices:

1. Add a little grit to the water (i.e., "boiling stones").
2. Scratch up the pot.

What we did at the Valero plant in Houston was to reinstall the old bundle. The rough surface area of the corroded and pitted tubes permitted heat transfer flux rates in excess of 17,000 Btu/hr/ft^2. Then I understood the 18,000 flux rate I had observed in 1966 at the Amoco refinery in El Dorado, Arkansas.

If you don't have roughened surface areas of the tubes, you have two choices:

1. Lightly sandblast the tubes before use.
2. Have the tubes coated (electrically) with a sintered metal coating on the process side. This technology was licensed by Union Carbide (Linde), but I suppose their patents have long since expired.

Forced-Circulation Refrigerant

I have applied my experience on reboilers to refrigerant heat exchange. The refrigerant is just like the process fluid, and the material being chilled is analogous to the heating medium. The refrigerant is often a pure component (Freon, NH_3). However, in a refinery, the cheapest refrigerant is propane.

In the experiment I conducted in El Dorado, Arkansas, the refrigerant was not quite a pure component:

- 60% propylene
- 35% propane
- 5% butane

The heating medium (i.e., the reactor effluent) was on the shell side of a vertical heat exchanger. The refrigerant was on the tube side. I had retrofitted the unit with a large refrigerant circulation pump with the expectation that the heat transfer coefficient would be enhanced with a high rate of refrigerant circulation. When I started circulation with the new pump, I began at a minimum rate so that 100% of the refrigerant would vaporize in the tubes. This was basically the old mode of operation. When I doubled the circulation rate to achieve a refrigerant flow sufficient to reduce the evaporation rate to 50%, the heat transfer coefficient increased by 60%. However, further increases in the refrigerant circulation rate did not noticeably increase the exchanger heat transfer rate, and even seemed to hurt a bit.

I imagine that maintaining sufficient excess flow of the pure component that will be vaporized—so as to keep all the tubes full (tube side), or submerged (shell side), in the heat exchanger—is important. But further increases in the flow rate apparently suppress nucleate boiling. Note that I have not yet mentioned heat transfer **film resistance**. When dealing with a pure latent heat transfer (condensation or vaporization) coefficient, I do not believe that film resistance is a factor. However, boiling and condensing heat transfer resistances are usually small compared to the heat transfer resistance of the fluid on the other side of the exchanger, or compared to the effect of fouling.

Film Resistance in Sensible Heat Transfer

If the primary component of the reboiler duty is sensible heat transfer, then film resistance is the limiting factor. I'll illustrate this principle with a troubleshooting example from a refinery in Durban, South Africa.

The process fluid was on the shell side of this horizontal reboiler. The heating medium was 400 psig saturated steam. The malfunction was low heat transfer coefficient:

- Design = 95 Btu/hr/ft^2/°F
- Actual = 30 Btu/hr/ft^2/°F

The initial concern was shell-side fouling. But the exchanger was cleaned with little observed benefit. The next thought was steam condensate backup. The 400 psig steam condenses at 448°F. I checked the steam condensate temperature draining out of the channel head (i.e., tube side). It was about 440°F. If condensate backup had been the problem, then the water draining out of the channel head would have been subcooled, which was not the case. Note that condensing steam has a heat transfer coefficient of over 1,000 Btu/hr/ft^2/°F. Hence, I ignore the steam side as a limiting factor in heat transfer problems, provided condensate backup is not an issue (see Chapter 14, "Steam Condensate Collection Systems").

By process of elimination, the malfunction was poor heat transfer efficiency on the shell side. One possibility that accounts for poor shell-side heat transfer is excessive clearance between:

- The tubes and the holes in the tube support baffles.
- The support baffle OD (outer diameter) and the shell ID (inner diameter).
- The outer row of tubes and the shell ID.

A field check showed that all shell-side clearances corresponded to Tubular Exchanger Manufacturer Association (TEMA) clearances. Apparently, the malfunction was then excessive shell heat transfer film resistance. Film resistance is a function of:

- Viscosity of the fluid
- Velocity of the fluid
- Density of the fluid
- Equivalent diameter of the flow path
- Thermal conductivity of the fluid
- Surface tension of the fluid
- Direction of fluid flow

This last factor is of critical importance. When liquid flows along the length of the tube, film resistance is high. But if liquid flows crosswise, perpendicular to a tube, **vortex shedding** results.

Effect of Vortex Shedding on Reboiler Duty

Did you ever watch a river flow rapidly past a tree stump? Do you recall the swirls of water around the dead tree? That turbulence is vortex shedding. Vortex shedding disturbs the stagnant liquid film

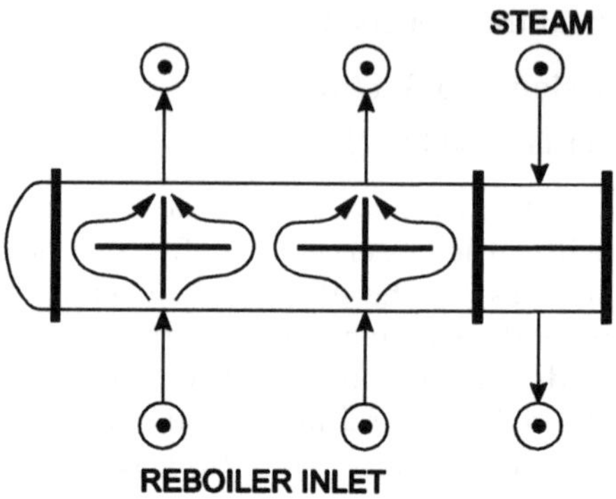

Figure 9-1 Liquid flow parallel to tubes is bad for heat transfer.

around the exterior surface of heat transfer tubes. This greatly diminishes shell-side heat transfer resistance. The greater the component of **cross-flow velocity** around the tubes, the more intense is the vortex shedding. Cross-flow velocity means the component of fluid flow running at 90° to the length of the tubes. Flow that is parallel with the tube length has a zero component of cross-flow velocity, and thus does not promote beneficial cross-flow velocity.

Figure 9-1 shows the internal shell-side baffles of the poor-performing South African reboiler. There are dual inlets and outlets. The baffle configuration is called a "double split-flow." There are four parallel liquid pathways. Other than the two vertical baffles used to divide the shell-side flow, there are no other vertical tube support baffles. The objective of such an arrangement (the four parallel passes, without intervening vertical baffles) is to minimize the shell-side delta P.

The exchanger data sheet specified a maximum delta P of 0.5 psi. The delta P measured in the field was essentially zero. The available **thermosyphon driving force** was equivalent to several psi, but the reboiler design did not use this available delta P.

The main feature of the reboiler shown in Figure 9-1 is that the shell-side flow is parallel to the length of the tubes, rather than perpendicular to the tubes. Also, the component of the parallel flow velocity was less than 1 foot per second. The reboiler configuration resulted in very little vortex shedding. The lack of vortex shedding would not have been of any particular consequence if the main mechanism of heat transfer in the reboiler had been nucleate boiling due to latent heat transfer. But in this reboiler, due to the large temperature rise of the process fluid, most of the heat transfer duty was in the form of sensible heat transfer. Hence the lack of cross-flow velocity, vortex shedding, and low velocities combined to produce the observed low heat transfer coefficient of 30 Btu/hr/ft^2/°F.

Effect of Additional Baffles

To induce vortex shedding, I designed a new tube bundle with eight sets of additional segmental baffles. In the new bundle, most of the tube surface area would be devoted to a cross-flow velocity of about 3 ft/sec. I've discussed the basis for achieving a minimum cross-flow velocity of 3 ft/sec in my book, *Process Design for Reliable Operations*. In practice, it's too difficult to add baffles to an existing tube bundle. So a new and rather expensive bundle was purchased by my client. When the reboiler was returned to service with its new bundle, the heat transfer coefficient was more than doubled.

I explain this story in my troubleshooting seminar as it illustrates the importance of vortex shedding when dealing with high-heat-transfer film resistance and sensible heat transfer. The importance of this concept increases for services with high viscosity and low **Reynolds numbers**.

Lost Thermosyphon Circulation

Most reboilers in process plants operate as natural or thermosyphon circulation reboilers. Figure 9-2 illustrates the origin of the pressure driving force that creates thermosyphon circulation. The ΔH dimension

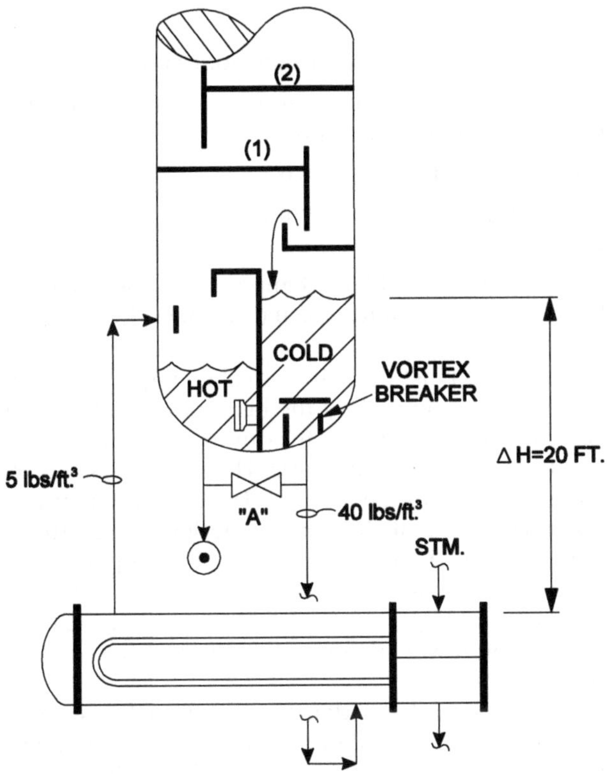

Figure 9-2 ΔH determines the thermosyphon driving force.

shown in the figure is multiplied by the density difference between the liquid flowing to the reboiler (40 lb/ft^3) and the mixed-phase vapor–liquid mixture leaving the reboiler (5 lb/ft^3). Then the available thermosyphon driving force in psi is:

$$20 \text{ ft} \times (40 - 5) \text{ lb/ft}^3 \div 144 \text{ (in)}^2/\text{(ft)}^2 = 4.8 \text{ psi}$$

I've used the parameters from the Durban, South Africa, reboiler in this example. An available thermosyphon driving force of 4 or 5 psi is quite large and not typical. If the frictional loss through the reboiler, nozzles, and piping is less than the available thermosyphon driving force, then the liquid level on the cold side of the tower will fall until the available thermosyphon driving force balances out with the required thermosyphon driving force.

But suppose the opposite happens. Perhaps the reboiler begins to foul on the shell side and the delta P on the shell rises; or the piping is restricted with coke; or some object partially plugs the vortex breaker, covering the tower nozzle that feeds the reboiler. The level on the cold side of the reboiler (see Figure 9-2) will increase until the liquid from tray #1 overflows onto the hot side of the tower. And as a consequence of this:

- The hot (or product) side of the bottom of the tower cools.
- The reboiler outlet temperature gets hotter, and may actually turn into superheated vapor.
- The reboiler duty goes down.
- The bottoms product, having partly bypassed the reboiler, gets lighter.
- At lower flows and higher temperatures, the reboiler will start to foul at a greater rate, which just makes the problem worse.

The malfunction I just described is called **vapor lock**. The rate of reboiling is no longer limited by the heat transfer film resistance or fouling, or condensate backup, or maximum flux rate due to nucleate boiling, or a lack of surface roughness. The problem is lack of process flow into the shell side of the reboiler. The evidence that proves this sort of malfunction is that the reboiler outlet temperature is rising and the bottom product outlet temperature is dropping.

Of course, the same symptoms will occur if tray deck #1 is damaged, and liquid bypasses the reboiler by dropping directly into the hot side of the tower bottoms (i.e., the left side of the vertical baffle in Figure 9-2).

Fouled reboiler shells or damaged bottom trays are common malfunctions that result in loss of thermosyphon circulation, or what operators call vapor lock. How, though, to discriminate between the two problems? There are two methods:

1. If you have a level indication on the hot side, see if this level matches with the top of the baffle. If it does, there is a hydraulic restriction in the exchanger, piping, or nozzles.
2. If you open up the startup line (shown as valve A on Figure 9-2) and the vapor lock is broken, then the problem is with tray #1. Or perhaps some evil person during a turnaround has left open the internal manway on the vertical baffle in the bottom of the tower.

Reboiler Delta P Causes Tower Flooding

I was working in Mumbai (formerly Bombay) on a debutanizer flooding problem. The tower started up at design capacity and worked fine. But as time progressed the tower started to flood. Tray delta P went up and fractionation efficiency went down. Increasing reflux and reboiler duty made fractionation worse—a sure sign of flooding.

The tower was taken offline to clean the trays. However, the trays were not found to be particularly fouled. When the tower was restreamed it flooded just like before the turnaround. Also, the problem became progressively worse.

Reboiler capacity had also declined, but was not a limiting factor, as the steam inlet valve was never more than 50% open. Therefore,

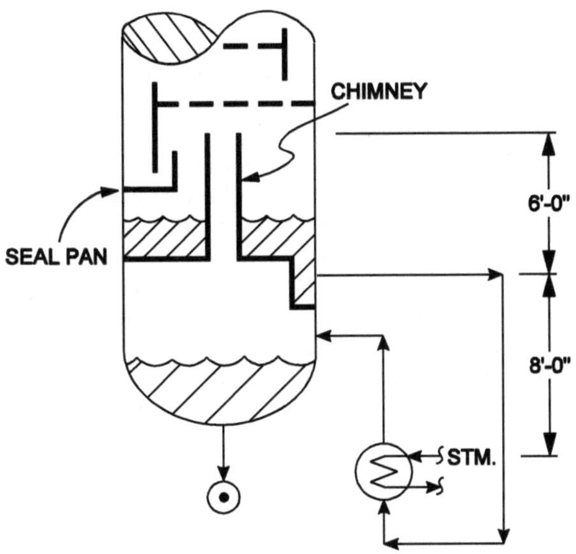

FIGURE 9-3 High reboiler delta P causes tower to flood due to lack of provision for internal overflow from chimney tray.

the reboiler shown in Figure 9-3 was not considered to be part of the problem and was not cleaned.

When I studied this problem in Mumbai, I noted three things that made me worry and wonder:

1. The reboiler feed draw-off nozzle on the tower was only a few feet above the top of the reboiler shell.
2. The measured pressure drop across the reboiler (after correcting for elevation) was about 3 psig.
3. The **total trap-out chimney tray** did not have provision for internal overflow.

Let's first calculate the required thermosyphon driving force, assuming the same fluid densities as in the previous example. Hopefully you have calculated, based on the previous example as a guide, 12½ feet. If not, I've repeated the calculation procedure as shown below:

$$(3 \text{ psi delta P}) \times (2.31) \times [62.3 \div (40 - 5)] = 12\tfrac{1}{2} \text{ ft}$$

where The 3 psi is the exchanger delta P.

The 2.31 factor converts psi to feet of water.

The 62.3 ÷ (40 − 5) factor corrects the weight of a column of water to the weight of a column of hydrocarbon with a density of 40 lb/ft^3.

I've subtracted 5 from the 40 to correct for the density of the mixed phase in the reboiler effluent riser line of 5 lb/ft^3.

Next, compare this to the elevation between the top of the reboiler and the draw-off nozzle. This dimension was only 8 feet, not 12½ feet. So the liquid level on the chimney tray would need to back up by 4½ feet (12½ feet minus 8 feet).

Actually, this kind of worked. The top of the chimney was 6 feet above the draw-off nozzle. So the liquid level backed up the chimney tray until the required thermosyphon driving force equaled the available thermosyphon driving force. Then as the reboiler fouled with time, and the shell-side delta P increased, the liquid level on the chimney tray was pushed up an inch or two each month. But circulation through the reboiler was maintained.

But you will notice on Figure 9-3 that the downcomer and seal pan from the bottom tray extends far below the top of the chimney. As the liquid level was pushed up on the chimney tray by the fouling reboiler, the level in this downcomer was pushed up, until the level on the bottom tray started to back up. And since flooding progresses up a tower, the whole tower flooded.

To stop the flooding, the operators had to reduce the reboiler pressure drop, by reducing the reboiler duty. Reducing the reboiler duty reduced the reboiler delta P by reducing vapor traffic. But of course, reduced reboiler duty necessitates less reflux. Which impairs fractionation efficiency.

What to do?

The fundamental malfunction was not reboiler fouling. This was a fouling service, and fouling was unavoidable. Certainly, we shut down to clean the reboiler. But the real fix was an 8-inch overflow pipe. I designed the pipe so that its top edge was in line with the top of the seal pan. While bypassing the reboiler due to overflowing my new 8-inch pipe diminishes thermosyphon circulation, this is not as serious a malfunction as flooding the entire tower.

The Indian engineers were truly shocked. They could not understand how the tower was designed without the overflow pipe in the first place.

"Mr. Norman, this tower was designed by a famous American engineering company. How could such a fundamental design error be made?" Kumar asked.

As I put aside my slide rule to answer his question, I noticed something rather odd. It was 9:30 p.m. I was working late to finish the design that evening. Kumar, Ashok, Shiny, and the entire HPCL Refinery Tech Service group were standing behind me, watching my calculations.

"Maybe," I thought, "that's the answer. Maybe that's the difference between American and Indian process engineers." But you hate to start mentioning such things at that time of night, and so far from home.

Incidentally, the bottom of the new 8-inch overflow pipe had to extend down below the liquid level in the bottom of the tower to maintain a liquid seal. Otherwise, rising vapors could have interfered with the downflow of liquid through the overflow pipe.

Kettle Reboilers

I have always disliked kettle reboilers because they are dirt traps. In clean services, they are fine. For example, they are a cheap way to build an evaporator with the clean, circulating refrigerant on the shell side and the process fluid to be chilled on the tube side. The only contaminant in the NH_3 or propane refrigerant is lube oil from the refrigerant compressor. This is readily drained off from the bottom of the kettle.

But in my world—amine systems, naphtha splitters, debutanizers, aromatic fractionators—there are no truly clean systems. Scale, corrosion products, and salts will accumulate behind the overflow baffle shown in Figure 9-4. Note that in a kettle reboiler, unlike the thermosyphon circulation–type reboilers previously discussed, there is, by design, no liquid circulation. Only vapor flows back to the tower. Thus, dirt gets trapped behind the baffle.

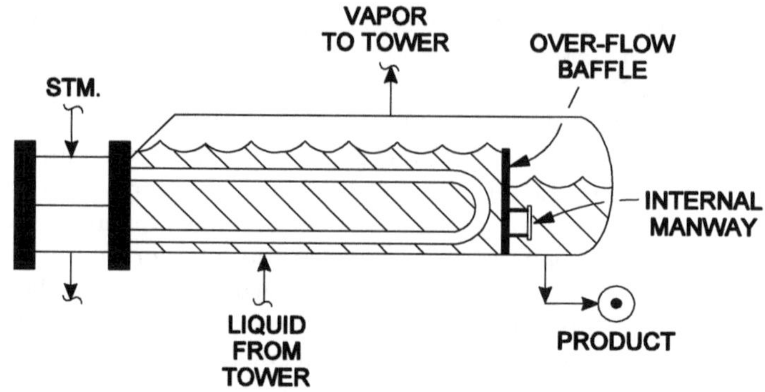

Figure 9-4 Kettle reboilers are dirt traps.

The consequences of this dirt buildup are:

- Tube failure due to under-deposit corrosion.
- Loss of heat transfer surface area.
- Increased shell-side delta P.

Unfortunately, the increased shell-side pressure drop has exactly the same result as in my previous example. That is, liquid backs up the tower to overcome the increased kettle reboiler **head loss** or pressure drop, which doesn't hurt anything until the liquid level in the bottom of the tower backs up above the reboiler vapor return or the seal pan from the bottom tray (whichever is lower). Then the tower will flood—just like in the previous story from Mumbai.

Cutting Holes in Kettle Reboiler Baffles

Most of my clients deal with this malfunction by cutting a large hole in the base of the overflow baffle. This reduces the kettle reboiler pressure drop and lowers the liquid level in the associated tower. But if you will study Figure 9-4, would you not agree that this is the same as bypassing some feed around the reboiler directly to the bottoms product? And bypassing feed around any heat exchanger will always have the same results:

1. Heat exchange efficiency is impaired.
2. Lower velocities promote higher rates of fouling.

In this case, as the kettle reboiler is contributing to fractionation, the fractionation efficiency of the tower itself is diminished.

Yet I myself, to overcome the inherent fouling nature of kettle reboilers, have had openings cut in the overflow baffles to relieve tower flooding due to liquid backup. The real answer to this sort of malfunction is not to build kettle reboilers in fouling services in the first place. However, if one does have a design with kettle reboilers in such a fouling service, then keep the tube support baffles, and hence the tubes themselves, elevated (maybe by 6 inches) above the bottom of the shell.

I have often cut out two rows of tubes in the bottom of a kettle reboiler to create such an open area where dirt can accumulate without restricting the shell-side flow (see my book, *Process Design for Reliable Operations*).

Tower Stab-In Reboilers

The only stab-in reboiler I have ever specified was for a delayed coker blowdown recovery quench tower. In this case, level control was not an issue. As I discuss in Chapter 16, "Level Control Problems," the use of stab-in reboilers creates a host of level measurement problems. Handling reboiler tube leaks is also a bigger problem for a stab-in reboiler. However if you do have such a reboiler, the most common malfunction occurs when too low a level uncovers tubes. The uneven heating of the tubes can cause mechanical stresses due to differential rates of thermal expansion of individual tubes. Try increasing the tower level. If this also increases reboiler duty, then the tower bottoms level was too low to start with.

Reboiler Floating Head and Tube Leaks

There are four types of leaks that may occur in a shell-and-tube reboiler:

- Ordinary tube leak
- Roll leaks in tube sheets
- Floating head gasket leaks
- Hole in floating head

The latter two modes of failure are not applicable to U-tube bundles. However, U-tube bundles are best not used anyway (see Chapter 10, "Shell-and-Tube Heat Exchanger in Sensible Heat Transfer Service").

Roll leaks are common but, at least the ones I've seen, are pretty small. Tubes are sealed in the tube sheets, not by welding, but by forcefully expanding the tube's outer diameter up against the holes drilled in the tube sheets. Roll leaks are found by pressuring up the shell side with water. The water will seep out around the OD of the tube, on the outside face of the tube sheet. Roll leaks are best repaired by rerolling the ends of the tubes rather by than welding up the leaking roll joints.

Ordinary tube leaks are also observed by pressurizing the shell with water. Only the channel head cover has to be dropped to find leaking tubes. I have a vast amount of experience in leaking reboiler tubes. Of course, corrosion on the steam side due to carbonic acid, and corrosion on the shell side due to sulfur, HCl, and weak H_2SO_4, are common.

I have discussed these subjects elsewhere in Chapter 10. But for now, let me tell you all something you will not read in any other book. Most tube leaks in reboilers observed on startup, after a unit turnaround, are caused by your maintenance people. Tube bundles are constructed from thin-walled (0.1-inch) tubes that are easily bent or broken if not handled with care. These bundles must be lifted and reinserted in the reboiler shell carefully to avoid damage. Especially for amine regenerator reboilers in Aruba, and coker stabilizers in Texas City.

Large tube leaks can cause even larger floating head cover leaks. If the shell-side pressure is much greater than the tube-side pressure, then the shell-side fluid can blow out against the floating head with great erosive force. Especially if the shell-side fluid contains corrosive components, a large hole can be eaten through the floating head. I had such a hole in an alkylation unit depropanizer in Texas City. It was 4 inches in diameter and the floating head was an inch thick!

However, most floating head leaks are less dramatic. They are gasket leaks at the closure between the floating head itself and the floating head tube sheet. It's a consequence of sloppy rebolting of the floating head and should never occur. Unfortunately, the leak cannot be found by hydro-testing by pressuring-up the shell. That just forces the floating head tightly up against the floating head tube sheet. But during operations, if the tube-side pressure is larger than the shell-side pressure, then the floating head will be pushed away from the tube sheet and establish a leak.

Maintenance-induced malfunctions are just part of the job. We process people have to accept that our coworkers in the maintenance division are not perfect. However, they ought to be held accountable for their poor workmanship, just as a consul operator is for operational errors.

Effect of Leaks on Process Operation

If the pressure of the steam or the hot oil side of the reboiler is greater than that of the process side, then other than contamination of the tower's bottom product, there is no real effect on the operation. But if the pressure of the tower or process side is greater than that of the steam or hot oil side, the malfunction is a lot more complex to troubleshoot. For example, in 1975, I was reboiling a butane splitter with 30 psig steam. The butane was on the shell side. The condensing 30 psig steam was on the tube side of this horizontal thermosyphon reboiler. The shell-side pressure was 180 psig.

The exchanger developed a very small tube leak. The liquid butane flowed into the tube, flashed, and accumulated in the channel head. The butane vapors were trapped in the channel head because vapors cannot flow through a steam trap. The purpose of a steam trap is to pass liquid but retard the flow of vapors. I knew that butane vapors were trapped in the channel head because I opened the ¾-inch vent on the head and observed hydrocarbon vapors being vented. These vapors filled the upper rows of tubes and thus restricted the rate of steam condensation. Opening the vent on the channel head increased my reboiler duty back to normal. But even in 1975, even in Texas City, I was not permitted to operate with a ¾-inch vent blowing hydrocarbon vapors continuously to the earth's atmosphere at grade level.

The cause of the tube failure was weak sulfuric acid corrosion in carbon steel reboiler tubes. The H_2SO_4 originated in my sulfuric acid alkylation unit reaction loop.

Another example, also more recently from Texas City, occurred on a debutanizer reboiler. Again, the process fluid was on the shell side of a horizontal thermosyphon reboiler. In this case, the tube-side heating medium was hot oil. The hot oil was circulated by a large centrifugal pump with a normal discharge pressure of 160 psig.

The debutanizer pressure varied from 140 to 180 psig. At 140 psig, the tower operated in a stable, controllable fashion. Above 160 psig, the flow of hot oil to the reboiler became erratically low. The malfunction was a tube leak. That is, the volatile debutanizer bottoms leaked into the hot oil return line and then:

- The debutanizer bottoms product flashed in the 6-inch return line.
- The evolved vapor increased the delta P in the return line.
- The hot oil centrifugal pump discharge pressure increased.
- The pump was pushed up on its performance curve.
- The flow of hot oil to the debutanizer reboiler (and to all the other reboilers served by this pump) decreased.
- The debutanizer reboiler duty dropped, and consequently the level in its reflux drum also declined.
- Reducing the debutanizer pressure 20 psig below the hot oil pump discharge pressure restored stability. But at the lower pressure, the tower overhead product could not be condensed.

The reboiler tube leak had occurred right after startup. Investigation proved the bundle had been damaged due to improper handling by the maintenance crew. My bitter comments at the time were not particularly appreciated by the plant management in Texas City. But that was a long time ago, and I've completely forgotten the entire rotten incident.

Steam-Side Problems

One of the main problems we have with reboilers is not on the process side, but on the heating medium side. While many towers are reboiled with circulating hot oil systems, pumparounds, or hot reactor effluents, most columns are reboiled with condensing steam. The resulting condensate (water) has to be recovered and not drained to the deck or sewer. And this creates a big, very complex problem, which degrades the capacity of many reboilers, due to condensate backup or blowing the condensate seal. This is the subject of Chapter 14, "Steam Condensate Collection Systems."

CHAPTER 10

Shell-and-Tube Heat Exchangers in Sensible Heat Transfer Service

We have met the enemy and he is us.

—*Pogo* comic strip

Heat exchanger fouling, blown trays, leaking pump seals, and faulty instrumentation are the main malfunctions I have had to contend with. I recall a refinery in Louisiana that had a malfunctioning refrigeration loop. The refrigerant compressor discharge pressure was excessive and was reducing the flow of the NH_3 refrigerant. The problem was poor performance of the compressor discharge NH_3 condensers.

The high cooling water outlet temperature and low cooling water flow indicated severe tube-side plugging and fouling. We do not ordinarily measure the cooling water flow. It's a calculated value:

Step 1—Determine the process side duty in Btu/hr.

Step 2—Measure the cooling water temperature increase in °F.

Step 3—Divide step 1 by step 2 to determine the cooling water flow in pounds per hour. You can divide by 500 to convert to GPM (U.S.).

I had one of the parallel condensers taken offline for cleaning. The cooling water was on the tube side, and the NH_3 was on the shell side. The bundle construction was of the U-tube type shown in the lower portion of Figure 10-1. To clean tubes, we use a water lance.

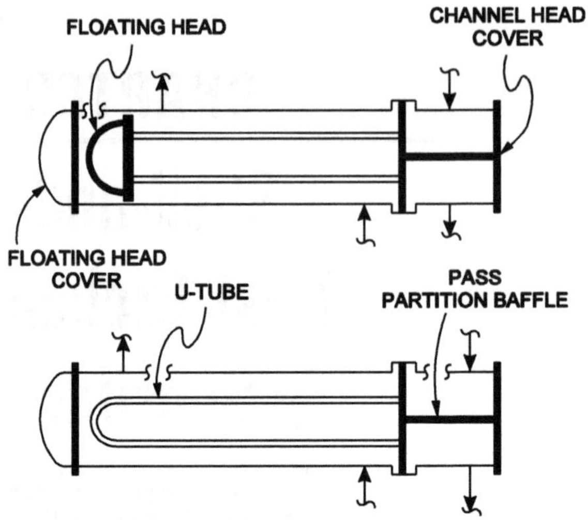

FIGURE 10-1 Floating head bundles are preferred over U-tube bundles, which cannot be inspected for clean tube side.

High-pressure water is injected into each tube using a lance constructed of 10 to 20 feet of ½-inch OD stainless tubing. The lance is then pushed through each tube separately. This is called "lancing the bundle." It is typically done by unskilled contract labor.

After the NH_3 condenser was lanced, its performance did not improve. My client was not impressed with my recommendation but nevertheless continued the cleaning program on a second condenser. This time, I decided to observe the lancing process myself, even though the work was done on the night shift.

As the cleaning progressed, I noted that the worker lancing the bundle was not "clearing the tubes." This term means that the injected water blows out of the far side of the tube being lanced. If the bundle has a **floating head** configuration as shown in the upper half of Figure 10-1, the water blows out the opposite end of the tube bundle being lanced. If the bundle has a U-tube configuration, the water blows out the same end of the exchanger being lanced. The problem was that on the NH_3 condenser being cleaned, water was only clearing half of the tubes.

So I asked the worker using the water lance, "Hey! How about the tubes you're not clearing? They're still plugged."

"I'll get them later. I got to put me on a special tool to make that turn in the U-bend."

"Yeah, but how do you know which tubes you have to redo," I asked. "This bundle has a thousand tubes."

"Well, I got me a super-good memory. That's how come I kin remember which of them there tubes I got to go back and reblast. But say, who the hell are you, anyway?"

Now there are two possibilities. One is that this guy really can remember which 500 of the thousand tubes he has to relance. The other possibility is that he isn't going to do it. And after he is done, how can I see if the tube bundle has really been cleaned? It's quite impossible to make any meaningful visual inspection. On the other hand, for a floating head-type bundle, the tubes can be easily inspected by looking through each tube individually from either end of the bundle.

U-tube bundles represent inferior engineering design and should not be used, as they cannot be visually inspected on the tube side after cleaning.

Advantages of U-Tube Exchangers

It is likely that 30% to 40% of the shell-and-tube heat exchangers I work with do employ U-tube bundles. You would then think they must have some sort of advantage over floating head bundles. Perhaps it's cost? The floating head alternate is 10% to 20% more costly than the U-tube bundle. But that's a small factor in a major project. The real advantage is the propensity of the floating head to leak. A leaking floating head allows the tube-side fluid to pass into the shell side fluid, provided the tube-side pressure is greater than the shell-side pressure.

The reverse is not true. If the shell-side pressure exceeds the tube-side pressure, no leakage is typically observed. The reason is that the floating head is forcefully pushed into a tight position by the differential pressure between the shell and tube sides. Certainly, there is no legitimate reason why we should have internal floating head leaks. It's just sloppy maintenance. But it happens all the time in process plants.

Effect of Surface Roughness

As discussed in Chapter 9, surface roughness can be quite beneficial in latent heat transfer. The pits and crevices provide sites that promote nucleate boiling. But these same pits and crevices also provide sites for the growth of fouling deposits. Especially in sensible heat transfer applications (no boiling or condensing), surface roughness due to corrosion ordinarily promotes fouling and retards heat transfer efficiency. The metallurgy of tubes is typically selected for a 10-year life. Selecting alloy tubes rather than carbon steel cannot be economically justified on this basis. But the alloy tubes will remain smooth, and hence they will resist fouling and the loss of heat transfer.

When my clients retube carbon steel (c.s.) bundles with alloy tubes, they create a potential problem in galvanic corrosion. For example, a tube bundle in Aruba was retubed with 316 s.s. (stainless steel), but the old c.s. **tube support baffles** were not replaced.

The high-alloy tubes were in direct physical contact with the c.s. support baffles. Electrons flowed between the dissimilar metals. When I had the bundle extracted from the shell, the tubes were fine, but the baffles were mostly corroded away. That's galvanic corrosion in action.

Yet ESSO, who owned the refinery when the bundle was retubed, must have known all about this common malfunction. They addressed the issue by installing a **sacrificial anode** on the shell side of this exchanger in this hydrodesulferizer reactor effluent service. I'm quite sure of this, as I saw the anode when we pulled the bundle. So what went wrong?

One of my colleagues explained the malfunction to me. The anode must be wired up correctly to the baffles for the anode to be sacrificed instead of the c.s. baffles. If these electrical connections come loose, then the anode cannot work. He also noted that the use of sacrificial anodes was poor engineering practice, as anode replacement is often neglected during turnarounds. Correct engineering practice is to construct the tube bundle out of the same or similar metals in the galvanic series. For example, as a refinery engineer, it's nice to know that 9 chrome steel is okay to use in direct physical contact with carbon steel components.

Fouling Due to Laminar Flow

Why do rivers cut bends in their channels? Because of erosion due to high velocity. The outside bend of the channel moves faster than the inside bend, as it must flow farther. The higher velocity and turbulence erode the outside of the bend faster than the inside of the bend. Hence, the river is forever moving to create bigger loops.

Inside the tubes of a shell-and-tube exchanger, we need to have a reasonably high velocity to keep the dirt eroded off the outside surface of the tubes or to prevent the dirt from sticking to the inside of the tubes.

What is a reasonable tube-side velocity to retard tube-side fouling? It rather depends on several factors:

- Surface roughness
- High viscosity
- High shell-side temperature

These factors all promote tube interior fouling. Based on my long experience, old American Oil design guidelines, recommendations in the TEMA data book (Tubular Exchanger Manufacturer Association), and Donald Kern's book, *Heat Transfer Fundamentals*, I've developed the following rules of thumb:

- For reasonably clean, low-viscosity fluids exchanging heat at moderate (100°F–400°F) temperatures with smooth tubes, use a tube-side velocity of around 3 ft/sec. By low viscosity, I mean less than 10 centistokes.

- For dirty, high-viscosity, hotter tubes using pitted carbon steel tubes, use a tube-side velocity of around 6 ft/sec.
- Regardless of other circumstances, keep the tube-side velocity below about 10 ft/sec, to minimize tube metal erosion.

Combating Tube-Side Fouling

Let's assume the material being heated is on the tube side. If it's a fouling service, you will have noticed that not all the tubes are uniformly restricted. What happens is that some tubes start to foul, and flow through these tubes gradually becomes restricted. The reduced flow through these tubes causes them to run hotter and foul faster. This then causes the tubes to run even hotter and to develop progressively lower tube velocities. The problem feeds upon itself until that particular tube is entirely plugged. Other tubes that are still flowing are running cooler. A symptom of this problem is that when the bundle is pulled, many tubes look bent and twisted. That's because the floating head shown in Figure 10-1 (top) cannot accommodate differential rates of thermal expansion between the hotter and colder tubes. If half the tubes plug off, then the tube-side delta P will increase by a factor of four (delta P varies with velocity squared). This differential pressure exerts a huge force on the pass partition baffle in the channel head, which not uncommonly starts to leak or fails entirely.

Probably the simplest thing to try to suppress such fouling is to blow nitrogen (or air if water is in the tubes) for a few minutes a day into the channel head inlet nozzle. I presume this helps because the tube-side deposits are disturbed before they can solidify their position. Of course, in cooling water service, the standard back-flushing procedure would be followed (see Chapter 13, "Cooling Water: Towers and Circulation").

A more aggressive method of disturbing the tranquil growth of the fouling deposits in hydrocarbon service is **online spalling**. Proceed as follows:

- Open the tube-side bypass.
- Block-in one tube-side valve.
- Leave the hot shell side flowing.
- Wait 10 minutes and restore the tube-side flow to normal.
- Retighten any leaking flanges.

Some of my clients follow this procedure but add a few barrels of solvent into the channel head inlet to dissolve tougher deposits. I've only actually used this method myself on crude exchangers upstream of the desalter, where tube-side temperatures never exceed 300°F.

Ultimately, the best fix is to correct the fundamental problem—low tube velocity. My usual method is to increase the number of the tube-side passes. For example, I could double the tube-side velocity in a bundle by going from two passes to four passes. This requires two off-center pass partition baffles in the channel head and a new pass partition baffle in the floating head. This is an expensive modification. You will have to have both tubesheets remachined to accommodate the three new pass partition baffles.

Strange to say, the single most effective way of combating overall heat exchanger fouling is a crash shut-down and rapid restreaming of the process. This is a lot more effective than the use of expensive anti-foulant chemicals. Naturally, most of my clients would not approve of this method, but they use it inadvertently anyway.

Avoiding the Root Cause of Fouling

A few basic suggestions are:

- For cooling water systems, keep hydrocarbon leaks out of the circulating water. The hydrocarbon is food for the bugs that foul cooling water tubes.
- For systems with H_2S, minimize the HCl content. When H_2S reacts with steel in the presence of wet HCl, the end product of corrosion is water-insoluble $Fe(HS)_2$ (iron sulfide), rather than water-soluble ferric chloride.
- For hydrocarbon systems, do not allow thermally degraded and cracked materials to contact air. The result will be gums that cause particulate matter to stick to metal surfaces.
- For refineries, do not mix chemical slops and oil recovered from sewers into the feed to the crude distillation unit.

Shell-Side Fouling

Once a tube is plugged by fouling deposits or solidified hydrocarbons, it will remain in that state until hydroblasted by a water lance. An area of the shell side where flow has been lost might be cleared by a localized high velocity flowing across an adjacent tube, but this potential for shell-side flow restoration will probably be lost due to "shell-side bridging." When a tube bundle is pulled, areas which have experienced shell-side bridging are rather obvious. The dirt is packed tightly into these areas. The root cause of the bridging is the tubes distorting and moving too close to each other. Once the fouling deposits on adjacent tubes touch, bridging has resulted. Low or zero flow will now occur in the region of the bridged tubes. Dirt will pack into these stagnant areas. Localized areas of the exchanger tubes will now be unavailable for heat transfer.

One of the causes of tube bridging I've described is plugging off of individual tubes due to low flow and fouling. The plugged tube can cause a strain on adjacent tubes, as it will be at a temperature that is different from the surrounding tubes.

Perhaps the main cause of tube bridging is improper tube support. Too few tube support baffles may be used in an effort to reduced the shell-side delta P. If the baffle spacing is increased from 10 to 20 inches, the calculated shell-side pressure drop will decrease by a factor of eight. The correct way to reduce the shell pressure drop, and at the same time discourage tube bridging with dirt, is to use the following tube bundle design criteria:

- Use 1-inch OD tubes; not ¾-inch OD tubes.
- Space tubes on 1½-inch centers, not on 1-inch centers.
- Place tubes on a rotated square **pitch**, not on a triangular pitch.
- Design the tube bundle for a shell-side velocity of 3 to 5 feet per second flowing between the tubes at the edge of the tube support baffle (this is called the shell-side cross-flow velocity).

Replacing an existing tube bundle with these superior features will cost about 10% of the installed cost of the existing heat exchanger. However, the new bundle may have only 65% of the surface area of the old bundle. So what could we do to an existing bundle to retard shell-side fouling?

Probably the best option is to use "seal strips." Look at the exchanger drawing or at the bundle after it is pulled. Visualize the pathways around the bundle where liquid can bypass the tubes. These bypass areas are created when the tubes are not placed close enough to the shell ID, or when tubes are omitted to accommodate the **impingement plate**. The seal strips are typically 2 to 4 inches wide, and ¼-inch thick. They extend down most of the length of the tube bundle, as close as reasonable to the shell's ID. The objective is to close off to liquid flow these bypass areas and thus encourage the fluid to flow across the tubes. You will have to cut slots in the existing tube support baffles to accommodate the new seal strips.

Caution! Make sure you have not trapped the flow between the seal strips and the impingement plate. To avoid this, truncate the seal strip opposite the baffle cut and above the impingement plate.

I deliberately did not provide any sketches of seal strip arrangements. You can only understand the problem by looking at the detailed mechanical sketches for each individual heat exchanger, or better yet, at the bundle itself after it is pulled for cleaning and repair during the unit turnaround.

Exchanger Tube Leaks

Corrosion is undoubtedly a major cause of heat exchanger tube leaks. Some of the other causes, however, are more easily preventable. My least favorite one is tube bundle damage by the maintenance department. When tube bundles are reinstalled during a unit turnaround, they are sometimes handled roughly, and tubes are damaged.

Damage to tubes immediately below the shell-side inlet is caused by erosion from the shell-side inlet fluid. Perhaps the impingement plate is too small or is missing.

If tubes have become thin and fail as they pass through the tube support baffles, that's due to tube vibration. This complex problem is a result of the harmonic vibration of the tubes and vortex shedding. "Dummy baffles" are required. I discussed this difficult malfunction in my book, *Troubleshooting Process Operations* (PennWell Publications, 4th ed.) in the chapter, "Unusual Noises and Vibrations."

Tubes are also subject to roll leaks. Tubes are not usually fixed to the tubesheet by welding. The tube is expanded against the hole in the tubesheet by inserting a tool in the end of the tube and forcefully expanding the end of the tube. Roll leaks are best fixed not by welding, but by rerolling the leaking tube ends. When you pressure-test the shell with water, roll leaks will be observed as water leaking out around the OD of the tubes on the face of the tubesheet.

One of the more frustrating tube leakage incidents I've witnessed was caused by tubes being too thin. It happened on a crude distillation unit in Texas City. The tubes had been subject to attack by wet HCl. Rather than attacking the root cause of the corrosion problem—poor upstream extraction of $MgCl_2$ salts in the crude—the overhead condenser was retubed with titanium (Ti) tubes. However, Ti is very expensive ($30 per pound). To obtain management approval for the project expense, the bundle was retubed with 16 gauge tubes. The bigger the gauge number, the thinner the tube wall. The tubes were impervious to corrosion, but they failed due to a combination of the mechanical issues described earlier. Worse yet, the operators, having been told that the tube bundle corrosion was solved, ignored the salt extraction process step. The old carbon steel overhead vapor line holed-through. Luckily, the resulting hydrocarbon vapor cloud did not detonate and destroy the Amoco Texas City refinery.

The more general problem of tube leaks due to corrosion is discussed throughout this text, where I have discussed corrosion in different process areas and equipment.

Finding Tube Leaks

This is a task on which I have had lots of practice. Let me give you a few examples:

- Cooling water leaking into a turbine exhaust surface condenser. To verify the leak, sample the recovered steam condensate for hardness deposits or chemicals (phosphates or chromates) used in the cooling tower treatment.
- Hydrocarbons leaking into a steam heater. Sample the gas vented off below the pass partition baffle. Submit for GC (gas chromatograph) analysis and compare to the heat exchanger feed composition.
- Steam leaking into a water stripper or amine regenerator. Here it's best to use a leak detection kit. Sulfur hexafluoride or lithium bromide is standard. I use Freon 22 refrigerant, used in air conditioner systems, because it's more readily available. Regardless, the technique is the same. Inject the tracer gas into the steam supply and check the vents on the process side with the detector. My Freon 22 detector (cost me $700 U.S.) is sensitive to a few parts per billion!
- Feed-effluent leaks on hydrotreaters. This technique was in widespread use in the Whiting, Indiana, refinery in 1964. Inject a dye such as that used to color diesel or leaded gasoline into the feed. Dye will be destroyed in a hydrotreating reactor. If the dye appears in the reactor effluent, then the feed-effluent exchanger was leaking.
- Adjust shell and/or tube pressures. Ideally, you can balance the pressures between the tube and shell sides. If altering pressures influences the level of contamination, that proves that the exchanger is leaking.
- Hydrocarbons leaking into cooling water. I hold the world's record for the largest isobutane leak into cooling water that went undetected for a period of 48 hours. 5,000 BSD! John Brundrett, the Texas City East Plant maintenance manager, said he never saw such stupidity in 30 years in the plant. I detected this leak by observing a geyser of water shooting 20 feet above the cooling tower distribution deck.

A less dramatic detection method is to check for hydrocarbon vapors above the cooling tower distribution deck with an ordinary **gas test meter** used by the operators to issue entry permits to vessels during a turnaround. If any hydrocarbons are found, check the cooling water return for hydrocarbons from each individual water cooler.

- For elevated cooling water exchangers, the water outlet is under a vacuum. To sample the water effluent, attach a hose to the outlet bleeder and run the hose down to grade. Insert the end of the hose in a pail of water. Fill the hose with external water to start with, and then open the sample bleeder. You can obtain your cooling water sample from the pail after 5 minutes have elapsed. The hydrocarbon vapors will bubble out of the pail of water.

- A vacuum tower feed versus whole crude exchanger was leaking. The crude was the higher-pressure side. A sample of the vacuum tower off-gas indicated the ratio of propane to propylene was 10 to 1. If this gas was only formed due to cracking in the vacuum tower heater, the ratio of propane to propylene would have been about 60% to 40%. In crude, there is only propane and no propylene. Hence, the crude was leaking into the vacuum tower feed. To prove my theory, the engineer on the unit equalized the exchanger tube and shell-side pressures; the vacuum tower off-gas rate dropped, as did the vacuum tower operating pressure.

Finally, there are radioactive tracers that may be used to detect leaks. I've never used such methods myself, so I'll withhold comment. I'm really only interested in methods that I can use by myself.

Effect of Viscosity on Laminar Flow

One frequent cause of poor heat transfer performance is laminar flow on the tube side of an exchanger. The problem is a low Reynolds number of less than 2,000:

$$\text{Re} = \frac{\text{Density} \times \text{Tube Diameter} \times \text{Velocity}}{\text{Viscosity}}$$

Once the viscosity rises above 40 or 50 centipoise, it's rather difficult to keep the Reynolds number out of the laminar flow range and in turbulent flow. If you observe that heat transfer efficiency (or heat transfer coefficient) shows a marked and immediate improvement with increased tube-side velocity, then the controlling resistance to heat transfer is laminar flow on the tube side. You can be almost positive this is the malfunction if the fluid viscosity on the tube side is high (above 30 centistokes or 150 sabolt seconds universal).

Ordinarily, the high-viscosity fluid is placed on the tube side, because it is most prone to fouling and it's easier to clean the tubes than the shell. However, switching the high-viscosity fluid to the shell side, even at the same velocity as that which was calculated on the tube side, can increase heat transfer by a factor of four to five. Of course, the pressure and temperature ratings, nozzle sizes, and baffle spacing must be checked before such a switch is contemplated. The reason for such an improvement is that laminar flow cannot actually exist on the shell side due to a **vortex shedding** flow regime that will develop across the tube bundle. (See *A Working Guide to Process Equipment* for details on calculating Reynolds number using American units.)

Excessive Exchanger Delta P—Tube Side

If the problem with an exchanger is low Reynolds number and laminar flow on the tube side, why not just increase the number of tube passes from two to four by modifying the pass partition baffle configuration (see Figure 10-1)? The new four-pass configuration will have twice the velocity. This is fine, but delta P will increase by a factor of eight. Delta P varies with velocity squared and linearly with the doubling of the flow path length. At a Coastal refinery, I modified several crude preheat exchangers from two- to four-pass without properly accounting for the high viscosity of the crude imported from Venezuela. The operators found my modified exchangers were restricting charge rates. So they opened the bypasses around the exchanger. An exchanger modification that results in bypassing defeats the purpose of the project.

On another occasion, I retubed a bundle with thicker tubes. That is, I changed the ¾-inch tubes from a wall thickness of 0.083 to 0.109 inches. The calculated effect is:

$$\text{Old tube ID} = 0.75 - (2 \times 0.083) = 0.584$$

$$\text{New tube ID} = 0.75 - (2 \times 0.109) = 0.532$$

$$\text{Increased delta P} = [0.584 \div 0.532]^5 = 1.60$$

That is, the small change in tube thickness increased the delta P by 60%. Pressure drop varies with velocity squared times the inverse of tube diameter. Or, for a constant flow, pressure drop in a tube or pipe varies with velocity to the fifth power, not to the fourth power.

Excessive Delta P—Shell Side

I was working for a small asphalt production plant in Benicia, California, preparing a revamp process design for their vacuum tower that produced paving asphalt. Included in the report I issued was the retubing without change of a vacuum condenser. The bundle was to be retubed because 10% of the tubes had been plugged due to leakage on past unit turnarounds.

When the unit was restreamed, the delta P across the shell side (i.e., process side) of the exchanger had increased from 5 to 50 mm Hg. As a result, the vacuum tower pressure was so high that paving asphalt specs could not be achieved. My client was quite angry (see Chapter 27, "Vacuum Surface Condensers and Precondensers").

In a way, it was my fault. When the bundle was sent out for retubing, the original bundle manufacturer drawings were sent with it. The original bundle was designed with a large impingement plate. The impingement plate is underneath the inlet nozzle. It protects those tubes adjacent to the shell-side inlet nozzle from the erosive force of the potentially high velocity of the shell-side inlet flow. Very often, in their zeal to extend this protection over a wide tube area, mechanical engineers make this plate too large. This chokes off the inlet flow and leads to excessive shell-side delta P. In this case at Benicia, the machine shop that retubed the bundle restored the bundle to its original design configuration. This included the very large impingement plate, which had been removed 20 years before because of excessive shell-side delta P. But neither I nor anyone else in the plant had been around 20 years before.

The moral of this story is that while impingement plates may be important, their size (i.e., diameter) must be moderated to minimize their contribution to shell-side pressure drop. The moral of the next story is that anything that can go wrong, will go wrong in a process plant.

Figure 10-2 is the exchanger configuration of several overhead crude tower condensers in Aruba. I had these bundles pulled for shell-side cleaning due to excessive delta P. When the condensers were restreamed, delta P had more than doubled from 4 to 10 psi. So we pulled the bundles again, to see what had happened.

The two bundles involved were identical, except for the location of their impingement plates. If you will study Figure 10-2, you can see the malfunction for yourself:

- The bundle that was supposed to be installed in the bottom position had been installed in the top position.
- The bundle that was supposed to be installed in the top position had been installed upside down, in the bottom position. In this orientation, the impingement plate was covering the outlet nozzle and restricting the outlet flow. This resulted in excessive shell-side pressure drop.

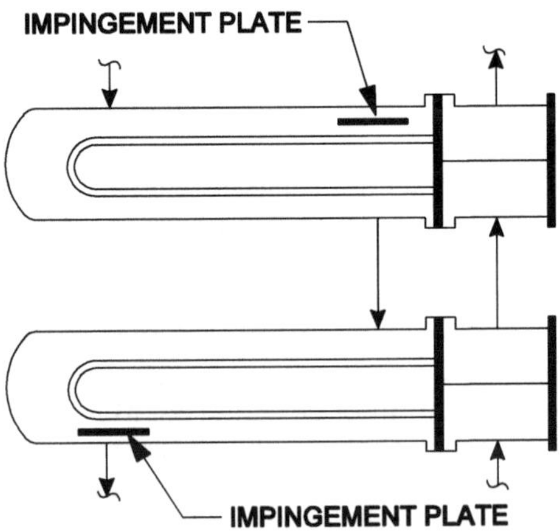

FIGURE 10-2 Impingement plate restricts flow out of shell due to bundle installation error.

Excessive Nozzle Delta P

I had been working in Lithuania on expanding a crude unit's capacity. The unit's feed limitation was 100,000 BSD, due to a relief valve pressure setting of 200 psig on the preheat exchanger train inlet. There were a dozen heat exchangers downstream of this relief valve that created a high backpressure of 180 psi, due to excessive pressure drop. Checking with my gauge, I found a 50 psi delta P across one set of exchangers. The calculated delta P through the tubes was only 5 psi. My client had been aware that the pressure drop across this exchanger was excessive. A year before, they had pulled the bundle and cleaned the tubes with no improvement.

The excessive pressure drop was not across the bundle, but due to small nozzles on the channel head. To calculate the nozzle losses, I used the following formula:

$$\text{Delta P} = 0.0002\,(D)\,(V)^2$$

where Delta P = psi per nozzle
D = density, lb/ft^3
V = velocity, ft/sec

The piping was much larger in diameter than the nozzles. The exchangers were stacked, as shown in Figure 10-2. Therefore, the delta P calculated above was multiplied by 4, as the flow had to be accelerated four times (i.e., once for each of the four nozzles in series). The calculated delta P through each nozzle was 10 psi, or 40 psi for all four nozzles. The channel head nozzle sizes were increased to full line size (4 to 8 inches). Delta P through the exchangers dropped from 50 to 12 psi. The crude charge rate increased from 100,000 BSD to 112,000 BSD as a result of this single simple change to the channel head.

Too often I have found that heat exchanger bundles are replaced to reduce delta P, but the designer forgets to check the pressure drop through the existing shell- and tube tube-side nozzles. Incidentally, heat exchangers are **coded vessels**. New nozzles must typically be **post-weld heat treated** to stay within the ASME Boiler Code for pressure vessels.

Two-Pass Shell Benefits

The objective of the two-pass shell is to produce an exchanger flow pattern that is consistent with true countercurrent flow. If you will consider the flow pattern of an ordinary exchanger, the flow pattern (see Figure 10-1) is truly countercurrent on the lower half of the shell, but co-current on the top half of the shell. This reduces the calculated Log Mean Temperature Driving Force (LMTD), which determines the rate of heat transfer in an exchanger. Note that in Figure 10-1, the shell inlet and outlet nozzles are on the opposite sides of the exchanger. However, in Figure 10-3*a*, the shell inlet and outlet are on the same side of the exchanger, next to the channel head. What then keeps the shell-side flow from jumping right across the shell side and bypassing the entire tube bundle?

It is the shell-side **horizontal pass partition baffle** that fulfills this function, as shown in Figures 10-3*a* and 10-3*b*. This baffle is part of the tube bundle and is not attached to the shell. I've shown the baffle for a U-tube–type bundle construction, but it is also used just as easily for a floating head bundle, shown in the top half of Figure 10-1. If you will now consider the flow pattern in the exchanger, which is called a

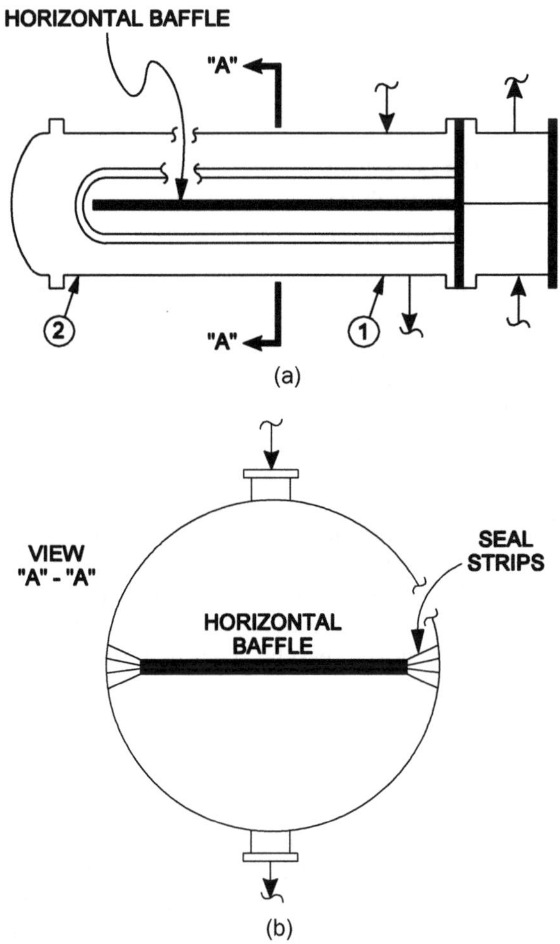

Figure 10-3 (*a*) Horizontal, shell-side, pass partion baffle creates a two-pass shell-side flow pattern. (*b*) Cross section of shell shows horizontal seal strippers which are subject to leakage between baffle and shell ID.

two-pass shell exchanger, shown in Figure 10-3*a*, you will see that the flow on the shell side is truly countercurrent to the flow on the tube side. Thus, the calculated LMTD can apply without any reduction due to non-true countercurrent flow.

For those of you who were absent the day they taught how to calculate LMTD:

$$\frac{(\text{DTH} - \text{DTC})}{\text{LN}\ (\text{DTH} \div \text{DTC})} = \text{LMTD}$$

where DTH = Temperature difference between the tube and shell flow at the hot end of the exchanger

DTC = Temperature difference between the tube and shell flow at the cold end of the exchanger

LN = Natural log

Two-Pass Shell Malfunctions

Two-pass shells are not good design practice. You can identify a two-pass shell exchanger in the field by observing that the shell inlet and outlet are on the same side of the exchanger, as shown in my Figure 10-3*a*. These exchangers work well when new, but often, after the first time the bundle is pulled for cleaning, their performance is degraded. The problem lies with the seal strips shown in Figure 10-3*b*.

The horizontal pass partition baffle, being mechanically part of the bundle, has to be sealed along its entire length to either side of the shell. This is done by what I call a "crush fit." These are thin metal strips that are attached to the long edges of the horizontal baffle. The seal strips are half as thick as a soup can lid. Six to 10 are used together. Then, when the bundle is inserted into the shell, the seal strips are crushed up against the ID of the shell. This retards bypassing on the shell flow from the inlet to the outlet. Bypassing inside the shell has the same effect as opening an external bypass line.

When such a tube bundle is reinstalled during a turnaround, the single-use seal strips should be removed and renewed by the maintenance department, using the metallurgy of the original seal strips.

To identify defective seal strips, proceed as follows:

- Let's assume that the hot material is on the shell side of the exchanger. Use your infrared temperature gun to check two temperatures.
- Skin temperatures at point 1 and point 2, shown in Figure 10-3*a*.
- The temperature at point 1 ought to be colder than at point 2.
- If point number 2 is colder than point 1, you have a serious internal shell-side leak across the seal strips, shown in Figure 10-3*b*.

The problem that I describe was so pervasive at Amoco Oil in the 1960s that the process design department was instructed never to specify such two-pass shells for new refinery units. Perhaps 5% of the exchangers I see are of this type. Many seem to be working quite well. So I suppose they are a good design, if we remember to renew the seal strips during unit turnarounds.

Fouling Due to Gum Formation

One of the most prevalent sorts of fouling in refineries and chemical plants with cracking processes is diolefin polymerization to gums. It works like this:

- Diolefins are hydrocarbons with two double bonds—that is, two olefin bonds.
- At about 350°F, the diolefins, in the presence of oxygen, react to form a free radical and water. That is, O_2 extracts a hydrogen ion from the diolefins.
- The diolefins then polymerize to form gums.
- The gums form a binder that causes particulates to adhere to the heat exchanger surface area and foul the exchanger's tubes.

When I just wrote 350°F, I was not actually referring to the bulk liquid temperature, but to the film temperature. Or, the temperature of the heating medium.

I've been working on a shell-side fouling problem with Hess Oil in New Jersey. The unit is hydrodesulferizing cracked naphtha from a fluid cat cracker. The feed preheat exchanger is fouling at 300°F. I just replaced the phone on its charger unit:

"But Norm, there's no air in the feed," said Bill and Rob, the unit engineers.

"Are you sure?" I asked.

"Yeah, Norm! All flow comes direct from run-down. Anything spilled to tankage does not get rerun on the desulferizer," Bill insisted.

"Are you sure there's no air exposure?" I asked again.

"There's no possibility of air contamination," Rob yelled into the phone.

"So it's all direct run-down?" I asked for the third time.

"Yeah, Norm. Don't you ever listen to us? All 22,000 BSD runs down directly from the FCU cracker to the desulferizer unit through a water wash vessel."

"A what?" I asked.

Rob explained. "A water wash vessel. To remove any residual NH_4Cl salts."

"And the source of the wash water?" I asked in my nicest voice.

"Service water," my colleague replied.

"Gee, Norm," Bill observed, "I guess the service water might contain some air. It's only filtered water from the Raritan river. Maybe we should have used de-aerated boiler feed water or clean steam condensate?"

"Maybe you should think about what you're doing before and not after you design process equipment," I added.

"Don't be nasty, Norm," Rob said.

"And how much am I being paid for this bit of consulting?" I asked.

"Norm. Don't be that way. We're all friends. Did you watch the game last night? The Saints were great."

It's true. The Saints won the Super Bowl in 2009. It's also true that if diolefins contact oxygen extracted from service water, gums and fouling will result.

CHAPTER 11

Condenser Limitations

Man proposes, but God disposes.

—Count Leon Tolstoy

How would one know that a process facility, such as a distillation tower, is limited by condenser capacity? The normal answer is that the system pressure increases. It's also true that a vessel downstream of a condenser may experience a loss in liquid level. This drop in level can be due to a lack of heat input or a lack of condenser duty. If the lack of heat input is the malfunction, then the vessel pressure will fall. If the lack of condenser duty is the malfunction, then the vessel pressure will rise. The idea is that "Heat Makes Pressure!"

The malfunctions that limit condenser capacity are:

- Blow-through, also called "blowing the condensate seal."
- Condensate backup, or subcooling.
- Noncondensable accumulation, also called vapor binding or vapor lock.
- Fouling.
- Problems on the cooling water or the air cooling side of the exchanger.
- Nonsymmetrical piping of parallel condensers.

Blow-Through

I was working for the Fina refinery in Bridge City, Texas. The project was to debottleneck their sulfuric acid alkylation plant. The unit was limited by the flow rate of isobutane refrigerant. The refrigerant was circulated by a motor-driven (i.e., constant speed) centrifugal

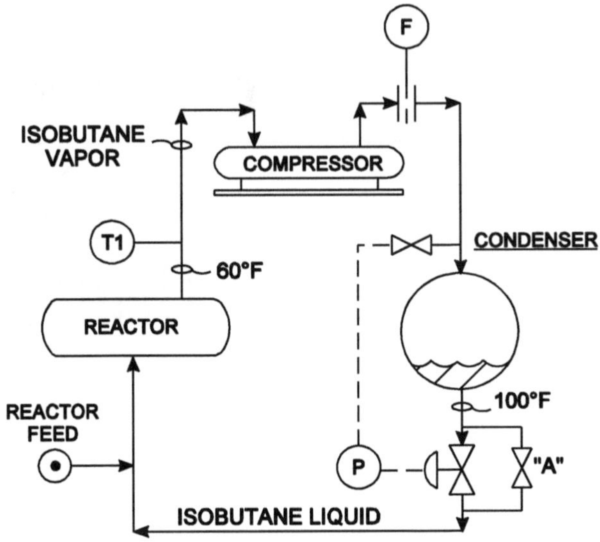

FIGURE 11-1 Setting the compressor outlet pressure controller too low will result in blowing the condensate seal. Too high results in excess condensate backup.

compressor, as shown in Figure 11-1. Centrifugal compressors, just like centrifugal pumps, run on a performance curve. Increasing the compressor discharge pressure is certain to reduce compressor capacity and in this case isobutane refrigerant circulation. The reduced refrigerant flow had increased the reactor temperature above 60°F, which degraded the product quality. The operators would then cut the reactor feed to reduce the exothermic heat of reaction so as to reduce the reactor temperature below 60°F.

My question to the Fina operators was, at what pressure did they set the compressor discharge controller shown in Figure 11-1?

"Lieberman, we hold 80 psig."

Now, the vapor pressure of pure isobutane at 100°F is only 58 psig, or 73 psia at sea level. This should have represented the condenser pressure and the compressor discharge pressure. So the 80 psig backpressure seemed rather high to me. But, perhaps my observation about the excessively high (i.e., 80 vs. 58) compressor discharge backpressure might be quite wrong. After all, if there were 10 mole percent propane in the circulating refrigerant, then the vapor pressure of the liquid at the current 100°F condenser outlet temperature would be 70 psig (85 psia), rather than my calculated 58 psig for pure isobutane.

Also, perhaps there was a 10 psi delta P across the condenser. Well then, 70 psig (for the 90%/10% isobutane/propane mixture vapor pressure at 100°F), plus 10 psi condenser pressure drop, would equal the 80 psig. So perhaps the operators' set point was correct. So I asked in my most polite tone:

"Why 80 psig?"

And the Fina operators responded with open hostility. "Because we always hold 80 psig, Mr. New York Engineer."

Dear Reader, you have to understand that Bridge City, Texas, is not too far from the San Jacinto Monument. There, 11 Texans defeated 16,000 Mexicans to gain freedom for the Lone Star State. Also, that I have a really strong Brooklyn accent. So I said, "Let's all try to be open-minded. Let's try to optimize the compressor discharge pressure. How about if we reduce the set point slowly a couple of pounds."

"Okay, your royal Yankeeship," said the panel operator, as he changed the compressor discharge controller set point from 80 to 70 psig.

"Hey, I didn't say 10 psi all at once. I said a couple of pounds slowly," I objected.

"Look, Lieberman. We just figured we would do what you want quickly. The quicker we do what you want, the quicker we'll get rid of you."

Then I thought, "Maybe this is all for the best. Maybe the optimum pressure is close to the 58 psig I just calculated. Maybe even the 70 psig is too high."

But then I thought, "No, these guys really dislike me. They probably have tried to change the compressor discharge pressure before. Probably they have seen that something bad will happen if they suddenly drop the compressor discharge pressure set point too far and too fast."

At first, the reactor temperature at T-1 slipped below 60°F, to 55°F, which was fine. But my worst fears were suddenly justified. The reactor outlet temperature started to climb above 65°F. The compressor suction and discharge pressures started to increase. The refrigerant circulation rate increased as well. As the compressor discharge pressure was rising, the pressure control valve holding back pressure on the refrigerant condenser opened. But this just seemed to make matters worse. Worse, in the sense that the reactor temperature as measured at T-1 increased to 70°F.

I looked around me. The operators were passing around a can of Copenhagen. That's finely ground chewing tobacco. "Hey, Mr. Norm! You all want a dip of snuff?" the stillman, John Hunter, called out happily. The panel board operator also smiled broadly as he spat on my boots. It's nice to make people happy in their work.

Now what? I was pretty sure what had happened. We had blown the condensate seal in the condenser. I call this **blow-through**. Let me explain in detail what had transpired:

Step 1—The panel operator had suddenly lowered the set point on the compressor discharge pressure by 10 psi, from 80 to 70 psig.

Step 2—The pressure control valve located downstream of the condenser (see Figure 11-1) opened rapidly from 70% to 100%.

Step 3—The entire isobutane refrigerant liquid level in the condenser drained out.

Step 4—Without a minimum liquid level in the condenser, refrigerant vapors blew out of the condenser outlet into the reactor. That's blow-through, which most operators call blowing the condensate seal. Especially as applied to steam condensation on the tube side of process reboilers.

Step 5—The uncondensed isobutane vapors raced through the reactor without extracting any appreciable heat from the reactor.

Step 6—The increased vapor flow to the compressor raised the compressor's suction pressure. This also raised the reactor temperature, as isobutane evaporates at a higher temperature, at a higher pressure.

Step 7—The higher vapor flow from the compressor raised the compressor discharge pressure. This in turn caused the compressor discharge pressure control valve to open further, which made the blow-through worse.

Thus, a positive feedback loop was created. I've written a lot about positive feedback loops in my book, *Troubleshooting Process Plant Control* (Wileybooks.com).

Optimum Drainability Point

Whenever you are caught up in a positive feedback loop, your first reaction must always be the same:

- Switch from auto to manual.
- Manually move the control valve in the reverse direction from its current direction.

This should be the first lesson process control engineers learn.

Having switched the compressor discharge control valve to manual, I closed it rapidly to 50%, to stop the blow-through of the uncondensed refrigerant vapors. At the same time, I cut the reactor feed rate in half. This reduced the T-1 temperature (Figure 11-1) back below the maximum 60°F target. As my objective was to increase rather than decrease reactor feed and alkylate production, this was not so good.

"Could be, Lieberman, you best go back to the admin building and leave us dumb operators to run this here alky unit. Could be we should put the pressure controller back on auto, at 80 psig," the panel board operator said with a sarcastic smirk.

It's true that at 50% valve position, the compressor discharge pressure was higher than before I began my plant test. It's also true that both the refrigerant isobutane flow and the reactor feed rate were lower than before my ill-fated experiment. I knew that my apparent failure was viewed by the Texas operators as a partial compensation for the Confederacy having lost the Civil War.

"Look," I said, "I'm going to try to optimize the compressor discharge pressure control valve on manual. When I find the optimum discharge pressure by experiment, then we will switch it back to auto. My experiment will determine the new pressure set point."

"How you gonna do that?" asked Leroy Johnson. Leroy was a young trainee operator, not qualified yet to work on the panel. "How you gonna know if this here speriment is workin, Mr. Engineer?"

"Well, Leroy, let me explain. You can have too many girlfriends or too few girlfriends."

"Ain't that the God's honest truth."

"The position of the control valve downstream of the condenser is like that. We can open it too much or too little."

"We just done seen what'll happen if you open it too much," said the stillman John Hunter. "Them compressor vapors just blow right through that condenser without liquefying."

"Right," I said, "We've just done that."

"But what'll happen Mr. Engineer, if we don't open that control valve enough?" asked Leroy.

"Well, that's the situation right now. We're suffering from condensate backup." And I drew a sketch similar to Figure 11-2. (Note that the refrigerant is on the tube side of the condenser and the level shown is in the channel head, and not the shell.) "You guys can see from my drawing that about 40% of the tubes are filled by the condensed isobutane. Those tubes are not available to help condense the vapor. In effect, that shrinks the size of the condenser, as far as the vapor is concerned."

The panel operator edged up closer. "That's all true. But I reckon that having 40% of the tubes cooling the liquid off from 120°F to 100°F makes the refrigerant take a lot more heat out of the reactor. Colder refrigerant is better than warmer refrigerant."

"That's true. Can I explain with some engineering calculations:

- Cooling butane off by 20°F will take 10 Btu/lb of heat out of the reactor. The specific heat of butane is 0.5 Btu/lb/°F. And 20°F times 0.5 equals 10 Btu/lb. That's sensible heat removal.
- But when a pound of butane evaporates, it absorbs an additional 130 Btu/lb. That's called the latent heat of evaporation of butane.

Thus, the extra 10 Btu/lb or 20°F of subcooling only increases the heat removed from the reactor per pound of refrigerant circulated by

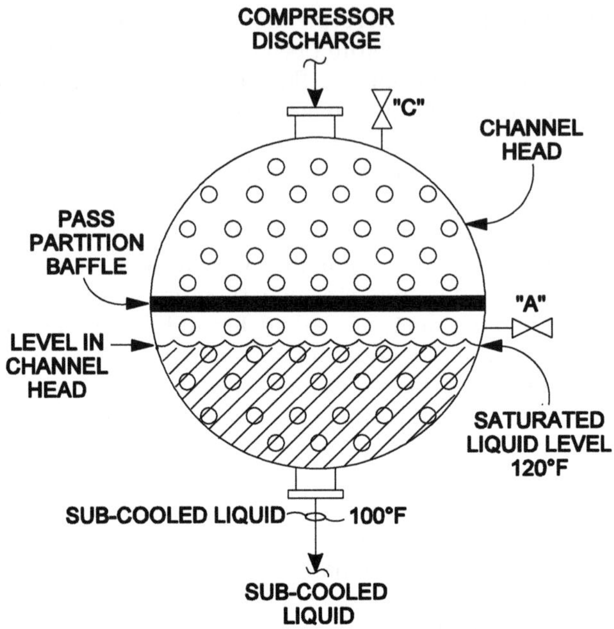

Figure 11-2 About 40% of condenser surface area is lost due to condensate back-up and subcooling of the isobutane refrigerant. Note that refrigerant is on the tube-side.

around 7%. That's not much compared to the 40% of the condenser surface area lost due to the isobutane condensate backup."

"Excuse me, Mr. Norm," asked Leroy, "But how do you all know where that liquid level really is in the condenser channel head? You got X-ray Superman-type vision?"

Identifying Condensate Backup in the Field

"Leroy, run your fingers around the channel head. Can you feel where it starts to get cooler?"

Leroy, the stillman John Hunter, and I had walked out together to examine the condenser. I could see the new bridge to Port Arthur shining in the distance.

"I surely can. The colder part starts just below the middle. I guess that's the butane getting cooled after it done condensed."

"Right! So what do you think we ought to do next, Leroy?"

"Well, that there condensate butane liquid is stealing the tubes away from the vapor. So I reckon we ought to lower the liquid level by opening up that control valve on the outlet. Make the condenser drain out faster."

John Hunter then cracked open valve A shown in Figure 11-1 (the control valve bypass).

"Say, Mr. Engineer," observed Leroy, "I can feel that warm level droppin in the channel head. Mr. John, you need to keep on with openin that bypass valve. But real slow, like. If you open it too much, we gonna blow-through that vapor out the condenser before it got's time to liquefy itself."

"Leroy, that's called blowing the condensate seal," I added.

But Leroy was ignoring me. He was seeking, with his fingers running up and down the channel head, to determine the **optimum drainability point**. That is, he was trying to minimize condensate backup without blowing through. As John Hunter opened valve A:

- The condenser outlet temperature went up.
- The compressor discharge pressure went down.
- The refrigerant flow went up.
- The reactor temperature cooled.
- And the still-hostile panel board operator increased the reactor feed flow, which was my ultimate objective.

The compressor discharge pressure was 73 psig. The compressor discharge control valve was switched to auto at this set point, and Mr. Hunter shut the bypass valve A.

"You know, guys, that you're going to have to repeat this procedure every time the cooling water temperature changes, or if the amount of propane in the circulating isobutane increases," I explained.

"Lieberman, why didn't you dumb New York City engineers design the condenser with a boot? Then we could have level-controlled out of the boot. This would have kept from blowing through, but also avoided condensate backup."

"Mr. Hunter! Better to light one small candle than curse the darkness," I responded.

Methods to Achieve the Optimum Drainability Point

The need to prevent blow-through, yet still minimize condensate backup, is one of the fundamental principles of process operations. The problem was recognized almost at the start of the Industrial Revolution with the invention of the steam trap. The purpose of the steam trap is to prevent condensate backup and avoid blowing the condensate seal.

In a properly designed refrigeration system, the condenser drains into the refrigerant receiver vessel. If the condenser is elevated above the vessel, which is best, then the condenser drains by gravity into the refrigerant receiver. The flow from the refrigerant receiver is by liquid level control, which in effect prevents blow-through of the refrigerant vapors.

If the condenser is located below the liquid level of the refrigerant receiver, then a small amount of condensate backup is required. This condensate backup is required not so much to provide head pressure, but to subcool the refrigerant, or condensed liquid hydrocarbon, below its boiling point temperature and pressure.

Required Subcooling of Condensed Liquids

The condenser shown in Figure 11-3 is located below the receiver. Vent valve B is shut. Therefore, the fluid entering the receiver must not contain any vapor. If it did, then:

- The receiver pressure would rise.
- The level in the condenser would be pushed up to fill more tubes.
- The liquid would become progressively more subcooled, so that when it rose to the higher elevation in the receiver, it would not vaporize.

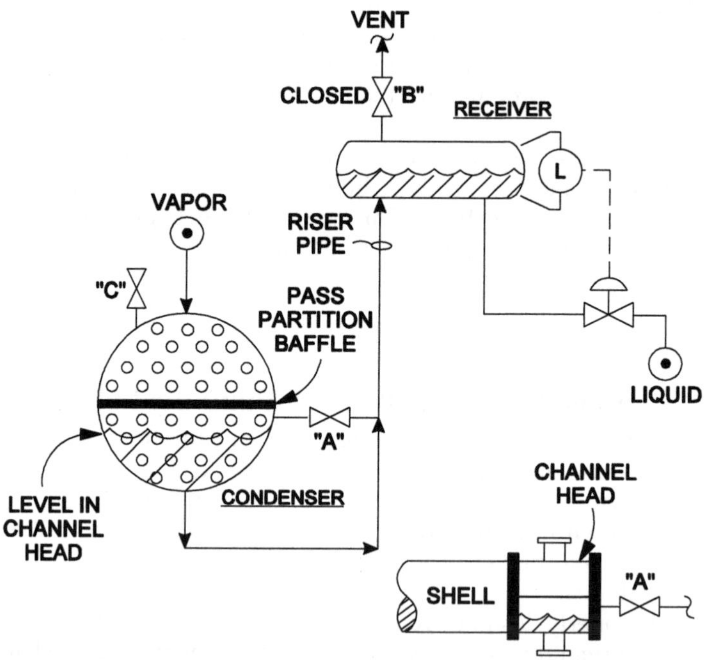

Figure 11-3 Noncondensable vent should be connected just below channel head pass partition baffle, when the condensing vapor is on the tube side.

But condensate backup is due to factors other than the rise in elevation. These are:

- Frictional loss in the riser pipe shown in Figure 11-3.
- Nozzle exit loss from the condenser channel head outlet.
- For refrigeration systems, heat gain from poor insulation of the riser pipe.

The effect of the preceding four factors must be added to predict the required amount of subcooling and condensate backup. And obviously, as seen in my story about the Fina refinery, if a control valve is located on the condenser outlet, its delta P will also contribute to condensate backup in the condenser channel head.

In summary, the concept of minimizing condensate backup while preventing blow-through is one of the most important concepts in process engineering. Think about a distillation tray downcomer. We're trying to minimize the downcomer backup, but without blowing the downcomer seal. How about the seal on a flare? We try to minimize backpressure on the flare header line, but without blowing the flare's water seal.

How about the condensate drain placed on a steam reboiler? Same problem. The loop seal from an absorber draining back into the feed recontact drum; the drain line from the interstage condenser draining back to the main condenser; the drain from a blow-down tower to a pressured condensate collection sewer. All the same problem.

Then we have the most famous blow-through incident of all. The BP Deep Horizon well blowout near my home in Louisiana. Also a matter of the loss of a liquid (drilling mud) seal. But that's another story.

Noncondensable Venting

Figure 11-3 shows the condensing vapors located on the tube side of the exchanger. I've shown it this way because half or more of process plant condensers are air coolers. Two-pass air coolers will also have the horizontal pass partition baffle shown in Figures 11-2 and 11-3. Especially on startup, noncondensables will accumulate just below the pass partition baffle on the channel head (for a shell-in-tube exchanger) or on the header box (for an air cooler).

The correct place then to vent off the noncondensable gas is through valve A (Figure 11-2). Venting through valve C will just vent the hydrocarbon vapor feed from the condenser, but not the noncondensable vapors. To separate the noncondensables from the vapor feed, the vapor must first be condensed. This occurs after, and not before, the vapor passes through the tube bundle.

If you have the condensing vapors on the shell side of an exchanger, locate the vent on the top of the shell, as far away from the vapor inlet as practical.

In Figure 11-3, I have shown the noncondensable vent connected beneath the pass partition baffle on the channel head and flowing to the riser line. Valve A and valve B should then be opened together, as required for noncondensable venting. After venting, the condenser pressure should drop. If that does not happen, you have vented off valuable vapors, and not noncondensable gas.

Leaving valve A on Figure 11-3 open, even with valve B closed, is still bad. Then you are bypassing some uncondensed vapors into the receiver vessel. This will result in an additional increment of condensate backup in the channel head and a loss of some condenser capacity. Then, closing valve A will reduce the condensate backup a little. Which is always a good thing.

The effect of noncondensable accumulation is called by the operators "vapor lock" or "vapor binding." After startup, air, nitrogen, or fuel gas that was used in startup for purging out air or vacuum breaking is the main problem. During normal operations, hydrogen or CO_2 can be produced as a by-product of corrosion or carbonate breakdown. Most so-called noncondensables are soluble in the condensate and thus do not have to be vented.

Exchanger Fouling

The three factors that control heat transfer rates in exchangers are:

- Viscosity
- Low velocity
- Fouling

The viscosity of vapors and condensed liquids is small and can be neglected. High velocities hinder, rather than aid, heat transfer in condensing services. Therefore, it's really only fouling that counts in process plant heat transfer in condensing service.

According to Donald Kern's book, *Heat Transfer Fundamentals*, the **heat transfer coefficient** for condensing steam is around 600 Btu/hr/ft²/°F. When I first read this, I thought Mr. Kern must be insane. I had never measured any coefficient in any service above 150 Btu/hr/ft²/°F. But then in 1990, I measured a coefficient of 400 in Aruba. It was on a steam surface condenser. The clean turbine exhaust steam was condensing on the shell side of a water-cooled condenser. The unit had been in service only a few days. There was zero condensate backup and no fouling—yet.

Within a year, the coefficient had fallen to about 200 Btu/hr/ft²/°F, due to water-side fouling. Let me explain:

- Let's say that the overall clean resistance to heat transfer is 0.0025 (the reciprocal of the 400 coefficient).
- A typical fouling factor for seawater on the tube side is at least 0.0025.

- Then, 0.0025 + 0.0025 = 0.0050. That is, I am summing up the resistances to heat transfer—film plus fouling.
- The reciprocal of 0.0050 equals 200 Btu/hr/ft^2/°F.

My point is that in the condensation of vapors, a relatively small degree of fouling, which might ordinarily go unnoticed, leads to a relatively great loss in heat transfer coefficiency in condensing service. When we are condensing steam or closed-loop refrigerant vapors, the fouling is almost always on the water or air side.

I say almost always. But I recall at a Chevron plant in El Paso, Texas, that they had a closed-loop refrigeration plant with lube oil fouling on the refrigerant side. The compressor seals were defective. I recall how an older and smarter engineer than me showed me that it was necessary to drain down the lube oil from the low point in the refrigerant receiver vessel, so as to maintain heat exchanger efficiency in both the evaporator and the condenser.

Use of Low-Fin Tubes in Condenser Service

I once made a serious mistake in retrofitting a sulfuric acid alkylation unit de-isobutanizer overhead condenser. To increase condenser capacity, I had the tube bundles retubed with Wolverine low-fin serrated tubes. The tubes' external surface area increased by a factor of 2.7. But since the controlling resistance to heat transfer was fouling inside the tubes (that is on the cooling water side), the effect of the water-side fouling also increased by a factor of 2.7. I had better explain. To calculate the overall heat transfer coefficient, we must multiply the two tube-side heat transfer resistances:

- Fouling factor inside the tubes.
- Plus the tube-side heat transfer film resistance.

by the ratio of the tube external area (including the fins), divided by the tube inside area. That is about a factor of 2.7 for the finned tubes.

So you may think I didn't gain very much from the finned tubes. Quite true. But you may be surprised to learn that I actually lost about half of the original condenser capacity. The corrosion products on the shell side lodged between the fins. Thus, the dirty fins acted as an insulating barrier between the cooling water and the condensing isobutane vapors. The refinery manager of the American Oil refinery, Colonel Norogaard, personally accompanied me as I inspected the fouled bundles.

But the project was a big success anyway. A success in the sense that I learned never to use low-fin tubes, unless the tube side and the shell side are both in clean services. I still recall that Colonel Norogaard, a legendary tyrant, was really very kind to a young engineer who had made a serious design mistake.

Cooling Medium Problems

For air-cooled condensers, please consult Chapter 12, "Air Coolers: Forced- and Induced-Draft Air Side Malfunctions."

For water-cooled condensers, please consult Chapter 13, "Cooling Water: Towers and Circulation."

For air coolers, the main problem is dirt accumulation between the fins on the bottom two rows of tubes. For water coolers, the main malfunction is carbonate deposits inside the final tube pass restricting the cooling water flow rate.

Parallel Condensers with Nonsymmetrical Piping

For condensers placed in parallel, two criteria must be met for the condensers to work properly:

- The condensers must be identical.
- The piping must be symmetrical.

One of the many examples of the consequences of the violation of these criteria occurred at the Texaco refinery in Convent, Louisiana. As in the first example in this chapter, the service was condensing the refrigerant isobutane effluent from the alkylation unit refrigerant recycle centrifugal compressor. Figure 11-4 shows two parallel identical condensers, A and B, with almost symmetrical piping. Almost, but not quite.

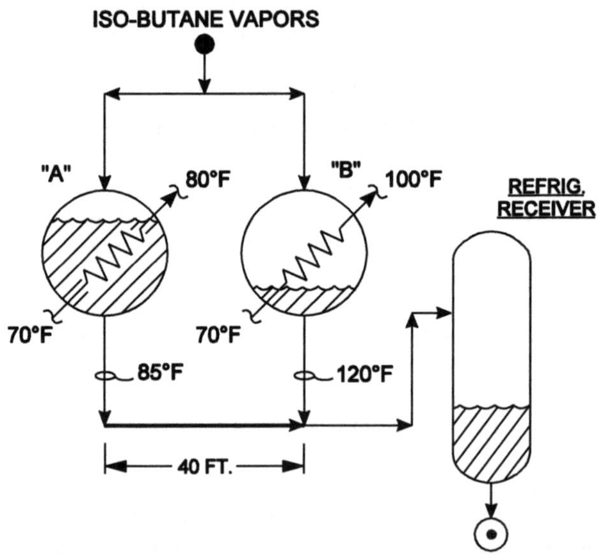

FIGURE 11-4 The extra 40-feet line shown causes excessive condensate backup in condenser A.

The nonsymmetrical pipe is shown by the 40 feet of heavy, horizontal line. The plot plan did not permit the refrigerant receiver vessel to be located between the two condenser shells. So it was located next to condenser B, and only 40 feet further from A. I calculated the delta P through this 40 feet of line as only 0.3 psi. But with a specific gravity of butane of 0.55:

$$(0.30 \text{ psi}) \times (28 \text{ inch } H_2O) \div (0.55) = 16 \text{ inches}$$

The 28-inch H_2O factor is the height of water equivalent to 1 psi.

The calculated 16 inches of frictional loss in the horizontal 40 feet of pipe will back up the liquid by an extra 16 inches in shell A, as compared to the liquid level in shell B. As the shell's ID was only 30 inches, the 16 inches of extra isobutane condensate backup also submerged well over half the tubes in shell A. This sounds really bad, but what proof did I have of this malfunction?

- First: The cooling water flow was symmetrical between exchangers A and B. Yet the observed water temperature rise for B was 30°F, versus only 10°F for A. This was happening because B was condensing three times as much of the refrigerant flow as was A.
- Second: I could observe, using my infrared temperature gun, that almost two-thirds of the height of the shell for condenser A was full of liquid. There was a distinct level I observed on the exterior of shell A where the shell's skin temperature dropped from about 120°F down to about 90°F. That distinct temperature gradient corresponded to the liquid level in the condenser's shell.
- Third: The effluent butane temperature from condenser A was about 40°F below its calculated saturated liquid or bubble point temperature of 125°F. The effluent isobutane temperature from condenser B was only about 5°F below its bubble point temperature. The only way to account for this was excessive condensate backup in the shell side of condenser A.

I have quite a similar story to relate from the Amoco refinery in Texas City, on my giant butane splitter. There the problem was my addition of a small new condenser piped up in parallel with a larger existing condenser. The smaller new condenser had a tendency to blow through. This caused extreme condensate backup and subcooling in the older and larger condenser shell.

Actually, I have dozens of similar stories which I will not relate. However, all these stories have one feature in common. They did not have a happy ending. I cannot recall even one instance where I succeeded in balancing the flows to the two non-symmetrical parallel condensers so that I could simultaneously prevent the dual malfunctions of:

- Condensate backup and subcooling.
- Blow-through and loss of the condensate seal.

That is, I could never achieve the point of "optimum drainability." Just ask Tiger Woods, the famous golf pro, about optimizing your number of girlfriends. It can't be done, unless you realize that the optimum number is one.

Combined Effect of Blow-Through and Condensate Backup

Let me refer again to Figure 11-4. Let's assume that condenser B was blowing through and thus had lost its condensate seal. Now note that there is no vapor vent from the refrigerant receiver drum. Thus it follows that the combined mixed effluent from both condensers A and B, at equilibrium, must be saturated liquid, or bubble point liquid, with zero weight percent vapor.

What then happened to the vapor flowing from condenser B? It must have condensed! It must have condensed in the subcooled liquid draining from condenser A. However, the only way the liquid leaving condenser A can become subcooled is for condensate to back up in the shell of condenser A.

So vapor blow-through in one of the two parallel condensers results in condensate backup in its sister condenser's shell, and a consequent reduction in net condensing capacity for both shells.

I've explained this concept in every one of the 700 plus process troubleshooting seminars I have presented since 1983. One time, an alert attendee asked the following pertinent question:

"Norm, wouldn't it be best, both from the aspect of condensatebackup and vapor blow-through, to put condensers in series rather than in parallel?"

The answer is certainly yes. My preferred method of expanding condenser capacity on existing units is indeed to add a new, low delta P condenser, upstream of the existing condenser. But engineering care must be exercised in the design.

First, if the new condenser is not elevated so that it is self-draining into the old condenser, the pressure stability of the upstream process vessels may be compromised. If there is a vapor–liquid mixture flowing uphill in a riser pipe, at a velocity of less than 20 to 30 ft/sec, phase separation will result in the riser pipe. The liquid will accumulate in the riser pipe and periodically blow clear. This is called **slug flow in risers**. As I write this, I'm returning from a BP refinery in Brisbane, where they have exactly this problem on their HF alkylation unit recycle isobutane condensers.

Second, placing equipment in series has the potential for a greatly increased pressure drop. If two equal exchangers are switched from parallel to series, the delta P might increase by a factor of eight. Thus, the new condenser must be of a very low delta P design. Also, it must always be placed upstream of the older exchanger and preferably elevated for gravity drainage to the existing exchanger. Which is exactly what BP failed to do on their alky unit in Australia.

CHAPTER 12

Air Coolers—Forced- and Induced-Draft Air Side Malfunctions

The future is unknown and unknowable. The past is dead and gone. But this moment in time is ours.

Air coolers are more expensive than water coolers. But water coolers need a source of cooling water. In the 1960s at the American Oil refinery in Whiting, Indiana, cooling water was provided free from Lake Michigan. Once we couldn't use lake water because of environmental constraints, we began to use air coolers.

There are two sorts of air coolers. Induced-draft fans are above the tube bundle. Forced-draft fans are below the tube bundle. Perhaps 70% to 80% of fans are forced draft. Not because of cost or efficiency; it's just that access for maintenance to the rotating components is easier below the tube bundle.

The tube bundle itself has aluminum finned tubes. The heat transfer film resistance of air is much greater than that of water. Therefore, we use finned tubes, which increase the exterior tube surface area by a factor of about 12. This offsets the low heat transfer coefficient of the air cooler due to high air heat transfer film resistance, as shown in the following equation:

$$Q = U \times A \times \text{Delta T}$$

where Q = Duty
U = Heat transfer coefficient

A = Surface area of tubes

Delta T = Process temperature minus air temperature

However, the use of these fins creates the potential for dirt accumulation between the fins. Also, the fins may be subject to corrosion.

Effect of Moisture on Aluminum Fins

At the Chevron plant in El Segundo, California, they have a large bank of air-cooled condensers adjacent to their delayed coker. Steam rises in dense clouds from the nearby coke pit perhaps 12 hours a day. Those tube bundles exposed to the steam have suffered severe fin corrosion. By severe, I mean I could break the fins off easily with my fingers. When I tried to hydro-blast (water clean) the upper rows of tubes, they broke off. Fins, in a corroded state, cannot efficiently conduct heat from the tube to the air. Might a change in aluminum fin type mitigate this problem? I'm not too sure. But what I am sure of is that it's a bad practice to continually spray water on the top of fin tube bundles. While this will dramatically improve cooling, with time it will deteriorate the fins and degrade cooling capacity.

Fin Mechanical Damage

The aluminum fins are quite easily crushed if someone walks across them. However, this does not particularly reduce airflow or even lower heat transfer efficiency. On the other hand, crushing the underside of the fins positively reduces airflow through the tube bundle and certainly inhibits cooling capacity. I have no real explanation as to why crushing the fins on top of the bundle, or straightening them, had no significant effect on cooling. But at the Good Hope refinery, I had several banks of air cooler fins on the top tube row straightened by two tough, quite heat resistant, young men, and there was no benefit. On the other hand, crushing of the underside of the fins on the lowest row of tubes, by excessive water wash pressure, reduced airflow by roughly 15%.

Measuring Airflow Through Bundle

Please refer to Figure 12-1:

Step 1—Note air temperature flowing to bundle. This may be hotter than ambient temperature. Say it's 90°F.

Step 2—Obtain a long-stem dial thermometer. Tape a small piece of wood to its tip. (Needed to keep the tip from touching the fins.)

Step 3—Check the air outlet from four to six locations above the bundle. Average these temperatures. Say it's 130°F.

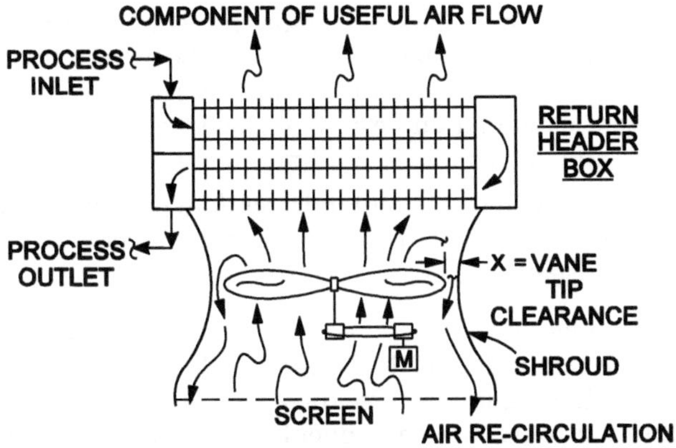

FIGURE 12-1 Forced-draft air cooler showing air recirculation. Tube side is two passes.

Step 4—Multiply 40°F (i.e., 130°F − 90°F) by 0.25 Btu/lb/°F (specific heat of air), to obtain 10 Btu/lb of air.

Step 5—Calculate the cooler or condenser duty from the process side. For example, we're condensing 10,000 lb/hr steam (212°F) to water at 212°F. The latent heat of condensation of the steam is 1,000 Btu/lb. Therefore the condenser duty is 10,000,000 Btu/hr.

Step 6—The airflow is then 1,000,000 lb/hr (10,000,000 ÷ 10 Btu/lb of air).

Next check the design airflow on the manufacturer's exchanger data sheet. If the design airflow is about 1,100,000 lb/hr, all is well. If the design airflow is about 2,000,000 lb/hr, then you need to read on.

Air Recirculation

Let's say dirt, oil, bugs, and moths have accumulated between the fins of the bottom two rows of the tube bundle, shown in Figure 12-1. The effect of this fouling will be:

- The airflow through the bundle will decrease.
- The total airflow discharging from the forced-draft, motor-driven fan will remain the same.
- The difference between these two airflows is airflow recirculation, which I'll explain in a moment.
- The air pressure discharging from the fan remains constant, because the fan runs on its head versus flow performance curve. If the flow remains constant, then so must the discharge pressure.

- It follows, then, that the airflow pressure drop through the bundle is also constant, even though the bundle is fouling. As a young engineer, I was always puzzled by the fact that severe exterior bundle fouling failed to raise the air inlet pressure to the bundle. But I'm smarter now.
- The amp load on the motor driver does not vary with fouling of the fins, because the flow and head developed by the fan are constant.

Air recirculation can easily be observed on all forced-draft air coolers. Note that in Figure 12-1, some airflow around the edge of the screen is flowing in a reverse direction. This is air recirculation. As the bundle fouls, the component of useful airflow, shown in Figure 12-1, is diminished, while the component of useless air recirculation increases. The area of the screen where a thin piece of cloth is blown off the screen surface will increase as the exterior of the finned tube bundle fouls. Washing these deposits off the fins increases the useful airflow, but it does not alter either the pressure drop of airflow through the cooler or the amperage load on the fan's motor driver.

Vane Tip Clearance

To some extent, the component of air recirculation is also a function of the vane tip clearance, shown as dimension X on Figure 12-1. This is the clearance between the shroud and the tip of the blade. When new, the clearance is a uniform ¼ inch. However, with time, the fan's **shroud** gets out-of-round. The fan tip blades rub against the interior of the shroud and thus wear down. Eventually you will find that:

- The clearance between the tip and shroud has grown to 2 to 4 inches in some areas.
- The clearance between the tip and the shroud has disappeared in other areas.

In order to reduce this tip clearance, you can install a vane tip seal. This is just a mesh material screwed inside the shroud. The fan blade, with a cutting tool temporarily attached, is used to cut a groove through this mesh. I've only used this mesh once, in Aruba, on a hydrogen plant effluent cooler, when the vane tip clearance was 4 inches. It helped effluent cooling, but I never rechecked the installation in the field. To observe the degree of vane-tip-to-blade clearance, spin the fan by hand. If the clearance is less than 1 inch, I'll ignore the problem. If the clearance is over 2 inches, I'll specify a new vane tip seal. I have no particular basis for this criterion. But, it's kind of like buying flowers for your wife for no particular reason. It doesn't cost much, but it's a move in the right direction.

Increasing Airflow for Forced-Draft Fans

On older fans, belt slipping was a big problem. In a modern installation, this can't happen, as the drive belts are notched. But if you think your belts are slipping, check the fan speed as follows:

- Place a bright piece of colored tape on the hub.
- You can now count the revolutions of the hub.
- Typically this will be 40 to 60 rpm, even though the motor is spinning at 1,800 or 3,600 rpm. This means the diameter of the wheel on the fan is 50 to 60 times bigger than the diameter of the wheel on the motor, assuming direct drive belts.

If you wanted to make the fan spin 10% faster, you would increase the wheel diameter of the motor by 10%. Both the torque on the fan blades and the amperage load on the motor driver will increase. Amps will increase by a factor of 34% (work varies with diameter or speed cubed, for rotating equipment). Don't increase fan speed without having some smart engineer verify that the blades are rated for the higher torque load or check with the manufacturer.

Changing the blades to new, lightweight fiber materials also can increase airflow. But this I've read in a book, and I don't have any firsthand experience. I have personally increased the blade angle or pitch on an air cooler from about 15° to about 20° to 25°. For me, plus or minus 5° is about as close as I could get. The net effect, based on the observed increase in amp load on the motor driver, was an increase in the airflow by around 10% (i.e., amps increased by about 10%). Still, it was only loosening two bolts per blade and the whole job took 30 minutes. Of course, you may be limited by the amperage rating of the motor before you increase the blade pitch or angle.

Where the tube bundle sits on top of its shroud structure is also an area where air can slip around the bundle. I routinely blank off such areas with sheet metal. How much good it does, I can't say. Like buying flowers for your wife, it couldn't hurt.

Cleaning Air Cooler Fins

I'm an expert in this area. It's true I've only done it three times with my own hands:

- Chevron—Salt Lake City
- Coastal—Aruba
- Secunda—South Africa

Still, that's three more times than most of the engineers my clients employ. So I'm relatively an expert. But, since people have been killed doing this job, I'll give you the step-by-step procedure to water wash the fins safely, but effectively:

Step 1—Determine that airflow is deficient and that the rate of air recirculation is increased.

Step 2—Electrically lock out the fan motor at the breaker. Meaning, put your own lock on the breaker.

Step 3—Remove the wire screen. This screen is only there for personnel protection and serves no process function.

Step 4—Tie off the fan blades with a rope. When I was cleaning a bundle in South Africa, the wind came up and started the fan blade rotating. That's when I found out it's best to take an extra step for safety.

Step 5—I have read that it's best to spray a detergent solution on the underside of the finned tubes and let it soak in for a few hours. I've never done that, but it seems like a good idea.

Step 6—You will need a source of clean water (boiler feed water is best). Don't use seawater. Water pressure at the required elevation should be about 30 to 50 psig.

Step 7—I use a section of ½-inch stainless tubing as a water gun. You will be cleaning the underside of the tubes. You will have to carefully spray the water between each of the bottom row of tubes to clean the lower half of the second row of tubes. If you hold your water gun too close to the fins, you will bend the fins. This mistake was made in Aruba, with a consequent 10% loss in airflow after the cleaning. Cleaning a large bundle will take 2 to 3 hours in this manner.

Step 8—Using a large-volume hose, water wash the top of the bundle for 5 minutes. This is mostly a waste of time, as the top of bundle is always cleaned by rain. But it's easy to do, and maybe it might help a little.

Removing Loose Deposits from Fins

If the fouling between the fins is dry dust or moths, there are two simpler methods that can be used. One way is just to brush the dirt off the fins with a broom. This method has another advantage. If the material being cooled is really hot, and if the cleaning is to be done on-stream (which is normal), the use of water washing can theoretically create a safety hazard. Both Chevron and Mobil (now part of Exxon) prohibited water washing air-cooled exchangers when process inlet temperatures were above 300°F. Their fear was that the sudden cooling of a tube might cause the tube to pull out of the header box. I never heard of this happening, but I can appreciate my clients' concerns. Brushing off the dirt for many air coolers removes the bulk of the restriction.

Another similar method is to reverse the polarity of the fan motor electrical connections. For a three-phase motor, the fan will now spin backwards. This sucks the loose dirt from the underside of the fin tube surface. Of course, this method is somewhat messy, but it is fast. Just keep in mind that none of these methods will do much if the dirt is greasy. If your driver is a single-phase, 1,500 or 1,800 rpm motor, switching the electric leads will not reverse the direction of fan rotation.

Fan Speed Adjustment and Louvers

Variable-speed motors are costly and are not normally required to control airflow. A three-phase electric motor runs at 3,600 rpm (3,000 rpm in Europe and Asia). A single-phase motor runs at 1,800 rmp (1,500 rpm in Europe and Asia). To switch from a three-phase to single-phase operation is not expensive and can be done with a single switch. Or, you can shut off the fan completely. Here then is my rough rule of thumb:

- Fan running at full speed: cooling capacity is 100%.
- Fan running at half speed: cooling capacity is two-thirds of full-speed duty.
- Fan when shut off: cooling capacity is one-third of that when running at full fan speed.

Especially if one has several banks of air coolers, this ought to provide sufficient flexibility in most applications for required cooling capacity.

I do not like louvers. A louver device sits on top of the forced-draft tube bundle and restricts the flow of air. It looks rather like a venetian blind with slats that are opened and closed. Process-wise, louvers are effective. And when new or properly maintained, they work fine. But the majority of the louver installations I've seen are no longer in working condition and are jammed in a partially closed position. With time, many of my clients abandon the louvers. They are discarded to increase airflow and cooler capacity.

Excess Amperage Load on Motors

It is natural for the amp load on the motor driver to increase in cold weather. The fan is like a centrifugal compressor. As such, both the airflow and the differential pressure increase as the ambient air temperature decreases. I've explained this in Chapter 32, "Centrifugal Compressors—Surge, Fouling, and Driver Limits."

If the fixed-speed motor trips off during cold weather, you have several options:

- Reset the amperage trip point. The motors we have at home trip off on a high winding temperature. At work, the motors trip off based on the **full limit amperage load (FLA point)** set on the motor. If you reset the FLA trip point 5% to 10% higher, the windings will run a bit hotter and the operational life of the motor between rewindings will be reduced. But many motor trip points are set too conservatively.
- A more conventional approach is to reduce the pitch on the air fan blades. This should probably be done anyway during the winter to save about 10% of the motor energy consumption.
- A more radical solution to excessive motor amps is to reduce the wheel (or pulley) size on the motor. Reducing the motor pulley diameter by 10% will reduce the amperage load on the motor by about 25%.

As I explained in the earlier section of this chapter, changes (cleaning or repairs) to the tube bundle will not affect, to any noticeable extent, the motor amperage load.

Induced-Draft Fans

In this type, the fan is located above rather than below the tube bundle. My experience with these less-common installations is limited. However, I can say that for such an induced-fan installation, the effects of fin fouling and vane tip clearance are completely different from those of the more common forced-draft variety. As the fins foul:

- Delta P across the tube bundle will increase.
- Airflow through the tube bundle will diminish, but so will the airflow to the fan.
- Airflow recirculation does not occur through the periphery of the screen.
- The amperage load on the motor driver will decrease as the airflow through the tube bundle diminishes.

I would imagine that an increased vane tip-to-shroud clearance will still reduce airflow through the tube bundle, as in the forced-draft air cooler. Certainly, everything I've noted pertaining to cleaning of the fins applies just as well to induced-draft and forced-draft air coolers.

Hot Air Recirculation

The sort of air recirculation I described earlier in this chapter does not affect the air inlet temperature, but only the airflow through the bundle. However, if adjacent air coolers are constructed at different elevations, then hot air recirculation is possible. At the Coffeyville refinery in Kansas, a new air cooler was built alongside an older air cooler, but about 10 feet higher. To some extent, a portion of the effluent warm air from the lower cooler was drawn into the inlet of the new, higher cooler. I do not believe this is possible when the adjacent air coolers are all at the same elevation.

I was working on a sour water stripper air-cooled overhead condenser in New Jersey last year. The air cooler was located atop a very tall platform. The top of the sulfur plant tail gas incinerator stack was essentially at the same elevation. Depending on the wind direction, the hot flue gas from the incinerator stack was blown directly into the airflow being drawn up through the air cooler. To a lesser extent, locating air coolers above shell-and-tube heat exchangers in hot service will result in the same type of malfunction.

Water Sprays with Air Coolers

Spraying water on top of air cooler tubes is highly effective in the short term in improving cooling. Heat transfer rates may double. But the continuous wetting of the tubes causes the aluminum fins to corrode. Then you can break the fins off with your fingers. A corroded fin retards, rather than promotes, heat transfer. Also, the salts and hardness deposits in the water will also, with time, degrade heat transfer efficiency.

An effective way to use a water spray is shown in Figure 12-2. This technique, which I learned about in a refinery in Lithuania, only works when the relative humidity is low. It works on the same principle as the swamp cooler. In New Orleans, it would be a waste

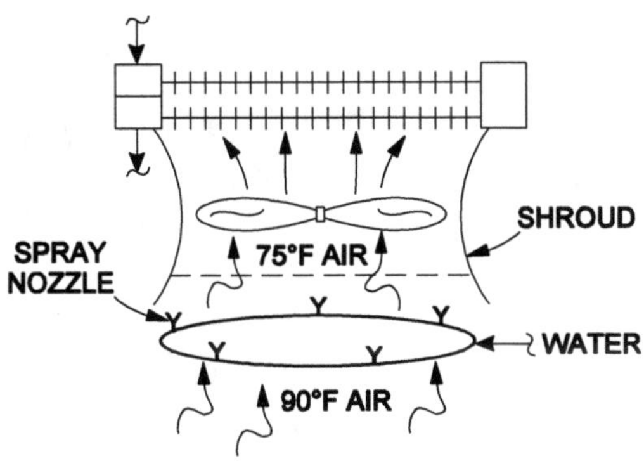

FIGURE 12-2 Humidifying will cool air flow 10°F to 15°F depending on relative humidity.

of time. In Lithuania, on hot summer afternoons, the humidity is low. Water is sprayed into the inlet (i.e., the underside of the screen) of the forced-draft fan through a distribution ring. The ring has six to ten mist formation-type spray nozzles. It's best to use demineralized water or steam condensate. A typical large air cooler fan might move 1,000,000 lb/hr of air. I would use roughly 15 GPM (U.S. gallons per minute) total water flow. In Lithuania, when the spray water was turned on, the process-side outlet temperature might drop 10°F to 15°F. In my daughter's house in California, her swamp cooler reduces her house temperature by 10°F. It depends on the relative humidity of the air. Quite a bit of water drips out of the bottom of the fan, and it does make a mess if used too often.

The key to success is proper selection of the mist formation nozzle. I pick out the appropriate nozzle from my Bete Fog Nozzle catalogue.[1]

Sloped Air Cooler Bundles

If you do not have sloped air cooler bundles (sometimes called A-frame) in your plant, don't bother reading this section. When I worked in Lithuania as a consultant, most air coolers were in this configuration. They all performed rather poorly. At first I thought that the large degree of air recirculation I observed was due to fouling. But then we installed some brand-new bundles, and they also suffered from severe air recirculation. It seemed as if the air had a tendency to bounce off the sloped angle of the bundles. Also, the tubes in the bundle had a tendency to pull out of the header boxes at elevated process temperatures.

The bundles themselves are totally conventional, long but narrow, air cooler tube bundles. I've shown the end view of the tube header boxes of four bundles in Figure 12-3. The purpose of sloping the header boxes and tube bundles is to reduce plot plan size for the air cooler support structure and to reduce the diameter of the fan. But something like 30% of the air in a new unit recirculates, with a resulting loss of airflow through the cooler itself.

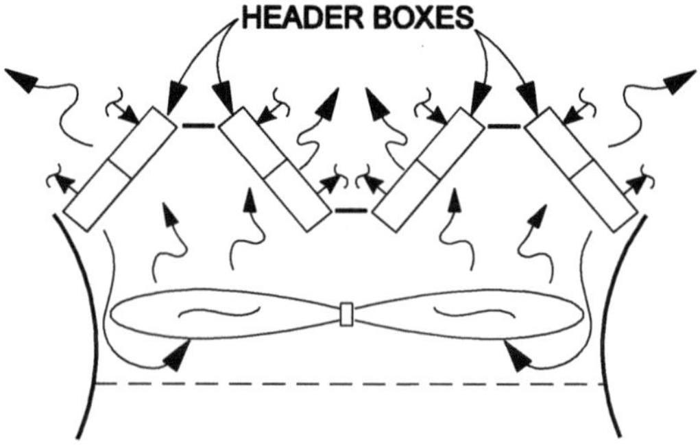

Figure 12-3 Slopped air cooler bundles suffer extreme air recirculation, even when new.

I found out from my Lithuanian colleagues that this type of sloped air cooler design was standard in the days of the Soviet Union. All these air coolers were produced at a factory in Estonia. So I did two things:

- First, I had the sloped air cooler bundles on the amine regenerator overhead condenser that were lacking cooling capacity laid out flat on a new and wider platform. This greatly reduced the air recirculation that limited cooling capacity.
- Second, I wrote a letter to the manufacturer in Estonia, requesting their experience about this malfunction. That was eight years ago. I'm still awaiting their reply.

One of the aspects of this story about the sloped air cooler bundles illustrates a general principle pertaining to many process equipment malfunctions. So often, the fundamental design is flawed. Yet no one questions the design, because it has become the standard design.

Reference

Bete—Manual No. 104.2. Bete Fog Nozzle, Inc. Greenfield, Massachusetts, U.S.A. Phone 1-800-235-0049.

CHAPTER 13

Cooling Water: Towers and Circulation

Now son, it's only a matter of time and determination.

—Chief Operator Leroy Wilkes, Texas City, 1975

Half the heat removed in most process plants is rejected to a circulating cooling water system. The heat is dissipated by water evaporation to air. For example, 120°F cooling water is contacted with cool, dry air in a cooling tower or cell to produce 80°F cooling water. The amount of sensible heat rejected by the water is:

$$(120°\text{F} - 80°\text{F}) \times (1.0\ \text{Btu/lb/°F}) = 40\ \text{Btu/lb}$$

The sensible heat is converted to latent heat as follows:

$$(40\ \text{Btu/lb}) \div (1{,}000\ \text{Btu/lb}) = 4\%$$

Therefore, if I'm circulating 100 pounds per hour of water, my makeup rate to my cooling tower is 5 pounds per hour. No! Not 4 pounds an hour. The extra pound of water is for blow-down. You see, there are salts or potential hardness deposits in the water supply to your cooling tower. If you just added 4 pounds of water, these salts would accumulate in the circulation water and eventually have to precipitate out of solution in the form of hardness deposits in your cooling water heat exchanger final outlet pass tubes.

If I am evaporating 4 pounds of water, but adding 5 five pounds of water, we say that the **cycles of concentration** are 5 to 1. Meaning, if the salt content on the makeup water to the cooling tower was 1,000 ppm, the salt content of the circulating cooling water would be 5,000 ppm.

Incidentally, when I write "salt," I do not mean NaCl. Mostly, I'm referring to carbonate salts. The higher the concentration of carbonate salts in the water supply to your shell-and-tube water coolers, the lower the temperature at which the hardness deposits will precipitate out of solution and foul your heat exchanger tubes.

In reality, in the plants I have run, excessive cycles of concentration were never a problem. On the contrary, due to losses and leaks in the cooling water system, my cycles of concentration have always been too low, and not too high. Meaning, I wasted a lot of cooling tower makeup water.

Hydraulic Losses in Cooling Water Systems

I was working at a refinery in Lithuania. My objective was to increase the capacity of their crude distillation unit. The bottleneck was the overhead condenser capacity in the summer. More to the point, the bottleneck was high overhead condenser cooling water outlet temperature.

The operators, based on their experience, had made the following correct observations:

- When the condenser cooling water outlet temperature exceeded 130°F, salts would start to precipitate in the final tube-side pass of the condenser's bundle.
- The salt accumulation would reduce cooling water flow to the condenser.
- Lower water flow increased the cooling water outlet temperature, which further escalated salt lay-down rates.
- The problem would feed upon itself.

So, as warmer weather approached, the operators would cut crude charge rates to keep the condenser water outlet below 130°F. The malfunction was clearly a lack of cooling water flow. The problem was not new. It had existed for two decades.

Therefore, I went to visit the Cooling Water System Control Center. It's really true. In this medium-sized refinery, they actually had a separate operating staff and control room just for their three cooling towers and water circulation pumps. They even had a dedicated operating superintendent, Comrade Vaidus, who served me tea and cookies.

"Comrade Lieberman. How good of you to visit us."

"Thanks for the tea. I've come to increase the cooling water flow to the crude unit."

"Comrade Lieberman. You're the first engineer from the main office to visit us. It's such an honor."

"Thanks, Vaidus. But can we turn on more pumps to increase the cooling water flow? I know that you'll have to increase the water

FIGURE 13-1 When operating on a flat portion of a pump curve, adding a second pump will not increase cooling water flow.

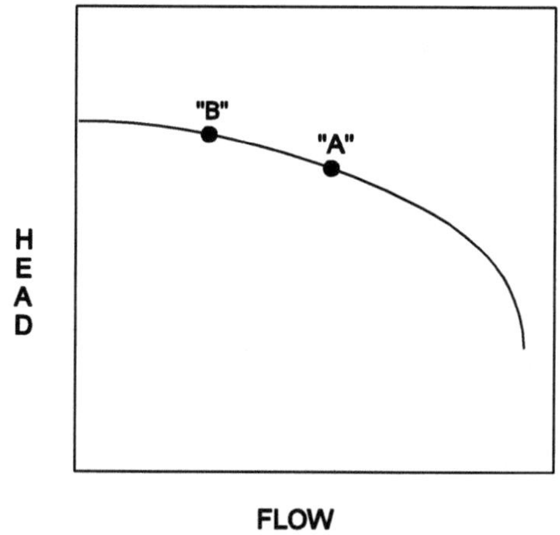

supply to the whole east side of the plant to get more water flow to the crude unit. But the other units would also benefit from more cooling water."

Vaidus looked across the huge and largely empty concrete pumphouse. "Mr. Lieberman, we expanded the entire cooling water system 2 years ago. We've always been short of water flow. Many times I have received unpleasant e-mails requesting more cooling water flow from the east plant units. But there's nothing I can do to help. Our expansion project cost $5,000,000—for the new pump and cooling tower—but it didn't help."

Regardless, I had Vaidus turn on another pump. Referring to Figure 13-1, you can see why the additional pump did not increase water flow. If we started at point A, I would be operating on the relatively flat portion of the pump curve. Then, starting up a second pump would split the flow of water between the two pumps. Both pumps would now be running at point B. But, because the pumps are both running on the relatively flat portion of their curve, only a small amount of additional head or pressure would be produced at the pump discharge. Let's assume the combined pump discharge pressure increased by 4%, as I commissioned the second pump. Thus, the delta P from the pump discharge to the cooling water return basin would also have increased by 4% as:

$$\text{Delta P varies with (Flow)}^2$$

The increased flow of water would only be equal to the square root of 1.04 or 2%. Hardly worth the extra electric power to run the additional pump. Evidently, to make any real progress on this problem of low cooling water flow, I would first have to reduce the pump's discharge pressure. Thus, I checked the cooling water circuit pressure profile, as shown in Figure 13-2.

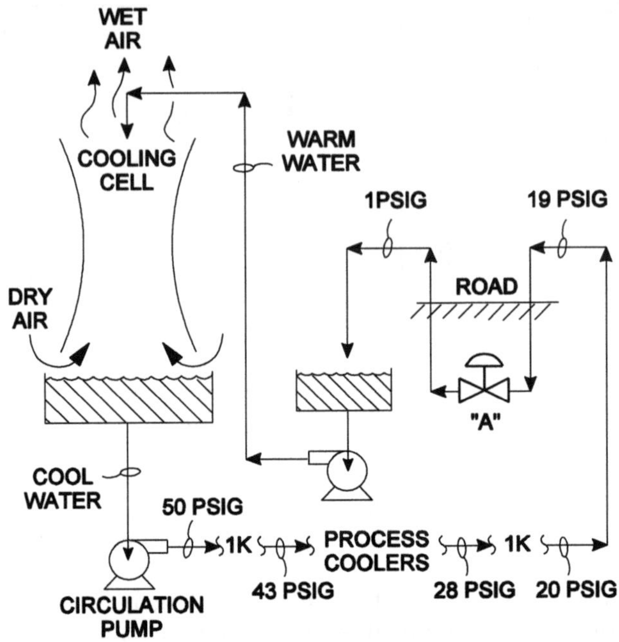

FIGURE 13-2 A valve buried underneath a road bottlenecked the cooling water supply to the entire refinery.

Note that the pressure drop in the water piping to and from the plant's process coolers was only 7 to 8 psi. That's about 0.2 to 0.3 psi per 100 feet of piping, which is quite a reasonably low pressure loss. Note also that the 15 psi pressure loss across the process coolers represented only 30% of the pump's discharge pressure. But the amazing feature of my pressure drop survey was the pressure loss across the road, as shown in my Figure 13-2. The distance between the two pressure points of 19 and 1 psig was only 100 feet. The pipe dipped down beneath a gravel road so as not to obstruct traffic. And when the pipe emerged from the other side, the cooling water had lost 18 psi. How could this be? Something must be constricting the pipe underneath the road! But what could that restriction be?

So the maintenance department dug up the road. And, as shown in Figure 13-2, they found a full-line-size isolation gate valve buried 10 feet beneath the road surface. And this buried valve was mostly closed! The carbon steel valve was badly rusted, was totally inoperable, and could not be opened any further.

The cooling water line was **hot tapped** on either side of the valve, and a bypass was installed. When the bypass was opened, the delta P across valve A declined from 18 to 8 psi. That is, the pressure drop through the cooling water system declined by 10 psi, or about 20%. With less resistance to flow, the water circulation pump moved out onto the steeper part of its curve (Figure 13-1) and flow increased. Now, when another pump was placed in service, there was a noticeable additional increase in cooling water flowing to the east side refinery process water-cooled exchangers and especially to the crude unit.

But Vaidus was angry. "Comrade Lieberman," he complained, "We've spent $80,000 on your American hot taps and bypass. And water flow has barely increased by 10%."

"But Comrade Vaidus," I explained, "You only eliminated 20% of the frictional loss through the system. And delta P varies with flow squared. Placing the additional pump in service has allowed us to maintain the same 50 psig discharge pressure at the 10% higher circulation rate. The extra 10% is all that can be expected."

"Still, I'm disappointed! So little results for so much money invested."

"How can you say that?" I answered. "You all spent $5,000,000 and 2 years on your new cooling cell and pump, and got nothing. I've spent $80,000 and a week, and get 10% more water flow."

But Vaidus was still angry. "The $5,000,000 were funds provided by the central government in Moscow. They have lots of money. The $80,000 came out of my operating budget. We won't be able to afford to buy tea, or cookies, or anything for many years, thanks to you and your reactionary, capitalist engineering."

Cleaning Water Circulation Lines

The pressure drop of cooling water in pipes may be approximated by:

$$\text{Delta P} = (0.15) \times (V)^2 \div (\text{ID}) \qquad (13.1)$$

where V = Water velocity, ft/sec
ID = Pipe inside diameter, inches
Delta P = Pressure drop per 100 equivalent feet, psi

When the observed pressure drop is twice the calculated delta P, then one should consider cleaning the cooling water lines. I've never done this myself, but I have investigated how this can best be done. I was advised by a water treatment chemical vendor to circulate a chelating solution through the cooling water system. This is an online procedure. The chelating solution should remove metallic ions (i.e., scale) from the interior of the piping. Thus the pipe's effective ID will be increased and the frictional losses in the pipe decreased.

Acidifying the system with HCl or H_2SO_4, as I often used to accidentally do on my alkylation unit at Texas City in the 1970s, essentially serves the same purpose. But it also created a host of cooling water leaks in my carbon steel cooling water circulation lines.

Changing Tube-Side Passes

Going from four tube-side passes to two tube-side passes will reduce the flow path length by a factor of two. This will reduce the frictional loss in the tube bundle because the cooling water will not have to flow as far through the tubes. Also, there will be twice as many tubes per pass. The combined effect of both factors will increase the water flow through the exchanger by:

$$(2) \times \sqrt{2} = 2.82 \text{ or } 282\% \tag{13.2}$$

The square root factor of 2 shown in this equation is a consequence of the reduction in flow path length and of the fact that the square root of the reduction in friction loss is proportional to the increase in flow. That is:

Flow is proportional to the square root of friction.

The "two" factor in Equation (13.2) is a consequence of the doubling of the number of tubes per pass.

Caution: I just returned from a job at Hunt Oil in Alabama. They retrofitted an exchanger, going from four- to two-pass, with disappointing results. The water flow barely increased. They had forgotten to change the channel head nozzle sizes and the water piping to accommodate the larger cooling water flow. I already detailed in Equation (13.1) how to calculate pressure drops in cooling water supply and return lines. The pressure drop in nozzles is:

$$\text{Delta P} = 0.012 \times (V)^2 \tag{13.3}$$

Delta P is in psi, and V is feet per second. Often piping and nozzle sizes are oversized, and this will not be a problem. But sometimes the piping and nozzles are not oversized, and one needs to carefully check the hydraulics of the entire cooling water flow system.

Fouling in Water Side

Factors that affect tube-side fouling in cooling water service are:

- *Excessive shell-side supply temperatures.* For instance, water cooling 300°F hydrocarbon vapors is okay. But water cooling 400°F diesel oil will likely lead to excessively hot tube skin temperature on the water side.
- *Excessive cooling water outlet temperature.* Some water fouls more readily than others. However, my general rule of thumb is if the water is less than 120°F, I'll not worry. If the water is over 130°F, I'll want to take corrective action.

- *Excessive cycles of concentration.* Perhaps four or five to one is a reasonable target. Certainly, ten to one is too high, and two to one is wasting the makeup water.
- *Hydrocarbon leaks feed the bugs.* This creates sludgy deposits on the cooling tower cell's distribution decks and inside the cooling water exchanger tubes. An increasing demand for the chlorine makeup rate needed to maintain your chlorine residual is a clear indication of a growing problem with hydrocarbon leaks.
- *Wood chips, small crabs, and paper cups* can restrict water flow into the first pass, by sticking into the tube sheet inlets. Back-flushing the exchanger once a week (see *Troubleshooting Process Operations*, 4th ed., PennWellBooks.com) will mitigate this problem.
- *Really high tube-side pressure drop* due to fouling (50 to 100 psi) can cause the channel head pass partition baffle to internally leak, or even collapse, against the channel head outlet nozzle. This cannot actually occur in most process plants, unless only a very few exchangers are connected in parallel on the discharge of the cooling water circulation pump, as delta P will not increase due to a single exchanger fouling in most systems.
- *Finally, dirty, sludgy, or high-salt-content cooling tower makeup water* should always be considered as a possible malfunction causing fouling.

Other Circulating Problems

- Sometimes, holding backpressure on the cooling water outlet increases water flow. This only happens for elevated coolers, where the water leaves the exchanger at subatmospheric pressures. Holding a very slight backpressure suppresses air evolution and thus increases water flow.
- Check the pump suction screen in the cooling tower basin. The water level upstream of this screen should be no more than 1 or 2 feet higher than downstream at the pump's suction.
- One method to increase water flow to a cooler, is to reduce the water flow to those coolers that have a small cooling water temperature increase—meaning, less than 10°F to 15°F—and hence are receiving too much water. If the cooling water circulation pumps are running on the flat portion of their curves, this will not help very much. However, if the pumps are operating on the steeper portion of their curves (see Figure 13-1), this may help quite a bit. I've done this several times, with varying results.

- For turbine-driven cooling water pumps, make sure the turbine is running at its maximum rated speed. Usually 3,600 rpm in North America and 3,000 rpm in Europe and most other countries.
- On startup, air may be trapped at the top part of a water cooler. Depending on the location of the water outlet nozzle, it may be necessary to vent the top of the channel head to eliminate accumulated air. This is especially true on those water coolers where the water outlet nozzle is not at the top of the channel head. Don't forget to check such vents for hydrocarbon vapors, indicative of a tube leak. Also, venting below the pass partition baffle on the channel head is very important.

Cooling Tower Efficiency

To gauge how well your cooling tower is working, you need to measure the water approach temperature to the "wet bulb" temperature, which is determined as follows:

- Tie a small piece of wet cloth around the end of a glass thermometer. Swing the thermometer for a few moments and read the wet bulb temperature.
- Measure the tower's water outlet temperature.

The difference between the two readings is the cooling tower's approach to the wet bulb temperature. It's best to use the same thermometer. My rough rule of thumb is that an approach temperature of 5°F or 6°F is excellent. An approach temperature of 12°F to 15°F is terrible. Parameters that increase the approach temperature are:

- *Low air flow*. See Chapter 12, "Air Coolers: Forced and Induced Draft; Air Side Malfunctions." More air flow will reduce the cooling water outlet temperature.
- *Bad water distribution*. Often caused by algae accumulation that plugs the cooling cell's water distribution decks. Shock treat the water with chlorine, and then maintain a 1 or 2 ppm chlorine residual. Water is not supposed to overflow the sides of the cooling tower. It's all supposed to flow down through the holes on the distribution decks.
- *Broken or missing fill or packing*. There are many different air-water contacting methods. Anything that diminishes the contacting efficiency is sure to increase the approach temperature. Especially on older, wooden cooling towers, the internal wood slats tend to break apart as the hot return water leaches out the wood's lignite content.

- *Uneven water distribution between parallel cooling tower cells.* At Texas City, I had a dozen cells. The hot return water was divided by manually operating the inlet valves on the top of each cell. The trick is to avoid introducing very much delta P on the return water, but to still have the water flow split evenly between all the 12 cells. If too much backpressure is introduced while you are adjusting these distribution valves, then water circulation rates to the process coolers will suffer.

Sometimes, cooling towers have a high water approach temperature, not because of any particular malfunction, but because they are overloaded, either with too high a water return temperature, or too much water circulation flow, or both. The cooling tower performance should correspond to a manufacturer's performance chart that specifies the expected cool water outlet temperature as a function of the wet bulb temperature, the hot water return temperature, and the water circulation rate. Based on this chart, you can predict the reduction in the cooling water return temperature as a result of a reduction in the cooling water circulation rate.

Thinking back over my long experience with cooling towers, if I had to specify one single action to improve water cooling for refinery process units, it would be to locate and repair the hydrocarbon leaks into the cooling water circulating system. Also, that's an environmentally friendly improvement.

Cooling Tower Fire Protection

In 1976, my cooling tower at No. One Alkylation Unit in Texas City burned down. Actually, this was no big loss. No. One Alky had been out of service for months and there was no plan to ever operate it again. However, Amoco Oil management suggested to me that it might be best not to burn down my cooling tower at No. 2 Alky—the world's largest sulfuric acid alkylation unit.

Cooling towers are subject to conflagration for several reasons. One such cause is hydrocarbon leaks. A sparking fan motor could be one source of ignition for the hydrocarbon vapors escaping from the cooling water return. Dry, older, wooden cooling towers can easily ignite when out of service for a while. When I shut down No. 2 Alky for a turnaround, I was worried about this problem. So I ran several hoses with lawn sprinklers to the top of each cell to keep the cooling cells wet. But a few days later we had two days of freezing weather in Texas City. Then I worried that the cells might collapse under their weight of ice. I had turned my cooling towers into fairyland ice castles.

Bypassing Water-Cooled Exchangers

If the overhead vapors from a distillation tower cannot be fully condensed, valuable product must be vented from the reflux drum. The inability to condense the overhead vapors is a combination of:

- Too low a reflux drum pressure.
- Too high a reflux drum temperature.

I was working on a debutanizer at the Coffeyville refinery that was subject to continuous venting from the reflux drum. The two-stage overhead condensers were:

- First stage—A large air cooler, oversized for its required capacity.
- Second stage—A small water cooler, undersized for its required capacity.

The process side skin inlet and outlet temperatures I observed with my infrared temperature gun were:

- Air cooler: Inlet = 160°F; Outlet = 75°F.
- Water cooler: Inlet = 75°F; Outlet = 78°F.

The 80°F supply cooling water temperature was heating the air cooler process effluent by 3°F! Perhaps you think that the 3°F, while bad, is of no great significance. However, I now also measured the delta P across the process side of the water cooler. It was 6 psi. This is not an excessive pressure loss under ordinary circumstances. But in this case, what was the reward for the 6 psi? Nothing!

So I opened the bypass around the shell (process) side and shut off the water flow to the tube side. The reflux drum pressure increased by 5 psi, its temperature dropped by 2°F, and the gas vent closed.

Author's Comments

In my books, seminars, and videos, I have been accused of "dumbing down chemical engineering." This could be true. It's not my fault. It's because I work at the interface between process equipment and engineering principles. I have a direct physical relationship with the equipment. When I try to describe this relationship to others, it often appears as if I'm relating the obvious. But that's not my fault. It's just hard to translate intimate thoughts and feelings into words. Process problems that seem difficult in the field appear obvious when described on paper.

CHAPTER 14

Steam Condensate Collection Systems

You can imagine how I've become so smart on this subject.

—Author explaining preference for total trap-out chimney trays in crude distillation towers

Probably more process heaters and reboilers malfunction because of steam condensate drainage problems than any other single cause. In most of the plants in which I have worked, steam condensate recovery is 30% to 40%, rather than 70% to 80%, because of this complex malfunction. The problem is that we operate between two extremes:

- Blowing the condensate seal
- Condensate backup

Both conditions result in a loss of reboiler capacity, regardless of whether the steam is on the shell side (vertical reboilers) or on the tube side (horizontal reboilers).

I began my career as a process design engineer in 1965 for the long-vanished American Oil Company. I recall the design of a stripper reboiler using 100 psig steam on the tube side. On my process flowsheet diagram (PFD), I showed 10,000 lb/hr of steam entering the top of the channel head of the reboiler. Draining out from the bottom of the channel head, I indicated a flow of 20 GPM of water at 340°F. The 340°F was the boiling point, or saturation temperature, of 100 psig water. My pipe sizing chart indicated that a 20 GPM flow of water required a line size of 1½ inches for a line velocity of 4 ft/sec.

In 1967 the new stripper was commissioned at the American Oil Whiting Refinery. I rode my bicycle out to the Cracking Complex to see my design in operation. All was fine except for one detail. The operators had diverted the steam condensate from the stripper's reboiler to the refinery sewer.

"It's okay, Lieberman," Stan Derwinski explained. "We always got's to drain reboiler condensate to the sewer. Not a problem."

"But Stan," I asked the stillman, "Aren't we supposed to recover the steam condensate and return it to the boiler house?"

"I reckon so, but it just won't go that way, Mr. Norm. So we just drain it to the sewer. You know that the flow must go on. We got the same problem with all our reboilers. The drain lines are too small. Maybe if you had made that skinny 1½-inch line a 4-inch line, it would have worked."

Let's see mathematically what Stan Derwinski was talking about. Let's presume that the pressure in the main condensate collection header was 15 psig. The boiling point of water at 15 psig is 250°F. As the steam condensate depressures from 100 psig and 340°F into the collection header at 15 psig, it cools down by 90°F, to 250°F. The specific heat of water is 1 Btu/lb/°F. Meaning about 90 Btu/lb of condensate is released. As the water cools, heat is liberated that goes into the vaporization of part of the 10,000 lb/hr of water. The latent heat of water is about 1,000 Btu/lb, so:

$$(10{,}000 \text{ lb/hr}) \times (90 \text{ Btu/lb}) \div (1{,}000) = 900 \text{ lb/hr}$$

So, 900 lb/hr of steam is generated. From my steam tables, at 15 psig and 250°F, I find that there are 14 ft^3/lb of steam:

$$(14 \text{ ft}^3/\text{lb}) \times (900 \text{ lb/hr}) = 12{,}600 \text{ ft}^3/\text{hr}$$

If I sized the line (which I did) just for water, the flowing volume would just have been:

$$(0.016 \text{ ft}^3/\text{lb}) \cdot (10{,}000 \text{ lb/hr}) = 160 \text{ ft}^3/\text{hr}$$

Actually, the real calculation is rather complex. To size the drain line correctly requires differential calculus or a line sizing computer program to account for the progressive flashing of water. In this case, the correct line size to account for the evolved steam due to depressuring the condensate was 4 inches, and not the 1½ inch line size that I had designed.

Consequences of Flashing in Condensate Drain Line

Figure 14-1 illustrates the malfunction caused by the evolution of steam vapor in a line designed for liquid water flow. To illustrate the problem, I'll make some simplifying assumptions that approximate the real situation:

- The steam condensate drain line is sized only for water (and not for any steam).
- There is no pressure drop on the steam side of the exchanger.

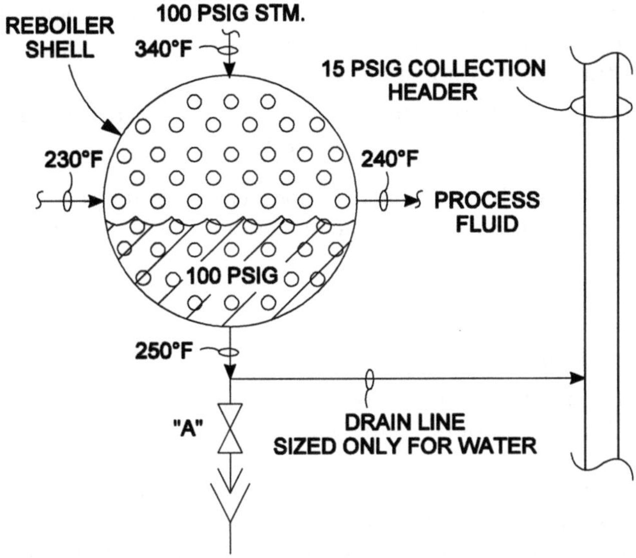

FIGURE 14-1 The result of flashing in condensate drain line is condensate back-up, subcooling of the condensate and loss of reboiler surface area.

- The pressure loss in the condensate drain line is 85 psig (i.e., 100 psig minus 15 psig in the collection header).
- Ambient heat loss in the condensate drain line is zero.

On the basis of these assumptions, the water draining out of the exchanger must be subcooled to 250°F. Because if it was not subcooled to 250°F, then the water would begin evolving steam before it entered the larger condensate collection header. The evolved steam would increase in volume by several orders of magnitude. The resulting expansion in volume would increase the delta P in the condensate drain line. And then:

- The condensate level in the exchanger would be pushed up higher.
- With more of the exchanger surface area devoted to subcooling water rather than condensing steam, the condensate would drain from the exchanger colder.
- The condensate temperature would diminish until the condensate was sufficiently subcooled so that at its lowest pressure in the condensate drain line, it would still be below its boiling point temperature.
- The reduction in exchanger surface area exposed to the condensing steam would reduce the exchanger heat transfer duty or capacity.

- The operators, refusing to accept this lost capacity in heat exchange capacity, would open valve A in Figure 14-1 to drain condensate to the sewer.
- This would lower the water level in the exchanger, and thus partially restore exchanger duty.

The problem that I've just described is not common—it's universal. It's the main cause of low steam condensate recovery in process plants.

Condensate Drum Elevation

There is another common variant of the same malfunction. I'll bet almost all of the readers of this text have seen it for themselves. It's depicted in Figure 14-2. I've encountered this problem in Baton Rouge, Aruba, Lithuania, and Mumbai. As the condensate drains through the steam trap, it's just water. The elevation of the exchanger above the steam trap is sufficient to retard vaporization due to loss of pressure in the steam trap. But, the steam condensate now must flow to a higher elevation. It must flow to the condensate flash drum set 34 feet above the steam trap. What, then, is the pressure loss of the water as it rises by the 34 feet? (In case you have forgotten—one atmosphere worth of pressure equals 34 feet of water).

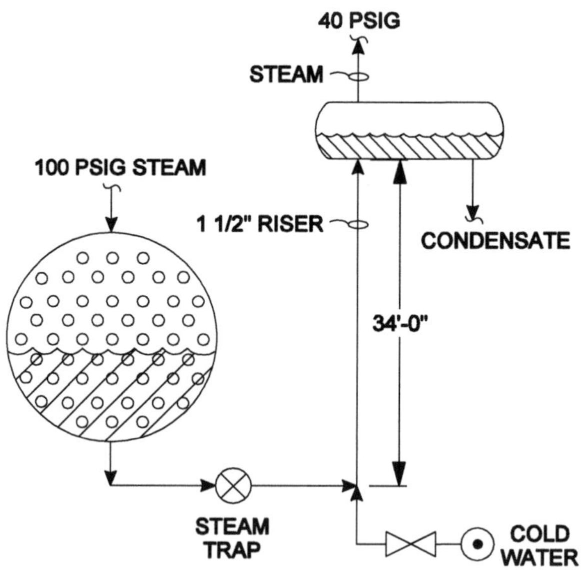

Figure 14-2 Condensate flashes as it looses head pressure and backs-up water in the exchanger.

The answer is not 14.7 psi. Here's why:

- The riser line to the drum was designed for water. The water leaving the steam trap is initially saturated water. As it rises to a higher elevation, it loses pressure and starts to vaporize. The evolved steam expands by an order of magnitude and pushes the water level up in the exchanger.
- The water becomes subcooled due to condensate backup, which suppresses some vaporization in the riser.
- However, the vaporization in the riser reduces the hydrostatic head loss between the exchanger and the drum.
- Depending on the diameter, friction factor, and equivalent length of the riser pipe, some steam vaporization in the riser will actually occur, as will some degree of condensate backup in the exchanger.

The one thing we can say about this problem is that you and I do not know how to do these calculations. Paul Dirac, or Heisenberg, or Max Planck probably could predict the degree of condensate backup and loss of heat transfer capacity. But for us mere mortals, the lesson is to locate the condensate drum below, not above, the exchanger.

Correcting Steam Condensate Backup

But this is a text about malfunctions and troubleshooting, not proper process engineering design practice. What did I do in Lithuania on their naphtha reformer depentanizer reboiler to correct their problem?

Step 1—I installed a bypass around the steam trap. This did no good whatsoever.

Step 2—I drained condensate to the sewer at the base of the riser. This helped, but my comrades in Lithuania said they didn't need an American to tell them to do that. They called this era Soviet technology.

Step 3—I removed the insulation from the 1½-inch riser pipe. This helped a bit.

Step 4—I lowered the pressure in the condensate drum from 40 to 30 psig. This also helped, but only by a few percent of the exchanger duty.

Step 5—I injected cold water at the base of the riser, as shown in Figure 14-2. This essentially corrected the problem and restored the heat exchanger duty to its required capacity. But the cold water had hardness deposits that contaminated the recovered steam condensate.

Step 6—The 1½-inch riser was replaced with a 4-inch riser. Pressure was still lost due to the change in elevation. But the riser was now large enough to accommodate the increase in flowing volume without significantly backing up the steam condensate in the exchanger.

How did I calculate the required line size of 4 inches? Well, I guessed, based on having 45 years of engineering experience during which I had undersized steam condensate drain lines.

Temperature Approach Pinch Point

Let's refer back to Figure 14-1. Note that there's an additional restriction, other than the loss of heat transfer surface area due to condensate backup: That limitation is the cold-end **approach temperature** between the 250°F subcooled steam condensate and the 230°F process-side inlet. The 20°F difference is approaching a practical limit. For example, if hydraulic constraints in the drain line required the steam condensate to back up to subcool the water effluent to 230°F, the exchanger heat transfer capacity would approach zero. Of course, the steam condensate flow would also approach zero, as would the delta P in the condensate drain line.

Limits of Process Engineering Design

So, what superficially appears to be a simple problem—draining water from an exchanger—is really quite complex. Very few process engineers even realize the scope of the problem. And no one that I've met correctly understands (including your author) exactly how to do the required calculations to size the drain line. But that's the true nature of process engineering. We combine field experience, engineering judgment, and simple calculations to guess at the answers. Unfortunately, if the engineer making the guess doesn't have the field experience, we wind up with a process malfunction.

So why not design equipment properly in the first place, by having experienced field engineers review detailed P&IDs and PFDs before construction? The answer to this critical question is that there is no answer.

Review Question

Let's refer to Figure 14-2 again. I began injecting cold water downstream of the steam trap. The cold water quenches the 1½-inch riser. But what happened to the temperature of the steam trap itself?

- Hotter?
- Colder?
- Same temperature?

The cold injection water reduced the need for subcooling. This lowered the unseen condensate level in the channel head of the depentanizer reboiler. Thus, the steam trap began to run . . . hotter.

I've said that the level in the channel head was "unseen." Not quite right. Using my infrared temperature gun, I can monitor the surface temperature on the **channel head cover**. Where this surface temperature drops is an indication of the water level in the exchanger channel head. (Note to Reader: Terms in bold are defined in the Glossary at the end of this text.)

Blowing the Condensate Seal

In Figure 14-2, I have shown a steam trap on the condensate drain. In this case, the steam trap serves no function, as the condensate in the bottom of the exchanger is subcooled. I've opened bypasses around such steam traps without any ill effects.

When I was 23 years old, I was working as a process design engineer for American Oil. Of all their 40,000 employees, I was the only person who did not know the purpose of the thousands of steam traps in the Whiting refinery. I was too embarrassed to ask. Eventually I decided that the steam trap serves the same function as the condensate drum shown in Figure 14-3. That function is to drain water from the channel head of the exchanger, and also to trap or prevent steam from escaping from the channel head.

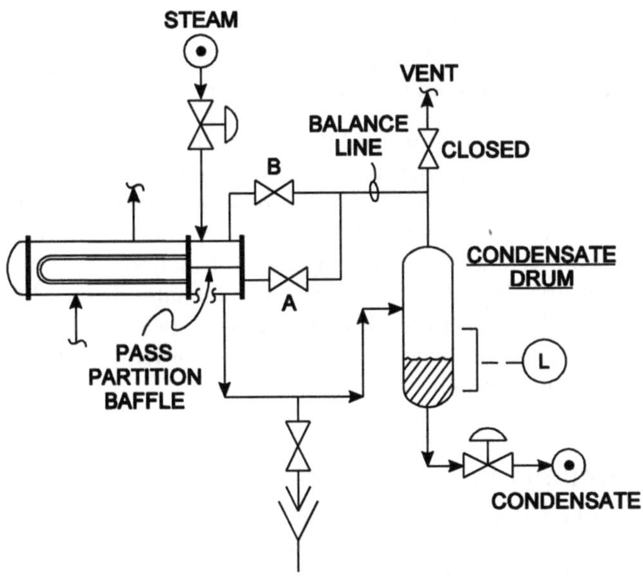

FIGURE 14-3 Condensate drum used instead of a steam trap.

Steam traps are typically used for smaller flows, condensate drums for larger flows. When to select a trap or a drum is something I do not know. I suspect that steam condensate drums with level control valves are more reliable than steam traps.

Let's assume that I have neither a trap nor a drum, but just a valve on the drain line from the channel head (i.e., the steam side) on my heat exchanger. I have conducted this experiment dozens of times. As I open my drain valve, the level of water in the channel head drops. More exchanger tubes are exposed to the condensing steam. The exchanger heat duty goes up. The temperature of the condensate increases and approaches the saturation temperature of the supply steam to the heat exchanger.

But if I open the drain valve too much, there will be a step-change decrease in the exchanger heat duty. Typically, I would observe a sudden drop of 10% to 30%, if not more, in heat exchange. I have just blown the condensate seal. Uncondensed steam is blowing through the tubes and out past my valve. The condensate drain is still at the saturation temperature of the steam supply, but the heat transfer coefficient (*U*) has dropped a lot. The problem is in the condensing service; high velocities appear to retard heat transfer rates, as I discussed in Chapter 9, "Process Reboilers—Shell and Tube."

The magnitude of the effect of the malfunction of blowing the condensate seal depends on the backpressure from the condensate collection header. For example:

- If you're using 400 psig steam draining into a 20 psig condensate header, a very large reduction in heat exchange efficiency will result.
- If you're using 50 psig steam draining into a 40 psig header, a small reduction in exchanger capacity will result.

If you have a steam trap, and the trap sticks open, the condensate seal will be lost. Restricting the flow from the trap with the downstream gate valve will then raise reboiler heat transfer rate, but diminish steam consumption. Sometimes, I've been told, hitting the trap with your wrench may correct this malfunction.

Loss of the liquid level in the condensate drum shown in Figure 14-3 will also lead to blowing the condensate seal. Correcting this malfunction is easier to observe, and less violent, than assaulting steam traps with wrenches.

Level Malfunctions in Condensate Drums

I have implied that the level in the condensate drum is indirectly controlling the level in the channel head of the steam-heated exchanger. However, for the level in the condensate drum to match the level in the channel head, the pressure in the drum must also match the pressure in the channel head. And this creates a problem.

There are two different pressures in the channel head: above and below the **pass partition baffle**.

Let's assume the pressure drop on the tube side of the exchanger shown in Figure 14-3 is 2 psi. Let's say the steam supply pressure measured at the top of the channel head connection B is 60 psig. Then the pressure measured beneath the channel head pass partition baffle is 58 psig (60 – 2). If we assume we want to maintain the channel head fully drained, but without backing-up liquid, then which pressure do we want in the condensate drum?

- 60 psig (through valve B)

or

- 58 psig (through valve A)

Certainly 58 psig. But most of the installations I've seen connect the condensate drum **balance line** in Figure 14-3 to the top of the channel head. The result of this error is to flood the channel head:

$$(2 \text{ psi}) (2.3 \text{ ft } H_2O \text{ per psi}) = 4.6 \text{ feet}$$

This suggests that the water level in the bottom half of the channel head will be pushed up by 4½ feet from the pressure in the top half of the channel head. Hardly likely! Because the diameter of the channel head is only 4 feet. So what really happens?

Field operators experiencing a reduction in exchanger heat capacity will drain the condensate to the sewer. The fix for this malfunction is to connect the balance line to valve A, rather than valve B. Then the pressure and level in the channel head and condensate drum will be identical. To summarize, the balance line must be connected below, never above, the pass partition baffle.

Steam and Water Hammer

One symptom of the dual problems I have just described, that is, blowing the condensate seal and condensate backup, is the often observed violent hammering of piping systems. I've never experienced an actual mechanical failure due to steam hammer. But based on the violent shaking and noise, I imagine that, if unchecked, such hammering would be quite destructive.

Steam hammering is most commonly observed when warming up cold steam piping systems. I well remember an evening during the startup of the Texas City sulfur plant in 1980. I was assigned to commission the 400 psig steam supply header. Bleeding live 400 psig steam into the main header produced the most awful clanging. Being startled by all the noise, I opened the steam supply faster on the theory that if something bad is about to happen, then let it happen quickly. This was not exactly the correct response to my problem.

Heating a cold piping system with steam is going to create condensate. The condensate, as it runs through the cold piping system, is going to become subcooled. The requirements for steam/water hammer are:

- Steam
- Subcooled water
- Mixed in a closed system of restricted volume

The steam, when contacting cold water (or cold steel piping), condenses rapidly. This results in areas of localized low pressure. Water races into these areas of low pressure, hits the piping bends, and produces the hammering sounds we are familiar with.

A condensate collection header in a refinery may have a hundred various contributors. If one reboiler is draining subcooled water and a second reboiler is blowing its condensate seal, then the result, depending on the geometry of the piping, may well be water/steam hammer.

The field operators will stop this hammering by dumping one or both condensate streams to the sewer. This leads to a loss of expensive steam condensate, a swelling of the refinery's effluent volume, and a waste of energy.

Correcting Steam/Water Hammer

If you read other texts, the stated cause of hammering is mixing steam and water in pipes. Not quite right. The correct statement is that hammering is caused by mixing steam and subcooled water in pipes. So to avoid this malfunction, all we really have to do is to operate our reboilers and steam heat exchangers in the manner in which they were originally intended to operate. That is, by avoiding condensate backup and blowing the condensate seal. If these dual criteria are met, then only saturated water will drain into the condensate collection header. Admittedly, high-pressure condensate will still evolve steam. But the residual condensate will be at its saturation temperature and not subcooled.

Unfortunately, if the vertical condensate drum in Figure 14-3 is run at a variable level to control the exchanger duty by condensate backup in the channel head, then subcooled water will always be flowing into the condensate collection header piping. In such a case, steam/water hammering may be unavoidable.

Corrosion Potential of Condensate

Water in general is corrosive to carbon steel, especially when it contains H_2CO_3 (carbonic acid). Steam supplied to a reboiler will contain variable amounts of CO_2. Residual carbonates from BFW (boiler feed water) will decompose in the boiler to CO_2. The CO_2 will accumulate in the reboiler or steam heater. If the steam is on the shell side of the exchanger, or if the reboiler is in a vertical configuration, then there is not much of a problem. To avoid carbonic acid corrosion, just open the high-point vent. Bleeding one-half of 1 percent of the steam supply continuously through a restriction orifice will eliminate CO_2 accumulation.

However, for the more typical situations, the steam is on the tube side of a horizontal exchanger. Then the CO_2 will not accumulate at the high-point vent of the channel head. If you inanely (as I did in Texas City from 1974 to 1976) vent from the ¾-inch bleeder on top of the channel head, you will just vent steam, not the accumulated CO_2. The CO_2 will accumulate not above, but beneath (see Figure 14-3) the channel head pass partition baffle, provided that you have not blown the condensate seal.

The CO_2 concentration will increase and dissolve in the condensate to form corrosive ($H^+ + HCO_3^-$) carbonic acid. If you have tube leaks due to corrosion originating on the steam side of an exchanger, especially if the leaks are in tubes located below the bottom channel head pass partition baffle, then this is carbonic acid attack.

H_2CO_3 produces a somewhat buffered solution. That is, even at rather high concentrations, the pH hovers around 6. For H_2SO_4, or HCl, or HNO_3, a pH of 6 would not cause much corrosion to carbon steel. But for H_2CO_3, it can cause quite a few localized tube leaks, as tube thickness is typically only 0.12 inches for a ¾-inch exchanger tube.

Most exchangers have two vents on the channel head, one on top and one just below the channel head pass partition baffle. The latter is the one to open. If the balance line to the condensate collection drum is properly located, as in Figure 14-3 (connection "A"), venting is even simpler. Just open the atmospheric vent in the top of the drum to bleed off the one-half percent of steam supply as I explained earlier. Just drill the appropriate-sized hole in the gate of the vent valve. For rough calculations, I assume a flowing velocity through the hole of 800 ft/sec.

But there is a completely different approach to this corrosive malfunction. Your chemical vendor will happily sell you a neutralizing chemical to react with the CO_2. This works! But unless you are getting the free duck hunting trips supplied by the chemical vendors, you may find my venting method to be somewhat more cost effective.

Shell-Side Condensate Backup

Examples I've cited so far for subcooling and loss of exchanger surface area have all been for steam on the tube side. But for most reboilers in the chemical industry or for refineries in Eastern Europe, the reboilers are of a vertical thermosyphon configuration, with steam on the shell side.

I was working on a gasoline splitter in Lithuania on such a reboiler. The condensate drain line was reasonably sized for steam

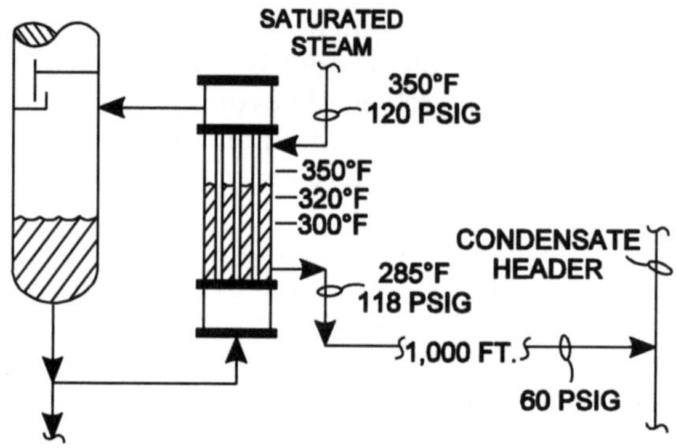

FIGURE 14-4 Effect of condensate backup is subcooling and loss of surface area.

evolution and the reboiler was elevated 20 feet above grade. But, I guess for circumstances not under the control of the process design engineer, the condensate drain line was far too long (see Figure 14-4). As a result, there was a huge delta P in the drain line due to the evolved steam. Condensate backed up. Fully two-thirds of the exchanger surface area was submerged. I checked the shell skin temperature with my infrared temperature gun, as shown in Figure 14-4, to determine the extent of the condensate backup level. The condensate was subcooled by 65°F.

My client felt that there was a positive aspect to this subcooling. That is, more heat was recovered from each pound of steam. That is true. But the extra heat recovered is small.

$$\text{Latent heat} = 900 \text{ Btu/lb}$$

$$\text{Extra sensible heat} = (350° - 285°\text{F}) \times 1.0 = 65 \text{ Btu/lb}$$

The 1.0 factor is the specific heat of water.

Condensate Recovery Pump

My first assignment as a process engineer was to stop condensate loss at VRU 300, in Whiting, Indiana. Over 200 GPM of steam condensate was being drained to the sewer at the C3-C4 splitter reboiler. This reboiler used 30 psig steam.

I started the assignment by measuring the pressure on the reboiler below the lower pass partition baffle. It was 25 psig.

Next, I measured the pressure in the 12-inch condensate collection header line, near the point that the 3-inch C3-C4 reboiler condensate drain line tied into the collection header. It was 38 psig. Obviously, this meant I had to pump the condensate out of the condensate collection drum shown in Figure 14-3 into the higher-pressure header.

The collection drum was several feet above grade. There was adequate clearance for me to install my new pump, which had the following design specifications:

- Flow = 200 GPM
- Head = 100 feet
- Required NPSH = 15 feet
- Power = 10 BHP (Brake horsepower)

I only required 30 feet of head. But I wanted my first project to work for sure. So I made sure my pump and motor driver were plenty big enough. A year after the pump was installed, I observed that 200 GPM of condensate was flowing to the sewer, and that the pump was not running.

When I instructed the operators to run the pump and close the 3-inch drain to the sewer, here's what happened:

- The pump cavitated for a while.
- The temperature of the water in the condensate drum dropped by about 8°F.
- The pump stopped cavitating and then ran fine.
- The reboiler duty on the splitter dropped by about 30%.
- The splitter reflux drum went empty and the reflux pump started to cavitate.
- The operators started draining the subcooled condensate back to the sewer. This lowered the unseen condensate level in the reboiler's channel head.
- My pump started to cavitate again and the operators shut the pump down.

And there, ladies and gentlemen, the idle pump still sits. A monument to a young engineer who was not acquainted with the concept that the **required NPSH** of a centrifugal pump must equal **available NPSH** of the system (see Chapter 29, "Centrifugal Pump NPSH Limitations: Cavitation, Seal, and Bearing Failures").

In practice, condensate recovery pumps are successful on vacuum systems and surface condensers, where the exchangers are running at subatmospheric pressures and it's impossible to drain steam condensate to the sewer. But if the operators can dispense with the condensate recovery pump operation, sooner or later they will. Then they will sewer the condensate and forget about the pump.

In summary, if you want to pump condensate, then you'll have to elevate the exchanger and drum well above grade, to provide adequate NPSH for your pump. Alternatively, you can create another malfunctioning monument, as did your author.

Summary of Malfunctions

When I'm faced with lack of reboiler duty or steam heater capacity, I'll proceed as follows:

- Place a pressure gauge below the bottom pass partition baffle, or on the shell, if the steam is on the shell side. Is the observed pressure within 5 or 10 psi of the steam supply pressure?
- Pinch back on the gate valve on the condensate drain line. If the exchanger duty goes up rather than down, your problem is a blown condensate seal.
- Open the condensate drain to the sewer. If the heat exchanger duty goes up rather than down, your problem is condensate backup (subcooling the condensate).

Condensate backup is the most common problem in heat exchangers using steam. It's caused by:

- Steam trap malfunctions
- Steam trap undersized
- Condensate flowing uphill
- Improper balance line connection
- Level control problems in the condensate collection drum
- High delta P in condensate drain line
- Back pressure from the condensate collection header
- Elevation of the condensate....!

Well, I'll stop. I guess you all get my point. This apparently simple problem of draining water from a heat exchanger is far from simple. Hidden malfunctions await the careless process technician or engineer. But if it was easy, they wouldn't need us.

Condensate Backup Due to Oversizing

A frequent problem in steam heaters and reboilers is failure to maintain condensate drainage. This occurs even when the steam supply pressure is substantially greater than the condensate collection header piping. In response, the operators drain the steam condensate to the sewer. As steam condensate has about 20 percent of the value of the steam itself (energy plus chemical costs), this represents a substantial waste of resources.

Often, the fundamental cause of this waste is that the steam exchanger has been vastly oversized for its current operating mode. Also, the control valve on the steam supply to the exchanger is on the inlet, and the condensate is drained via an outlet steam trap. Because the exchanger has excessive surface area, the steam inlet control valve operates in a mostly closed position. The resulting high pressure drop causes a low steam condensation pressure in the heat exchanger itself. As the pressure in the heat exchanger approaches the pressure in the steam condensate collection header piping, the steam trap will fail to function. It could even happen that water will flow out of the condensate piping, backwards through the trap, and into the steam heat exchanger. This is all going to cause heat transfer instability. If the heat exchanger is a distillation column reboiler, then the column operating pressure will become erratic due to this condensate backup malfunction.

To correct this problem, the steam inlet control valve should be opened 100%. A new control valve should be installed on the condensate drain line. This control valve will close to reduce the steam heat exchanger's duty as required.

The negative feature of this retrofit control scheme occurs if the steam is on the tube side of the exchanger. The cycling of the hot water level in the channel head tends to cause the closure between the shell and the channel flanges to leak hot water. I've put a ring-type clamp around these flanges to control this water leak. It's more of a housekeeping issue than a significant safety problem. But it is annoying.

At the higher exchanger heat duties, one should revert back to controlling the steam with the inlet control valve. I have dealt with this subject in greater detail in my book, *Troubleshooting Process Plant Control*, (Wiley, 2008).

Correcting Water Hammer

Water hammer is a result of mixing cold water and hot steam condensate. Or mixing cold water and steam in an enclosed piping system. The cold water, and the steam that is flashed out of the hot water, combine to form an area of low pressure. Water from a more distant portion of the piping system then rushes (i.e., accelerates) into the low-pressure region. The high-velocity water hits a piping elbow or tee-junction with great force. The result is water hammer. The problem is accentuated by steam traps that relieve intermittently into the condensate collection piping header.

One proven method to substantially suppress water hammer is shown in Figure 14-5. A cone-shaped insert, with orifice holes, is installed in the condensate piping. This insert acts as a dampening brake on the accelerating water. The more restrictive the insert, the more completely the water hammer is suppressed. The negative aspect of this cone insert shown in Figure 14-5 is that it increases the pressure drop in the collection header piping and possibly could plug with corrosion deposits.

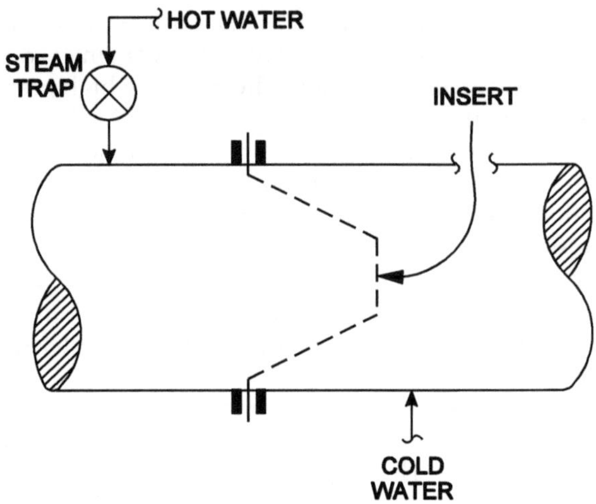

FIGURE 14-5 Insert used to supress water or steam hammer in a steam condensate collection piping header.

I learned about this bit of useful technology from one of my students in Qatar (Ras Gas). It was installed on his unit in India. It almost completely eliminated a rather serious water hammering malfunction in the unit's condensate collection header piping system. I've had 17,000 attendees in my seminars since 1983. The sum of the knowledge I have learned from them exceeds the net amount of knowledge I have imparted to my students.

CHAPTER 15

Steam Quality Problems

Norm! My girlfriend stole my new truck. What should I tell my wife?

—Shift foreman Zip, Texas City, 1976

Why worry about the quality of steam? Is this an important subject? It is, if you work with:

- Steam turbines that vibrate and lose efficiency
- Steam superheat coils that leak
- Vacuum jets that malfunction
- Steam strippers that have low efficiency
- Steam/methane reforming catalyst tubes in hydrogen plants that plug
- Spent catalyst steam strippers (i.e., FCU spent cat strippers)
- Overspeed trips on steam turbines that stick open

I could go on with this list, but perhaps I've given you enough to worry about. Certainly, the most common and serious result of poor steam quality is the effect of hardness deposits on turbine vibration and reduction in steam turbine shaft horsepower efficiency.

Steam Turbine Vibrations

The quality of steam refers to its moisture content. Dry steam, while possibly saturated, can have zero moisture. Very poor quality steam might have 10 wt% moisture. The origin of the moisture could be condensation in the steam supply piping due to ambient heat loss. Or perhaps more commonly, if the steam is produced from waste heat boilers, the moisture is due to carryover from the boilers.

The boiler carryover origin of moisture in the steam supply to a turbine is far worse than the ambient heat loss condensation problem. Let's say the waste heat boiler **blowdown** contains 1,000 ppm (0.1 wt%) of **total dissolved solids (TDS)**. That's rather typical for such a boiler. Then the entrained moisture will contain the same 1,000 ppm of solids, or potential hardness deposits, much of it in the form of silicates. If the steam quality is poor (assume 5% moisture), then the steam will contain 50 ppm of solids. If our turbine uses 10,000 lb/hr of steam, then we are introducing into the turbine case per day:

$$(10{,}000)\ (24)\ (50) \div 1{,}000{,}000 = 12\ \text{lb/day of solids}$$

Depending on a wide variety of factors, several thousand pounds per year of solids may stick to the turbine blades. In reality, though, that's quite impossible. Long before this amount of solid accumulation occurs, turbine efficiency will slip by 10% to 20%. And before the turbine efficiency can get much worse, the hardness deposits will begin to spall off the turbine wheels. The deposits do not break off evenly. Hence the turbine's rotating assembly becomes unbalanced. The turbine now trips off on high vibration. Attempts at washing a badly fouled turbine onstream usually just results in a shut-down. The deposits, if thick, break off and unbalance the rotor. However, cleaning a mildly fouled rotor is often successful (please consult my book, *Troubleshooting Natural Gas Processing—Wellhead to Transmission*, for a historical account of cleaning rotors to retard excessive vibrations).

Leaking Superheat Coils

A typical boiler may produce 600 psig saturated steam in its **radiant section**. The saturated steam is then superheated by an extra 100°F to 200°F in the boiler's **convective section**. If the saturated steam contains entrained boiler feed water, then the TDS in the entrained droplets will salt-out inside the superheat section tubes. The deposits accumulate over time. The more the entrainment and the greater the TDS of the boiler's blowdown, the thicker and faster will be the deposit accumulation. Ordinarily, carbon steel tubes can be used to superheat 600 psig steam to about 700°F without a problem. But if the tubes themselves get much over 1,000°F, the tubes (typically 4- to 8-inch ID) will likely fail. And then the boiler shuts down.

In a refinery in Spain, 10 bar (150 psig) steam was being produced on a small kettle-type waste heat boiler. The heating medium was vacuum tower bottoms product. The steam was being superheated in the crude furnace. As the steam was consumed as stripping steam on the crude tower, superheating the steam was of no particular benefit. Anyway, I had been hired to revamp the crude unit trays for better fractionation efficiency and not to optimize steam superheat.

However, due to improper level control on the waste heat boiler, there was a large carryover of entrained boiler feed water (BFW) to the superheat coil, which ruptured. During the subsequent startup, there was a pressure surge in the crude fractionation tower due to a slug of water in the stripping steam. This blew my new trays off their tray rings. Tower fractionation was worse than ever. The plant manager, Señor Gonzales, blamed me for the degraded fractionation efficiency.

Vacuum Jet Performance

Poor-quality motive steam to a vacuum jet contributes to three classes of jet malfunctions:

1. As discussed previously, salts in the steam deposit at the jet steam inlet nozzle and unfavorably change the nozzle's performance characteristics (see photo of a fouled steam nozzle in Chapter 26, "Vacuum Systems and Steam Jets").
2. The moisture in the steam partly evaporates as the steam expands into the jet's mixing chamber (see Chapter 26). The latent heat required for this evaporation robs energy from the expanding steam.
3. The droplets of moisture erode the steam nozzle throat. This allows an undesirably large amount of motive steam to flow into the converging section of the steam jet.

All of these problems will lead to a reduction of vacuum developed by the jets. The power generation and refinery industries likely lose billions of dollars every year because of these malfunctions. One of my clients claimed to have saved $14,000,000 per year (U.S.) by cleaning a single steam nozzle (Sasol-South Africa) on a surface condenser.

Steam Stripping Hot Oil

Stripping lighter components from a hydrocarbon liquid requires heat. The heat is needed to provide the latent heat of vaporization of the stripped lighter components. This latent heat does not come from the steam, but from the sensible heat content of the stripper feed. That is, the feed cools as it is stripped. As the feed cools, it becomes progressively more difficult to drive off the lighter components dissolved in the feed. If the steam is wet, the feed cools even faster. That's because the entrained moisture in the steam also requires latent heat. If the steam quality is 1% or 2% moisture, the effect is small. But, if the steam has 10% moisture, the stripper efficiency will be severely degraded.

Steam/Methane Reformers

The Coastal refinery in Aruba had three steam/methane H_2 production reformer plants. The reaction between the hydrocarbon feed (usually C_3/C_4 rather than methane) and the 450 psig steam took place in the catalyst-filled furnace tubes. The 450 psig reaction steam was generated in the convective section of the reformer furnace. There were level control issues in the steam drum used to separate the circulating BFW from the steam. The resulting periodic carryover of high-TDS BFW caused hardness deposits (silicates) to accumulate on the catalyst. Plugging of the catalyst caused only two of the three H_2 plants to be in service at any particular time. This single problem was the bottleneck of the entire Aruban refinery.

Plugging Steam Sparger

We sometimes use steam to strip volatile hydrocarbons from circulating aerated solids. My experience with this operation is limited to stripping 1,000°F circulating fluid catalytic cracking unit (FCU) spent catalyst with steam. The steam is dispersed in a pipe grid distributor with ½-inch distribution holes, called a **steam sparger**. Here, the solids content of the steam is not the issue. It's the moisture of poor-quality steam mixing with the catalyst that occasionally backs into the sparger. The moisture mixes with the hot catalyst, dries out, and temporarily plugs the sparger holes. The pluggage is localized. This causes poor distribution of the stripping steam and degraded stripping efficiency of the circulating, sand-like fine catalyst. As time goes on, the problem self-corrects itself, as the solidified catalyst clears the distribution holes, if the steam quality malfunction has been corrected. A significant percentage of the world's gasoline production is lost due to this moisture problem in the spent catalyst stripping steam.

Salting Out of Overspeed Trip on Turbine

From 1974 to 1976 I was the superintendent of the world's largest H_2SO_4 alkylation plant. I produced 0.1% of the entire world's supply of gasoline in Texas City. One day as I was making my morning round on the unit, I noted that my alky unit, turbine driven, centrifugal refrigeration compressor was running in a **tripped position**. The machine used 400 psig motive steam, exhausting to a 30 psig header. The **governor** speed control valve was holding the turbine speed at 3,600 rpm. But the spring-loaded trip valve was unlatched. Kind of like the sticking screen door of my garage. The trip valve should have been pulled shut by the big spring. But hardness deposits from the 400 psig steam had caused this valve to stick. So even when low lube oil pressure to the turbine bearings declined, the steam supply would continue. The low lube oil pressure had unlatched the trip valve, but steam hardness deposits froze the trip valve in an open position. I've told this sad story in detail in my book, *Troubleshooting Process Plant Control* (Wiley, 2008).

The 400 psig steam used to drive my alky compressor came from a 20-inch refinery wide steam header. How or where steam got into this header, I never thought about. My operators just said that the steam in Texas City was dirty and rendered our turbine trip valves inoperable with time. However, dirty steam is not an act of God. It's an act of man, and thus can be corrected without divine intervention. In 1974, it never occurred to me to inquire as to the origin of my poor-quality steam supply. Note that the 400 psig to my alky unit was somewhat superheated. How could superheated steam be dirty? Well, if superheated steam and wet steam are mixed together, then the resulting steam will be dry but will contain entrained solids. Anyway, superheating wet steam will not remove the TDS in the entrained water from the boiler.

Measuring the Moisture Content of Steam

To correct any process equipment malfunction, one should best begin by developing a method to quantify the extent of the malfunction. The solids content of entrained moisture may simply be assumed to be a little less than the TDS silicates of the boiler blowdown (not the BFW).

The correct way to measure the moisture content of steam is by the use of the throttling calorimeter. This is a standard piece of portable lab equipment that's described in the front of your **steam tables.** Now, nobody cares about the throttling calorimeter, so I won't describe it. I'll just explain my field method to approximate the quality of steam.

When steam expands (de-pressures) from 400 psig to a lower pressure, it cools as shown in Figure 15-1, for four reasons:

1. Lower-pressure steam has a bigger latent heat than higher-pressure steam. To supply that extra latent heat, the sensible heat content of the steam (i.e., the temperature of the steam) falls. You can read this effect from the **Mollier diagram** in the back of your steam tables.
2. Some of the enthalpy (heat) of the steam is converted to kinetic energy (velocity or speed) as the steam expands.
3. Ambient heat loss.
4. Moisture in steam flashing (evaporating) to steam. For each one weight percent moisture evaporating, a 17°F drop in temperature of the flowing steam will be observed.

Item 3 above can be neglected if piping runs are only a few feet in length.

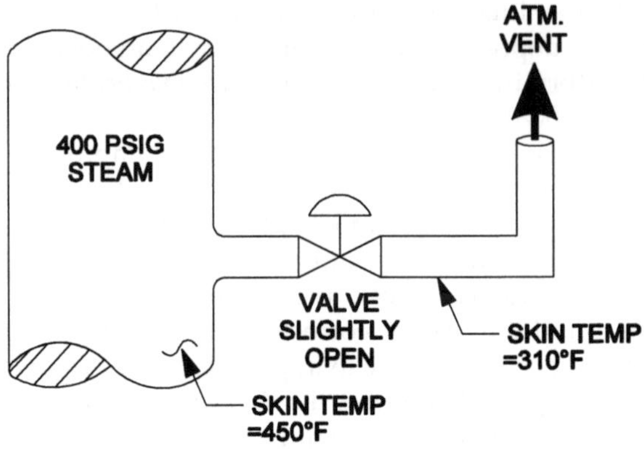

FIGURE 15-1 Measuring moisture content of steam using the temperature drop of the flowing steam through a vent.

The increase in the kinetic energy of the steam (item 2) must be dissipated. Just open the small gate valve shown in Figure 15-1 one or two turns. The velocity underneath the gate will be sonic, but the velocity in the downstream piping will be relatively small.

Referring to the figure, you would then measure the two skin temperatures shown, using your optical infrared surface thermometer:

$$450°\text{F} - 310°\text{F} = 140°\text{F}$$

However, from our Mollier diagram, we calculate that for an isoenthalpic expansion (i.e., a drop in pressure conducted at constant flowing steam velocity), the temperature drop should not be 140°F, but 90°F:

$$140°\text{F} - 90°\text{F} = 50°\text{F}$$

$$50°\text{F} \div 17°\text{F} = 3 \text{ wt\% moisture (see item \#4 above)}$$

The steam is said to be of 97% quality, which for most applications represents an excessive moisture level.

Appearance of Wet Steam

If the steam blowing out of a vent is clear or invisible, then the steam in the header is reasonably dry. If the steam blowing out of a vent is white, that's entrained water. Steam, like air, methane, and argon, is invisible when free of water. I can, with long experience, blow the steam against my wrench handle and approximate its moisture content. But that experience is a little hard to reproduce in this text. For the less experienced technician or engineer, wet steam appears white whether it has 1% or 10% moisture content. To discriminate between good steam (1%) and excessively wet steam (10%), one pretty much should rely on the quantitative method I just described. Or use a throttling calorimeter, which works on the same principle. That principle being thermodynamics.

Suppressing Entrainment

There have been a wide variety of devices invented since the start of the Industrial Revolution to remove water from steam:

- Steam trap—Drains water from low points that have already separated water from the steam.
- Steam filter—A somewhat efficient device that removes gross amounts of water from the steam. When not overloaded, I have found these devices to work surprisingly well, considering their small size and cost.
- Cyclone separator—Same comment as above, but I don't have much experience with this device.
- Sloped demister—The proper use of demisters is the correct way to minimize the moisture content of steam. Almost all steam drums or kettle-type waste heat boilers are so equipped. I have described how one designs such a demister in my book, *Process Design for Reliable Operations*, 3rd ed.

Regardless of how well the demister on a steam drum or boiler is designed, it doesn't matter if the cause of entrainment is high water level in the drum or kettle. I first noted this problem on a kettle-type steam generator on a delayed coker in Texas City. The boiler feed water makeup was on level control. While my level in the gauge glass was 2 feet below the top of the kettle, the evolved steam was very wet. So I lowered the level and the steam flow appeared somewhat less wet. It's true, as I kept lowering the level in the glass below the elevation of the tube bundle, the steam quality visibly improved. But it's also true that the steam generation rate and kettle duty went down as I pulled the level down below the midpoint of the tube bundle. To visualize the problem, refer to Figure 15-2. The difficulty, as depicted in my sketch, is that the external level (flat or still water) is being compared to the internal level (aerated or boiling water). The pressure head is the same in the kettle and in the glass. But the flat water in the glass is maybe 2, 3, or 4 times denser than the bubbly water in the kettle. That means the height of the bubbly water in the kettle is 2 or 3 or 4 times greater than the height of the flat water in the glass or in the **level-trol** (see Chapter 16, "Level Control Problems"). The density of the boiling water phase in the kettle is diminished by:

- Increasing the kettle duty.
- Increasing the TDS in the blowdown.

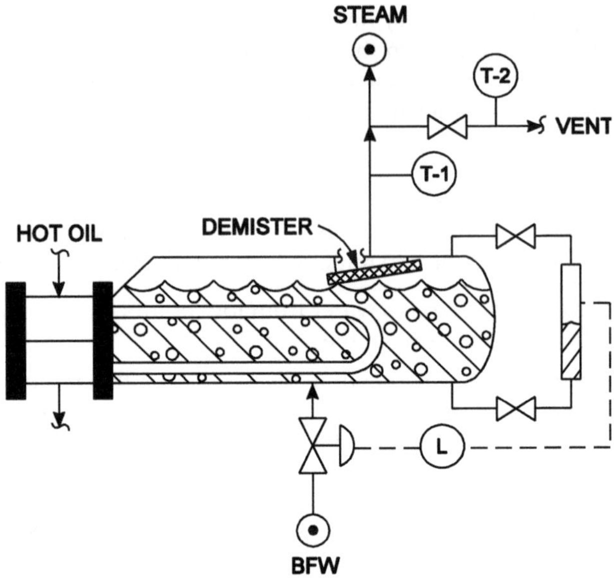

FIGURE 15-2 A kettle waste heat steam generator with external level lower than internal level. The problem is density difference.

- Increasing particulates inside the shell side of the kettle.
- Increased surface roughness on the outside of the tubes (which promotes nucleate boiling).

With the resulting variable density difference, how does the operator determine the correct set point for the kettle LRC? The answer is that there is no answer. And that's why we have dirty steam with salts and solids and steam of variable quality (moisture).

Delta T Level Control

Let's monitor the temperature difference of the two points shown in Figure 15-2:

$$\text{Delta T} = (T_1 - T_2)$$

If we lower the level to reduce entrainment, delta T will be reduced if there is actually a reduction in entrainment. On the other hand, as we lower the level too much below the top tubes of the bundle, the steam production will decline.

I have used this technique (just on manual control) at the hydrogen plant in Aruba and on the slurry waste heat boiler on an FCU spent cat stripper in Delaware City (the old Getty refinery). In both locations the malfunction due to poor-quality steam was corrected (see Chapter 16, "Level Control Problems," Figure 16-7, for details).

Moisture Due to Turbine Exhaust

When we expand steam, this may make the steam drier or wetter. Steam is always going to cool as it expands, either a lot or a little, depending on how we expand it. I'll discuss the three different ways we can expand steam. First, I can expand the steam and keep the heat content of the steam constant (isoenthalpic expansion). This also means that I don't change the speed of the steam. If I have 400 psig steam flowing in a 2-inch pipe and it flows through a letdown valve into a 4-inch pipe operating at 100 psig, the kinetic energy or velocity of the steam is the same. The steam might cool off by 30°F due to its bigger latent heat at lower pressure. But the saturation temperature of the lower-pressure steam drops from 440°F to 330°F. So the steam is actually 80°F superheated at the 100 psig pressure, even though it is 30°F cooler. The 80°F superheat would make the steam drier if it contained entrainment before expansion.

Second, I can expand the steam and let the velocity go up. Then I have to destroy the increased velocity by a combination of frictional losses and letting the steam flow into a larger piping system. This is just an alternative to the first method of expansion, with the same result of increasing steam superheat by 80°F.

The third method is to expand the steam at **constant entropy**. This is called an isoentropic expansion. That means I have expanded the steam through a nozzle designed to convert as much of the internal energy of the steam to speed or kinetic energy as possible. Now, the steam gets a lot colder and a lot wetter. Commonly 10% of the steam will condense to water. Of course, if you let down the cold, high velocity, 90% quality steam into a larger-diameter, low-pressure header, you would have the first case again. But if you allow the speeding steam to transfer its kinetic energy to a turbine wheel, the steam will slow down without heating back up. Now, if you exhaust the turbine into the low-pressure header, your steam supply will be contaminated with moisture.

Vacuum Jet Demonstration

In case you doubt my veracity, I invite you to prove it by observing the skin temperature on a steam vacuum ejector. The motive steam is quite hot, perhaps 350°F. After the steam enters the mixing chamber (see Chapter 26, "Vacuum Systems and Steam Jets"), it has expanded through the steam nozzle to sonic velocity. The mixing chamber is about 80°F. But then check the jet's outlet temperature. It's below the 350°F motive steam temperature and above the 80°F mixing chamber temperature. Perhaps 200°F. That's because the steam has slowed down as the diameter of the ejector has increased.

That's thermodynamics in action!

Ambient Heat Losses

If we start off with saturated steam, how much moisture can be attributed to ambient heat loss? The following formula applies whether the steam pipe is insulated or bare:

$$\text{LB} = (U \times A \times DT) \div 900$$

where LB = pounds per hour water formed in flowing steam
A = area of piping and/or vessel, in square feet
DT = the temperature difference between the outside measured skin or surface temperature of the radiating surface and the ambient air temperature, °F
U = the heat transfer coefficient between the outside pipe insulated or bare surface and the ambient air, in Btu/hr/ft^2/°F
900 = latent heat of steam, in Btu/lb

U for still air, if the radiating surface is a cool 100°F to 200°F, is about 1. For hot surfaces of 400°F, on a day when a 30 mph wind is blowing, *U* is about 4. You'll have to interpolate between these two extremes. Bare steam piping does radiate lots of heat because the exterior skin temperature is almost as hot as the saturated steam temperature.

The simple way to preserve steam quality for long piping runs is to steam trace and insulate the steam piping. The tracing does nothing to stop the energy lost due to ambient heat radiation. Its only purpose is to reduce moisture formation in the steam for the benefit of downstream process equipment. Increasing the insulation thickness on small, bare piping by more than a few inches does not help reduce heat losses significantly.

Effects of Superheated Steam

I once had a client in Australia who was designing offshore platform process facilities. The idea was to make diesel from crude, as it was produced for use on the platform. I got into a terrible argument with this client and have never heard from them again. Here's what happened:

To flash the diesel out of the crude, they needed a temperature of 550°F. Originally they were going to use an electric heater. But this turned out to be impractical as there are only 3,415 Btus in a kilowatt-hour. So they decided to use steam. The platform had 440 psig superheated steam. The steam was superheated from its saturation temperature of 460°F to a temperature of 600°F. The crude preheat exchanger data sheet they sent me read as follows:

- Shell inlet/outlet = 400°/550°
- Tube inlet/outlet = 600°/460°
- LMTD (Delta T) = 55°F (i.e., log mean temperature difference)

The sort of exchanger that they were going to use was a true countercurrent **double pipe** type of exchanger, so no correction factor was required for nontrue countercurrent-type flow (see Chapter 10, "Shell-and-Tube Heat Exchangers in Sensible Heat Transfer Service"). So that part of the calculation was okay. But what they had forgotten was that most of the heat is available only at 460°F, which is the saturation temperature of 455 psig steam. To quantify the division between the sensible heat and the latent heat of the steam:

$$\text{Latent heat} = 780 \text{ Btu/lb}$$

$$\text{Sensible heat} = (600°\text{F} - 460°\text{F}) \times (0.55) = 77 \text{ Btu/lb}$$

The 0.55 term is the specific heat of steam in Btu/lb/°F.

Only a very small portion of the available heat (less than 10%) is available above 460°F. Yet over 60% of the Btus needed to heat the crude from 400°F to 550°F are required above 460°F. Thus, it is quite impossible to use superheated steam in this service. As far as I can tell in the field, superheat in steam does very little to contribute to the temperature driving force. I just assume all the heat content of the steam (sensible and latent heat) is available at the steam's saturated temperature and pressure.

Some designers will spray steam condensate (water) into the steam to remove all superheat, on the basis that the superheated steam has a low heat transfer coefficient. In Aruba, on a naphtha stabilizer reboiler using 400 psig steam, I turned this de-superheating condensate flow on and off several times. The observed effect on the reboiler capacity was zero. Therefore, other than the extra Btus in superheated steam, it's best just to ignore the effect of the superheat on both *U* and delta T.

As for my Australian client, I tried to explain to them mathematically how their crude preheater exchanger could not work. That the condensation temperature of the 440 psig steam was too cold to flash diesel product out of the crude on their platform in the Indian Ocean.

They refused to accept my calculations.

Then I tried in a polite and tactful way to explain to them that they did not understand the essentials of calculating the exchanger LMTD. That's the last time I'm going to try niceness. I never heard from those guys again.

Boiler Feed Water Quality

Poor BFW quality results in the following malfunctions:

- Dirty steam containing salts (silicates) in the entrained water.
- Corrosion due to carbonic acid formation in the steam condensate, as discussed in Chapter 14, "Steam Condensate Collection Systems."

- Oxidation and tube leaks in the boilers themselves. This is largely a function of the deaerator. The function of the deaerator is to steam strip the dissolved air out of the BFW. Residual O_2 is removed by use of pyridine or other O_2 scavenger chemicals.

To reduce the amounts of solids in steam from entrained moisture, it's best to improve the operation of the demineralization plant. This is an ion exchange operation that removes cations and anions. An alternative is to increase boiler blowdown rates. Typically, 5% to 10% of the BFW flow is drained to the sewer to control the TDS as measured by a sample of water drained from the blowdown. Increasing blowdown above 10% is expensive, and it swells the plant's effluent water volume.

My way of reducing solids in steam is the best way. That's to improve steam condensate recovery rates. Most of the condensate lost to the sewer is a result of the condensate drainage problems I have enumerated in Chapter 14. Correcting these problems will unload the demineralization plant and directly reduce salts in steam. And as a consequence, improved process operation and reduced maintenance will result for:

- Steam turbine rotors
- Steam superheat coils
- Vacuum jets or ejectors (see photo in Chapter 26)
- Overspeed trips on steam turbines

Steam Contamination with Noncondensibles

I was using 60 psig steam to reboil a low-pressure tower in Texas City. One day, I rapidly lost most of the reboiler duty. This was indicated by a loss in tower pressure and by a loss in the reflux drum level. My immediate thought was that the steam trap draining the channel head side of the reboiler (steam on the tube side) had malfunctioned and was not opening properly. So I had an operator bypass the steam trap and blow out the steam condensate to the sewer. This restored my reboiler duty. But, looking at the condensate stream, I observed that it appeared somewhat gassy. That is, I could see hydrocarbon vapors flashing out of the steam.

Ordinarily, this would suggest a tube leak. However, in this service, the process side pressure (shell side) was lower than the pressure I had measured for the steam in the channel head. I managed to aspirate a small amount of the hydrocarbon vapors that had evolved from the steam condensate into an inflatable sample container. The lab gas chrome analysis showed:

- 20 mole% propylene
- 10 mole% propane

- 40 mole% butylenes
- 30 mole% butanes

This composition matched a de-ethanizer distillation tower bottoms product, located at the cracking unit. However, the cracking unit was on the other side of the road from my unit. Apparently, the 60 psig steam supply to my unit was contaminated with hydrocarbons from the cracking unit. Was there some sort of process piping cross-connection between the 60 psig steam system and the cracking unit de-ethanizer tower?

I followed the steam piping back through the cracking unit looking for such connections. At the bottom of the de-ethanizer, I found the culprit. It was not a process piping connection. It was a **steam-out** connection, used only during shut-downs and startups. Connected to the bottom of most process vessels there is a small (in this case a 2-inch) connection to purge hydrocarbons out of a vessel during shut-down, and to purge air out on startup. This connection is supposed to be **blinded** after startup. But I could not see the steel handle of the blind in the flange where it was supposed to be inserted. The blind in the bottom steam-out connection of the de-ethanizer had not been installed.

The de-ethanizer pressure was 200 psig. Should the isolating 2-inch gate valve leak, hydrocarbons would be pressured into the 60 psig steam supply. I use the term "leaking" in a diplomatic sense. Actually the valve was not fully closed. I imagine there was some dirt between the valve's gate and seat that blew out and started the hydrocarbon contamination of the 60 psig steam to my reboiler. Anyway, I tightened the valve with my wrench and restored the reboiler duty on my tower across the road. I had identified the malfunction by noting that the 2-inch steam-out valve was cold due to the flashing of the 200 psig liquid hydrocarbons down to the 60 psig pressure in the steam supply line.

Steam-out connections on the bottoms of process vessels are supposed to be blinded. When I worked at the Amoco Oil refinery in Texas City, we were supposed to remove a short piping spool piece between the steam-out connection and the steam supply line.

My boss told me essentially the same story. But this time the culprit was nitrogen, cross-connected to the steam header. The effect on the downstream reboiler was the same. That is, loss of reboiler heat transfer capacity due to accumulation of noncondensibles in the channel head of the reboiler.

On the other hand, CO_2 accumulation in steam reboilers does not often appear to interfere with heat transfer to a great extent. As I discussed in Chapter 14, CO_2 typically contaminates the steam supply to reboilers. However, the CO_2 does dissolve in the form of carbonic acid (H_2CO_3) in the steam condensate, more readily than either nitrogen or light hydrocarbons. The CO_2 is more of a corrosion problem due to carbonic acid attack on the carbon steel reboiler tubes than it is a problem in reduction of heat exchanger duty. Regardless, venting off the CO_2 from a steam system is always a good operating practice.

Using Low-Quality Steam

In practice, steam from some sources is likely to be of poor quality. For example, many units have small, kettle-type waste heat boilers that generate low-pressure steam and receive little operator attention. It would be best to avoid malfunctions in the first place by consuming this low-quality steam locally, rather than venting it into a main steam line. Some of the ways in which such variable-quality steam can be used without detrimental effect to process equipment are:

- Reboilers and steam heaters
- Steam tracing
- Purge steam, for example, on coke drums in Delayed Cokers
- Steam to the flare
- Cushion steam between valves (to prevent process leaks)
- Tank coil heating
- Heating buildings
- Open stripping steam on waste water strippers
- Steam used underneath relief valves to protect them from coking
- Velocity steam in heater passes

The main objective is to keep such low-quality steam away from turbines, vacuum ejectors, superheat coils, and catalytic processes. Flowing steam streams are rather like people. We need to segregate the good guys from the evildoers. We need to confine the poor-quality steam to those process services where it can do us little harm. As I write these words, I'm stranded in the Edmonton, Alberta, airport. I've been working here at the Suncor refinery. The biggest complaint I heard all week is about turbine vibration problems. Undoubtedly, they are caused by the low-quality steam from the process unit, kettle-type, waste heat boilers. Interesting enough, the guilty boilers are located on the very same process units that are suffering from the turbine vibration malfunctions.

CHAPTER 16

Level Control Problems

Señor Norman, the 80% level indication has not changed in 37 years. Good stripper level control, I think?

—Operator in Spain running in a "tapped out" level condition, 1993

If I look at a level in a gauge glass on the side of a tower, as shown in Figure 16-1, I will observe a level that's usually lower than the actual level in the vessel. The primary reason for this discrepancy is that the fluid in the bottom of the vessel is partially aerated. By partially aerated, I mean the vessel bottom is acting like a vapor–liquid separator. The liquid dropping into the bottom of the vessel has come from the bottom tray of a tower, from a reboiler outlet, or from a partial condenser. Depending on the fluid viscosity and vapor density, it takes a few minutes for the vapor phase to disengage itself from the liquid phase.

The gauge glass, however, contains a stagnant fluid and is hence denser than the fluid in the vessel. Thus, the observed level in the glass is less than the real level in the vessel. The lower level observed in the glass is then not a malfunction; it is normal and is unavoidable.

The glycol-filled chamber shown in Figure 16-1, on the opposite side of the vessel, is called a **level-trol.** Ethylene glycol is usually used for freeze protection and as a calibration fluid. By calibration fluid, I mean that the instrument technician knows the specific gravity of both the calibration fluid and the process fluid inside the vessel. If the process fluid has an SG of 0.60 and the glycol has an SG of 1.20, then 2 feet of glycol represents 4 feet of the process fluid inside the vessel.

The height of the glycol in the level-trol pot is measured by the delta P between P-1 and P-2 shown in Figure 16-1. But how does one know the SG of the process fluid inside the vessel? The instrument

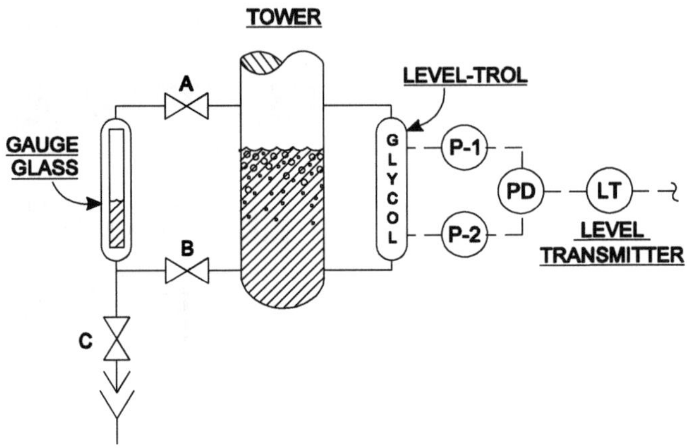

FIGURE 16-1 Valves needed to blow-out level taps to remove pluggage.

tech uses the standard process fluid density of a known hydrocarbon composition and corrects for temperature. But suppose the fluid in the tower is aerated. Then what? Well, the indicated level on the control console will be less than the true fluid level in the tower. Eventually, all this will lead to the vessel becoming **tapped out.**

Becoming Tapped Out

When the fluid level in a vessel rises above the top tap of the level-trol, three things happen:

1. The pressure at P-2 (Figure 16-1) goes up.
2. The pressure at P-1 goes up as well.
3. But the pressure differential (PD) between P-1 and P-2, and hence the output of the level transmitter, doesn't change.

If the SG of the fluid in the tower is 20% less than the density assumed by the instrument tech when she calibrated the level-trol, then the indicated level on the control console will read 80%, regardless of how high the liquid level rises in the vessel. If you think this is a minor problem, you're sadly mistaken. It was exactly this sort of malfunction that cost BP billions of dollars at the terrible fire and explosion at their Texas City refinery in 2004. I've described this incident in detail in *Troubleshooting Process Operations*, 4th ed. (PennWell Books, 2009).

In a sense, becoming tapped out is not a malfunction. Nothing is actually wrong. It's rather just a problem of perception, as to which fluid density is appropriate to use in calibrating the level-trol correctly at a particular time.

Other than the unavoidable foam inside vessels, there are a number of other common factors that cause the liquid level in the bottom of a vessel to be higher than the level generated from the level transmitter:

- The temperature of the tower liquid is 200°F hotter than the liquid in the level-trol or gauge glass due to ambient heat loss. Thus, the external liquid levels will be 10% lower than the internal liquid level in the vessel.
- The fluid in the tower is a two-phase mixture (water and lighter hydrocarbon liquid floating on the aqueous phase). Because of the location of the level tappings, the liquid in the gauge glass and level-trol is just water.
- The tower's bottom product is lighter than normal. Say you're starting up a naphtha splitter on total reflux. Then the bottoms liquid will be the same as the feed composition and not the normal heavy naphtha product. The SG of the bottoms liquid will be lower than normal, and thus the panel level indication will be biased toward a lower level. This is precisely what happened at the BP refinery, Texas City, naphtha splitter fire.

Plugged Level Taps

What are the results of plugged level taps? Let's make two assumptions in dealing with this question:

1. That the bottom of the tower is operating well above ambient temperature and hence the gauge glass, or level-trol, is cooler than the liquid in the vessel itself.
2. That the vapor phase, in contact with the liquid phase in the bottom of the vessel, is fully condensed at the gauge glass or level-trol temperatures.

Both assumptions are typically true for distillation towers, reboilers, and steam generators. If then either the top or bottom tap plugs, the indicated level will increase relative to the actual level in the vessel. If it is the top-level tap that has plugged, then a low pressure will develop (due to condensation of vapors) in the glass or level-trol. This rapidly draws liquid up through the bottom-level tap. If the bottom tap plugs, then condensation in the glass or level-trol slowly accumulates in the glass or level-trol. Either way, the indicated liquid level will go up relative to the real level in the vessel.

But suppose the vapor phase in the vessel is hydrogen or fuel gas, and the liquid phase is gas oil. Then, if the top tap plugs, the level in the glass or level-trol will be lower than the vessel level. If the bottom tap plugs, the level will either be stuck at a single value or will slowly increase relative to the real level in the vessel.

Blowing Out Level Taps

It seems from the preceding discussion that verifying whether level taps are unplugged is the first and critical step in troubleshooting a liquid level malfunction. Proceed as follows:

- Step 1 (see Figure 16-1) is to close valve B and open valve A and valve C. This clears the top tap.
- Step 2 is to close valve A and open valve B and valve C to clear the bottom-level tap.

Unfortunately, valves A and B may be **ball-type check** valves. Such valves have a steel ball inside the valve body, which seats in a closed position in the event of breakage of the gauge glass crystal. However, in normal use, this ball has a tendency to stick in a closed position, especially in fouling service. To unstick the ball, close the valve halfway. A stem inside the valve will push the ball off its seated position.

Safety Note *Blowing out H_2S-laden solvents in the above manner without wearing fresh air breathing equipment can, and has, killed the careless operator or engineer.*

Many level-trols are filled with glycol. If you blow out the glycol to clear the taps, ask the instrument technician to refill the level-trol with glycol. Refilling the level-trol with process fluid causes an erroneous low level reading on the control console. I've made this mistake on quite a few occasions.

Erratic Level Indication Due to Tray Flooding

I've just been on the phone with Gissel from Petro Canada. Her problem is an erratic liquid level in the bottom of her steam stripper. The level cycles on 3- to 5-minute intervals. She can see the erratic level indication locally in the sight glass shown in Figure 16-2, as well as on the control console level recorder control (LRC). A radiation neutron backscatter scan shows that the level in the bottoms of the stripper is actually erratic.

I asked Gissel to place a pressure gauge at the top of the gauge glass at position P-2, shown in Figure 16-2.

With the steam stripper top pressure fixed, Gissel noted that when the pressure at P-2 increased from 10.0 to 10.5 psig, the tower bottoms level would rapidly drop from 60% to 30%. The increase of 0.5 psi corresponds to 20 inches of liquid head, assuming a fluid SG of 0.70:

$$\frac{(10.5 - 10.0) \times 28 \text{ inches}}{0.70} = 20 \text{ inches}$$

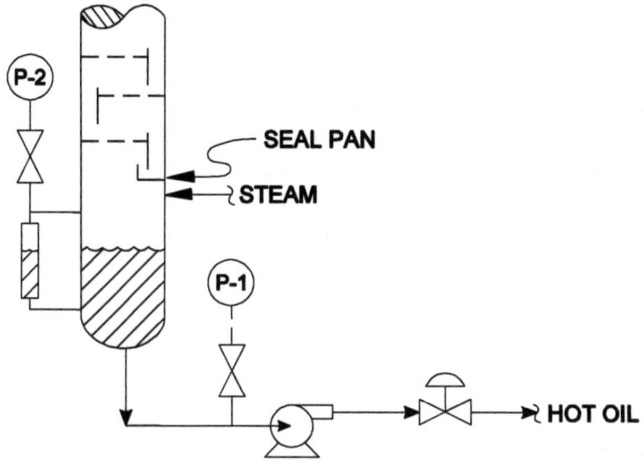

FIGURE 16-2 P-1 minus P-2 is a measure of liquid head.

The 28 inches factor is the conversion factor for changing 1 psi of head pressure into inches of water head pressure.

The overall span of the level-trol taps is 6 feet, or 72 inches. The observed loss in the liquid level in the bottom of the tower was also about 20 inches:

$$72 \text{ inches } (60\% - 30\%) = 21 \text{ inches}$$

What happened to this 21-inch drop in tower bottoms liquid level? The answer is that the stripping trays were **flooding**. That is, the steam flow rate was causing the liquid bottoms product to accumulate on the stripping tray decks.

To prove my theory, I asked Gissel to increase the stripping steam rate and observe the initial response of the stripper tower bottoms level. She reported that right after the steam rate was increased,

- The pressure at P-2 went up.
- The stripper level went down.

"But why," Gissel asked, "does the stripper bottom level oscillate rather than just keep falling, if the trays are flooding?

The reason is that if the stripping steam rate is not too high, the weight of liquid on the trays builds until the liquid head causes the liquid to dump down the tower. The unit operators had chosen to operate at a stripping steam rate that would permit the liquid to drain down through the trays.

But how, Gissel asked, did the operators know what steam flow to select? I explained that if they used a much greater steam rate, then the stripper's bottoms level would fall to zero, and the bottoms pump would have lost suction pressure and cavitated.

"Yes, Norm, you're quite right. We used to use 3,500 lb/hr steam and now we're only using 2,000 lb/hr," Gissel concluded. "But what's causing the stripping trays to flood?"

"Most likely dirt," I explained. "Probably corrosion products have accumulated in the seal pan (shown in Figure 16-2), or on the tray decks themselves. This promotes downcomer backup and tray deck flooding" (see Chapter 1).

There's a lesson here. Most apparent control malfunctions are usually process malfunctions. Trying to solve a control problem without reference to the process operation is counterproductive. I've written an entire book on this subject, *Troubleshooting Process Plant Control.*

Moisture in Level Taps

If you will look closely at Figure 16-2, you will observe that the steam inlet nozzle is located above and on the other side of the vessel from the top-level tap. However, on many such towers, the steam inlet is located just below the top-level tap. This latter location will produce an erratically high liquid level indication, both in the gauge glass and in the level transmitter output displayed on the control console. One additional symptom of the problem is that the gauge glass and level-trol will make a gurgling sound, especially at higher stripping steam rates.

The problem is that the gauge glass and the level-trol are going to be slightly cooler than the stripper. Thus, steam drawn into the top tap of the level-sensing device will partly condense. You can see the droplets of water formed from the condensed steam, dropping through the oil level in the glass. And when these droplets of water hit the hot oil covering the bottom-level tappings, the water flashes to steam. That's the gurgling sound. You can see the steam vapors pushing up the liquid level in the glass. It isn't only that the steam is pushing up the liquid that creates the high level indication. The steam is restricting drainage through the bottom tap, and thus a high liquid level is formed in the level-trol chamber and in the gauge glass.

Note that the actual level in the stripper is neither high nor erratic. It's just the level indication that is malfunctioning. To cure this problem, an inert gas purge (N_2 or fuel gas) is used in the top tap to prevent moisture entering through the external top tap. The details of this design are presented in my book, *Process Design for Reliable Operations.*

Suction Pressure Control

I was working in a refinery in Spain when I encountered another common malfunction in an erratic vessel bottoms level control. The problem was that the liquid hold time between the level taps was only about 20 seconds. A normal liquid residence or hold time would have been 10 times as long, or about 3 minutes. This small hold time of 20 seconds magnified the oscillations of the bottoms level by a factor of 10. Hence, the erratic bottoms level—both actual and indicated.

I have a method to overcome these problems. That is:

- Lack of adequate liquid residence time.
- The tendency for level taps to plug.
- Moisture in the level-sensing device.
- Levels becoming tapped out, as discussed above.

This method is to discard level control completely. I call this "suction pressure control," as shown in Figure 16-3. The idea is not new. It predates level control. It's just to control the flow rate from a pump to minimize the level in the upstream vessel, consistent with not losing the required NPSH (net positive suction head) pressure that is needed to keep the pump from cavitating due to loss of suction pressure.

There are a few problems with suction pressure control:

- The pressure-sensing point must be located downstream of the pump suction screen. If it is located upstream, when the screen plugs, the pump discharge control valve will open instead of closing, and the pump will cavitate. This necessitates having a pressure transmitter for both the primary and spare pumps.
- A significant percentage of the pump suction pressure must be due to static head of liquid, and not the operating pressure of the tower. For example:
    - A vessel is operating at 15 psig (or 30 psia) and is elevated 50 feet above the pump. The pumped fluid is saturated water (one atmosphere of pressure

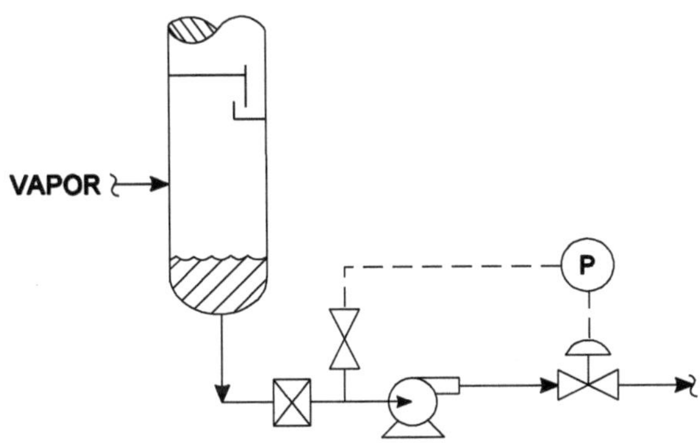

FIGURE 16-3 Suction pressure control eliminates level control malfunction.

equates to 34 feet of water). The pressure at the pump suction (neglecting frictional losses), is then:

$$34 \text{ feet} + 34 \text{ feet} + 50 \text{ feet} = 118 \text{ feet}$$

Therefore, 42% of the pump suction pressure (50 ÷ 118) is due to static head of liquid. In my experience, suction pressure control will work just fine.

- A vessel is operating at 75 psig (5 atmospheres) and is elevated 20 feet above the pump. The fluid is saturated water. The pressure at the pump suction is:

$$5(34 \text{ feet}) + 34 \text{ feet} + 20 \text{ feet} = 224 \text{ feet}$$

Therefore, only 9% of the pump suction pressure (20 ÷ 224) is due to the static head of liquid. My experience indicates that suction pressure control will not work consistently, due to small changes in the tower operating pressure.

In conclusion, in many low-pressure services, suction pressure control totally eliminates level control malfunctions, because there is no level to control.

Split Liquid Levels

An operator observes in the gauge glasses shown in Figure 16-4 a different level in each glass. This is called a "split liquid level." Naturally, this does not represent layers of vapor and liquid inside the tower. What it does represent is foam. The foam is originating from the bottom tray or the reboiler return nozzle. If the foam is

FIGURE 16-4 Split liquid levels indicate foam in vessel.

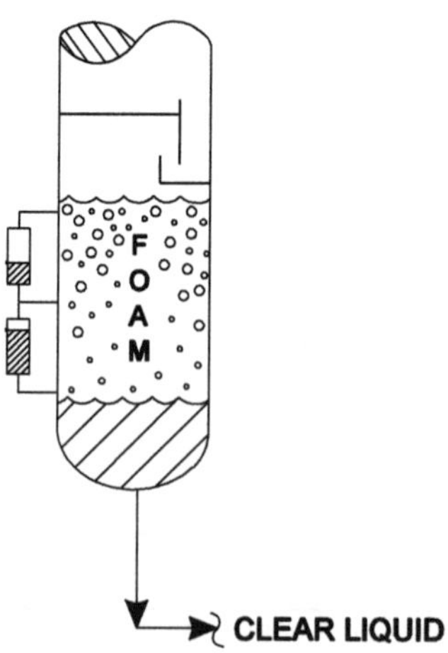

covering the top tap of the upper glass, then the liquid levels observed in the gauge glasses only reflect the density of the foam between the level taps and not any particular liquid level in the tower. You will observe that there is a higher level in the lower glass, and a lower level in the upper glass. This is only because the lighter foam in the vessel is floating atop the heavier layer of foam.

The liquid in the gauge glasses is relatively stagnant, compared to the liquid level in the vessel. Hence, foam is not observed in the gauge glasses. As there is always some amount of foam in the bottom of a distillation tower or vapor–liquid separator, most services are subject to the formation of split-liquid levels. The most common causes of foam in hydrocarbon systems are particulates resulting from corrosion. You can observe the effect of particulates on liquid levels in your home. Boil a pot of water. Then add some pasta. The starch particles on the pasta surface will cause the water level to foam up.

This all sounds rather bad if you are trying to monitor the real fluid level in a vessel subject—as many process vessels are—to foam formation. The conventional answer to this problem is to measure the foam density inside the vessel at a particular elevation by radiation. Neutron backscatter is the preferred method for hydrocarbon systems. A beam of neutrons is scattered from the hydrogen ions in the foam. Based on the intensity of neutrons reflected back to the radiation source, the hydrogen ion content, and hence the foam density, may be calculated. At an Ashland refinery crude preflash tower, the measured density using neutron backscatter ranged from 4 to 40 lb/ft^3, for an overall foam height of about 30 feet.

Insulating a Level-Trol

One of my towers on the alky unit at the Good Hope refinery was a propane–butane splitter. The tower ran fine, and I never paid it much attention. Unfortunately, I once observed that the level-trol was uninsulated. Unfortunately, I also forgot some wise advice from my Dad:

"Norman, just leave well enough alone."

So I had the maintenance department insulate the bare level-trol. The result was that the liquid in the level-trol became hotter and less dense. The reduced butane SG generated a lower differential pressure between the top and bottoms taps. The reduced delta P caused the level transmitter output to drop. This resulted in a lower measured tower bottoms level.

The tower bottoms LRC started to cut back and the tower became "tapped out." The level rose above the reboiler return nozzle, which eventually caused the tower to flood. Liquid hydrocarbons loaded up the tower overhead condensers. The tower pressure safety relief valve popped open, and we started to flare liquid propane–butane.

Yes it's true, "If it ain't broke, don't fix it."

External Skin Temperature

In general, the internal level in a vessel usually does not correspond to the indicated liquid level in the level-trol or in the gauge glass. Note that I do not even call the gauge glass a level glass. The gauge glass is just measuring the pressure difference between the two level taps, in terms of the SG of the liquid in the glass. Of course, if the SG of the liquid in the vessel is identical to the liquid in the gauge glass, then the two levels are going to be the same. But, having said this, it's usually the density of the fluid in the vessel that is lower. Thus, its level is higher than the level in the glass.

The instrument technician can calibrate the level-trol by using an assumed SG, or by reference to the level observed in the glass. Either way, the output from the level transmitter will usually be lower than the real level in the tower.

One of the more common difficulties that leads to this malfunction is shown in Figure 16-5. A hydrocarbon with an SG of 0.50 (i.e., propane) is floating on an aqueous phase with an SG of 1.0. If the ratio of the two phases in the vessel is known to be 1 to 1, and the actual level in the vessel was 8 feet, then a level in the gauge of 6 feet would be observed:

$$(1.0)\ (4\ \text{feet}) + (0.5)\ (4\ \text{feet}) = 6\ \text{feet}$$

But suppose I did not tell you the ratio of the phases. Then how can you use the observed liquid level in the vessel's gauge glass to determine the real level inside the vessel? The answer is that there is no answer.

However, there is a different technique to finding a liquid level in the vessel shown in Figure 16-5. Note that the liquid phase temperature is 160°F and that the vapor phase temperature is 100°F. Then, cut a series of 2-inch holes in the vessel insulation. Using your infrared surface temperature gun, you can locate the vapor–liquid interface based on the external temperature profile. A gradual change in the temperature profile is a sign of foam at the vapor–liquid interface.

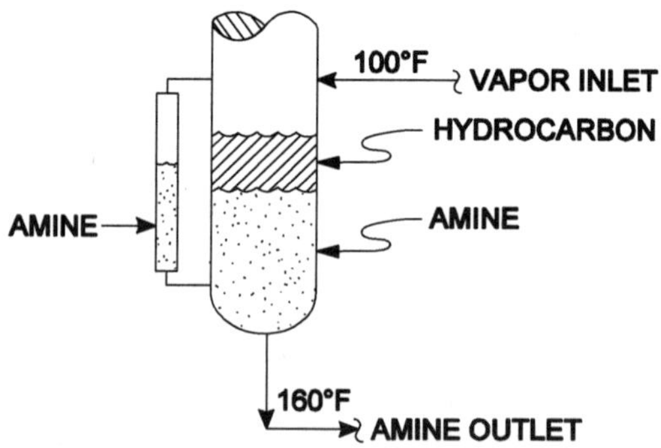

FIGURE 16-5 Effect of two phases in amine absorber.

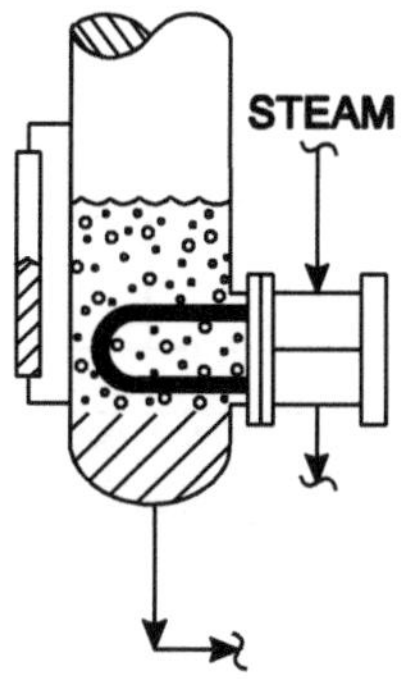

Figure 16-6 Stab-in reboilers produce levels in vessel higher than measured levels.

Stab-In Reboiler Level Control

I have never designed a tower with a **stab-in** reboiler as shown in Figure 16-6. One reason is that there is a nasty problem in measuring the tower bottoms level. The fluid in the tower is boiling, and thus it is partially aerated. Depending on the concentration of surfactants and particulates, the resulting vapor–liquid mixture might have a density of 50%, or 30%, or 10% of the clear, stagnant liquid in the external level-sensing chamber. Therefore, the external level might be 50%, or 30%, or 10% of the internal boiling liquid level. What's the solution to this problem? Short of radiation level detection, there is no answer. Well, not quite true. There is a good answer to this problem. Don't use stab-in reboilers without complex internal baffling to create a zone for internal vapor–liquid disengagement.

Level Control in Waste Heat Boilers

The reason I've introduced the subject of the seldom seen stab-in reboiler is to introduce a more common problem: measuring BFW (boiler feed water) levels in steam generators. By way of example, let's consider the waste heat kettle boiler shown in Figure 16-7.

This is not a malfunction unique to kettle type steam generators. It's a problem that occurs in all steam generators, or any vessel that contains a boiling liquid. Certainly, the boiling water inside the kettle will be less dense than the stagnant water in the external gauge glass or level-trol. The instrument tech has used an SG of 1.00 to calibrate the level transmitter. The SG of the boiling water will vary with:

- Particulates in the boiler.
- Size of the kettle.
- Rate of steam generation.

The level that is used to control the BFW makeup LRC will be lower by some unknown and unknowable amount. Uncovering boiler tubes may damage or distort the tubes due to local overheating.

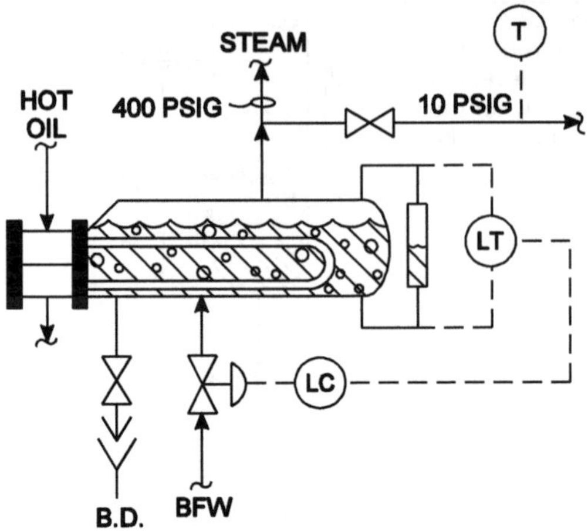

FIGURE 16-7 The level in the kettle will always be higher than the level in the glass.

Running with too high a BFW level will cause entrainment of TDS (total dissolved solids) in the steam. Solids in the steam (i.e., silicon hardness deposits) promote fouling in:

- Steam turbine rotors.
- Catalysts that use steam as a reactant, like steam-methane reformers that produce hydrogen.
- Steam superheat furnace tubes.
- Overspeed trips on steam turbines.

How then can one adjust the level in any boiler, or in any refrigerant evaporator or in any vessel in which a liquid is vaporized? I've tried this method at a fluid cracking unit in Delaware and at a hydrogen plant in Aruba. For details, refer to my books, *Troubleshooting Process Plant Control* and *Troubleshooting Process Operations*, 4th ed. I'll just summarize the basic principle here (refer to Figure 16-7).

- Allow a small amount of steam to depressure from 400 to 10 psig. If the steam is free of water, the temperature will drop by perhaps 70°F.
- If the level is increased high enough to promote entrainment, then the temperature of the 10 psig steam will begin dropping by more than 70°F, due to evaporation of the entrained moisture.
- If the level of BFW is decreased low enough to uncover tubes, then the steam flow rate for a given hot oil flow will go down. Theoretically, the steam should get a bit hotter, but I don't recall seeing that effect.

The BFW level controller should be set halfway between the two points determined above. Sadly, I was teaching a seminar for Repsol in Spain. An older instrument control engineer informed me that my innovative idea had been used on boilers for some time at her plant.

Observing Interface Levels in Clear Liquids

I've spent a long time in this chapter enumerating the limits of a gauge glass in observing levels in vessels. But still, there are many services where knowing the level in a gauge glass is critical. For example, finding the level in a hydrogen off-gas, alkaline water wash drum being used to extract HCl from hydrogen produced from a naphtha reformer. The problem is that both phases are transparent. To find the interface level, you have to find the meniscus. But to visually discriminate between the dirt streaks and scratches on the gauge glass surface and the meniscus is quite a challenge. The only real way to make this observation is to make the meniscus move. Even then it will be very difficult to observe the movement of the meniscus unless the **backing lights** behind the gauge glass are working and the gauge glass itself is reasonably clean.

I taught a short class in 1979 at the Chemical Engineering Graduate School of Northwestern University entitled, "Advanced Process Control." Short in the sense that I was fired by Dean Gold after one day. Probably because I taught that the most important aspect of advanced process control was to keep the backing lights working on gauge glasses. Without these lights, you'll need to hold a flashlight behind the glass to see the meniscus move.

To drop the meniscus, close-in the bottom level tap completely and just barely open the gauge glass drain. The meniscus will now start to drop. Typically, operators open the drain too much and the meniscus drops out too quickly to see. That's okay, because when you open up the bottom tap, you can see the meniscus rising back up to its correct liquid level. Hence, the other important aspect of advanced process computer control I advocated at Northwestern University was to keep the gauge glass drains operable and connected to a safe location.

Water-Hydrocarbon Interface

The malfunction that I have most frequently encountered, at least in refinery applications in maintaining a correct interface level between water and liquid hydrocarbons, is plugging of the level taps, especially in the aqueous phase. Most often the particulates that plug the level taps are corrosion products—iron sulfide in refineries, iron oxide scale in chemical plants. The bigger the level taps, the smaller the problem. Half-inch taps are too small. Use of 2-inch taps is great, but expensive and not typical. Of greater importance than the size of the level tappings is the elevation of the bottom tap. Ideally, it should be located 6 inches above the vessel bottom tangent line.

Regardless of these design considerations, I deal with these problems by connecting a clean water purge to the aqueous phase taps. The purge is operated manually for use on the gauge glass and flows continually through a restriction orifice for use on the level-trol. In the vessel separating H_2SO_4 and the hydrocarbon reactor effluent in a refinery sulfuric acid alkylation unit, I'll employ the same technique, but I'll use alkylate product instead of water as the level tap purge medium.

Measuring Levels of Solids

My experience on this subject is limited to measuring the height of coke in a delayed coker, coke drum, and heights of catalyst in fluidized beds.

The density of fluidized catalyst varies in a refinery **fluid cracking unit (FCU)** with such factors as:

- Coke on catalyst
- Vapors intermingled with the catalyst (steam, flue gas, combustion air)
- Type of catalyst

At the GHR Energy refinery, we switched to a denser, fresh catalyst and found that we needed higher indicated catalyst levels than before. Of course, the real catalyst levels had not changed. On the other hand, if catalyst density falls due to increased aeration, then a lower indicated catalyst level is required. I had this problem at the "Giant Refinery" in New Mexico. I wrote a paper on this complex subject, "Troubleshooting FCU Regenerator Slide Valve Problems" (*Hydrocarbon Technology*, Spring Edition, 1996). The level control malfunctions I described in this article also apply to Exxon-type fluid cokers.

Level indication in coke drums depends on neutron backscatter. That is, neutrons emitted by a radioactive source are reflected back from the hydrogen ions in the coke. This works really well when measuring coke levels. The problem arises when we fill the coke drum with water during the quench portion of the coking cycle. It rather seems to me as if the neutron backscatter device does not always discriminate properly between boiling water and the evolved steam, which is also rich in hydrogen ions. Thus, operators often underfill a coke drum with quench water. This malfunction can be corrected by the neutron backscatter level-sensing equipment vendor (K-RAY). It's just a simple recalibration that's required. I suppose the recalibration reduces the sensitivity of the reflected neutron receiver, so it no longer thinks that steam is boiling water.

Mislocated Top-Level Taps

Two stories illustrate these rather common design malfunctions. The first story took place at a chemical plant in Augusta, Georgia. The tower involved was fractionating between methanol and water. Fractionation was poor. I checked the delta P across the trays and

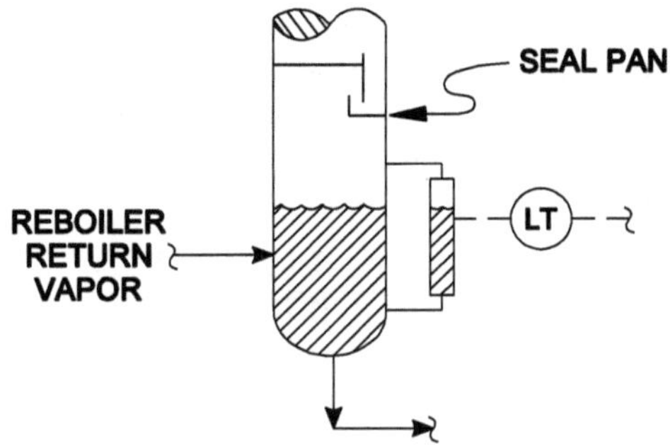

Figure 16-8 Tower floods because the level tap is located too high.

determined that the tower was flooding. I noted that the bottoms liquid level was holding dead-steady at 50% level. So naturally I thought the tower was tapped out. That is, I thought the bottoms liquid level was above the upper-level tap. But when I observed the top-level tap location on the vessel itself (Figure 16-8), I realized that the malfunction was not an operating error, but a design mistake.

The level span—that is, the distance between the taps—was 4 feet. However, the top tap was 2 feet above the reboiler return nozzle. Thus, when the liquid level rose a few inches above this nozzle, the force of the vapor blowing up through the liquid would cause liquid entrainment. The droplets of liquid would then be blown up against the bottom tray and cause the tray to flood. The combination of entrainment and bottom tray flooding would stabilize the level in the bottom of the tower, a few inches above the reboiler return nozzle. The lesson is to always design the elevation of the top-level tap one or two feet below the reboiler return or stripping steam inlet nozzle.

Another design malfunction is shown in Figure 16-9. I've made this mistake myself on a condensate drum. In this case, the bottoms level was in a crude fractionator with a tower pressure of 10 psig. The

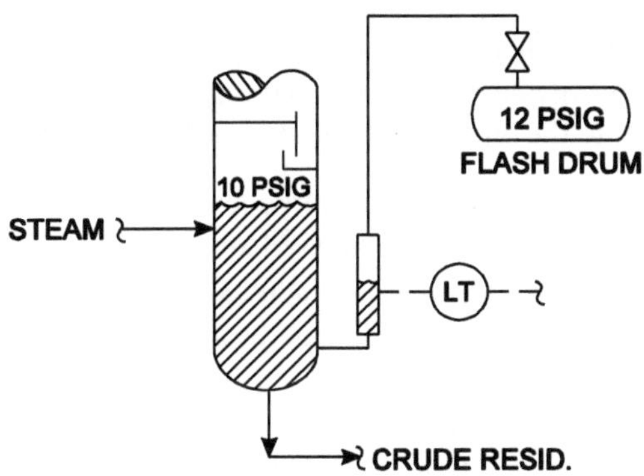

Figure 16-9 Tower floods because top-level tap is connected to a higher-pressure vessel.

top tap was connected to the 12 psig flash drum shown in the figure. The SG of the bottoms product was 0.70. One psig of water head pressure equals 2.31 feet of water. Then the elevation difference between the level in the tower and the indicated liquid level in the level-trol will be:

$$\frac{(12 \text{ psig} - 10 \text{ psig})\ (2.31)}{0.70} = 6.6 \text{ feet}$$

But if the tower liquid level is 6.6 feet above the indicated level, then the liquid will rise above the tower bottoms stripping steam inlet. This high liquid level will cause the tower to flood, due to entrainment of the crude resid.

In the case of my error on the condensate drum, the mislocated top-level tap resulted in the steam condensate backing up into the channel head side of the steam reboiler. This reduced the reboiler capacity by over 50%.

Redundancy Minimizes Effect of Malfunctions

There are four types of level indications that should be available to the operator who is controlling critical liquid levels:

1. The panel or console indication.
2. The output from the level transmitter displayed in the field, just below the transmitter.
3. The gauge glass.
4. The high and low level alarms.

At the Good Hope refinery FCU (i.e., Fluid Gas Oil Cracking Unit), I had all four of the above indications coming from a single set of level taps. A plugged level tap is then going to make all four level indications erroneous. Even worse: On the wet gas compressor suction KO drum, the high liquid level compressor trip was also connected to the same set of level taps! I wrote an extensive article on this subject of redundancy in level measurement, "Instrumenting a Plant to Run Smoothly" (*Chemical Engineering*, September 12, 1977). The reference is old, but my clients are still making the same mistake. Here's my rule to mitigate the effect of plugged level taps in critical services:

- Both the high and low level alarms should be located on taps *not* used by the level-trol.
- The output from the level-trol should be displayed locally, so that the outside operator knows what the console operator is seeing.
- Gauge glass liquid level taps can be shared with alarms and trips, but *not* with the level-trol.

- The drain from any level-sensing device should be connected to a safe location, so that the outside operator can safely blow out the level taps.

Consequences of Level Malfunctions

The following tabulation is based on results that I've personally observed on operating units:

- Fractionator trays damaged due to high liquid level above bottom few trays.
- Loss of reboiler thermosyphon circulation due to high liquid level rising above the reboiler return nozzle.
- Absorber floods due to high level above feed vapor inlet.
- Tower overpressured when the entire tower fills with liquid.
- Operators shut off stripping steam due to erratically high level indication.
- Erratic level causes downstream fired heater to trip off on low feed flow.
- Bottoms pump cavitates and damages mechanical seal due to low liquid level.
- High-pressure gas blows through bottoms LRC valve and overpressurizes downstream product storage tank.
- Hydrocarbon drained with sour water to sour water stripper feed storage tank, due to faulty interface level control.
- Operator killed due to wrong level indication on sulfur plant acid gas KO drum when H_2S vapors blew out of a water drain.
- Light liquid hydrocarbons released from relief valve on top of column when tower completely filled with naphtha. Resulted in 16 deaths, 400 injuries, and $2 billion (U.S.) in damages.
- Compressor wrecked when the high level trip fails to shutdown the compressor.
- Gas oil product to FCU turned black due to high level of resid in bottom of crude distillation tower.
- Concentrated H_2SO_4 overflows from an acid alkylate separator due to high acid level. Operators then drain the acid to a sewer from a downstream separator and acidify Galveston Bay in Texas. Your author is unfairly blamed for this fiasco, and as a consequence, is put in charge of the United Way Fund for the Texas City refinery. The 1974 United Way Fund program achieves its lowest level ever for percentage of employee participation. The plant manager asks, "What idiot put that fool Lieberman in charge of United Way…?"

Okay! I've gotten off my subject. But you get my drift. Level indication malfunctions are serious problems, and there is often no single or simple way of determining the real liquid level in a process vessel.

Stilling Well

Let's consider the fuel gas scrubber shown in Figure 16-10. Sour refinery fuel gas, rich in H_2S, is scrubbed with lean amine for H_2S removal. The amine has an SG of 1.0. The fuel gas is at its saturation dew point temperature of 150°F. Ambient conditions in St. Croix this day were 85°F.

The **stilling well** was 4 inches ID by 20 feet long. The purpose of a stilling well is . . . ? Actually, it serves no purpose. It's a common engineering error. I'll have to explain.

Because the saturated sour gas feed is 65°F hotter than ambient conditions, the stilling well, which is just a section of 4-inch pipe, will

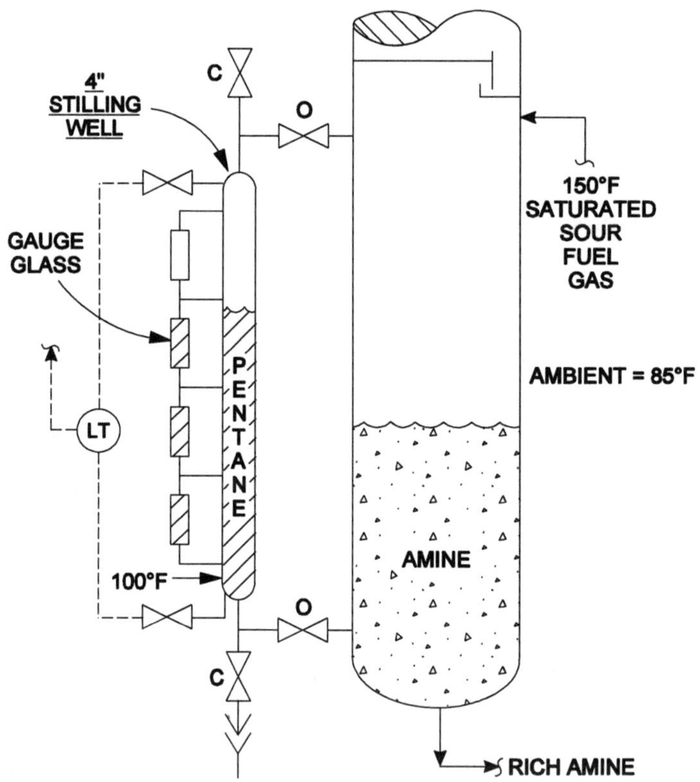

FIGURE 16-10 Connecting a level transmitted (LT) and gauge glasses to a stilling well is a bad engineering practice.

cool, due to ambient heat loss. The heavier hydrocarbon components in the sour gas (i.e., pentanes), will condense in the stilling well and accumulate. The pentanes will partly displace the amine. The SG of liquid pentane is about 0.5. Therefore, if we have 10 feet of pentane liquid in the stilling well, we'll only have 5 feet of amine in the tower.

Note in Figure 16-10 that the four gauge glasses are connected to the stilling well, and not to the tower. The instrument technician observes the level in the glass at 50% or 10 feet high. Using this visual level as a guide, she then recalibrates the level instrumentation to read 50%. But because the real fluid in the tower is amine, which has twice the density of the fluid in the stilling well, the level in the tower is only 25%, not 50%.

Stilling wells actually suffer from a more serious malfunction. Because the fluid in the stilling well is stagnant, dirt tends to settle out and accumulate inside the bottom of the well. With time, the dirt plugs off the bottom level-sensing connection to the level transmitter (the LT, shown in Figure 16-10). Then, the level shown on the control screen remains the same, even though the real level in the tower is changing quite radically. At some point, the operators decide to flush out the stilling well by:

- Closing both valves between the vessel and the stilling well.
- Attaching a water or steam hose to the top of the stilling well.
- Opening the bottom drain from the stilling well, and flushing it clean.

Now the operators return the stilling well back into service. But as they have displaced all of the 0.5 SG pentanes in the stilling well with 1.0 SG amine, they observe that the calibration of the level indication on the control screen has changed relative to the visual liquid level observed on the gauge glasses. So the output of the level transmitter is changed to force the gauge glass level to agree with the control center screen level.

But after a few hours or a few days, as lighter hydrocarbon components accumulate in the stilling well, the external level seen in the gauge glasses grows in proportion to the real amine level in the tower. Actually, as I write this, I'm becoming confused. And if I, who have worked on this problem for 46 years, have become confused, think about how this all appears to the operators. Remember that each operator only sees half the story at a time:

- The outside guys see only the gauge glasses.
- The inside guys see only the level transmitter output on their control screens.

In the Hovensa Refinery in St. Croix, the problem I've described had some nasty consequences. A high level caused the fuel gas scrubber to flood. Massive amounts of rich amine were carried overhead into the sweet fuel gas system. Or, a low liquid level caused gas to blow out of the bottom of the tower. The pressure in the downstream rich amine flash drum increased in 2 minutes from 10 to 50 psig. Which next caused the relief valve of the rich amine flash drum to pop open.

As I've explained in my book, *Process Design for Reliable Operations*, 3rd ed., the correct design for level indication is to:

- Avoid the use of stilling wells.
- Connect each individual gauge glass directly to the vessel itself.
- Connect the level-sensing taps for the differential pressure transmitter directly to the vessel itself.
- Same goes for level alarms and level trips. Especially for trips!

Why then the widespread popularity of the stilling well? If you use my design standard, you'll wind up with a dozen connections on the vessel wall. If you use the stilling well as a standard, you'll only have two such connections. And each connection increases the fabricated cost of the vessel.

When I produce a process vessel sketch for a client, I'll always show the orientation, elevation, size, and service for every connection on the vessel. Including all gauge glass connections, but never for a stilling well, because that is poor process design practice.

CHAPTER 17

Excessive Ambient Heat Losses

I travel my own path and I'm almost there.

—Irish folk song, "Peggy Morgan"

For a refinery crude distillation unit with a feed capacity of 200,000 BSD, ambient heat losses are relatively small. I've measured such losses in the range of 1% to 2% of the heat inputs to the units. On the other hand, I recall a 1,200 BSD crude topping plant in North Dakota where half of the furnace duty went to ambient heat losses. When I designed the 500 BSD White Oil Hydrofinishing Unit for American Oil in Whiting, Indiana, the project was plagued with heat losses.

Whiting Hi-Fi Unit

Did you know that cosmetics are made primarily from baby oil? Baby oil is vacuum-distilled lube oil base stocks that have been hydrotreated in separate stages for desulferization and for saturation of the aromatic rings. In 1968, I designed the Wax and White Oil Hydrofinishing (Hi-Fi) units for American Oil. The aromatic saturation stage was carried out at 3,000 psig, moderate temperature (about 600°F) in a hydrotreating reactor. My design for this reactor required a catalyst bed depth of 80 feet, divided into eight beds, each of a progressively greater length. The calculated reactor ID was 2 feet and the length was 120 feet tangent to tangent. Project engineer Andy noted, "Sorry, Norm. You can't have a 2-foot-diameter vessel. To install the internals—the interbed catalyst supports, interbed mixing, and redistributors—will require a minimum of 3 feet to permit a welder access to the inside of the reactor."

"Oh! I guess that's okay, Andy. The extra catalyst and lower **space velocity** will not impede aromatic saturation. The project incentives are so great, the extra cost of the catalyst is of no consequence."

"Really, Norm? Making white oil is that profitable?"

"Yeah, Andy. A 4-ounce bottle of face cream, which is 90% white oil, sells for $15. The white oil costs us 50¢ a gallon."

When the unit started up, the reactor failed to saturate the aromatic rings. The reactor was running too cold.

"Andy," I complained, "The reactor's too cold. The inlet temperature is 630°F. That's fine, but the outlet is only 530°F. That's too cold for the aromatic ring saturation with hydrogen."

"Your reactor is insulated. Don't complain to me," said Andy.

"No, Andy. The reactor's not properly insulated. There are nine 24-inch manways. Taking into account their thickness for the 3,000 psig service, and their bolt circle, each manway has an exposed surface area of 8 square feet. Including some of the uninsulated inlet flanges, that's 100 square feet of uninsulated metal surface. That's radiating 300,000 Btu/hr of heat to the atmosphere. You know, it's really windy by Lake Michigan. Also, the reactor shell itself, while properly insulated, is radiating another 100,000 Btu/hr to the atmosphere."

"Look, Lieberman. I'm really busy. What'd you want?" growled Andy.

"Insulate those damned manways and flanges!" I answered.

Andy looked out through his dirty office window at the ancient asphalt tanks and truck racks. "Look, Lieberman. I can't. I can't insulate over those bolt circles. At that 3,000 psi pressure, and with that metallurgy, those bolts will stretch. Then your manways and flanges are gonna leak, if the bolting gets too hot."

"But what am I going to do Andy?"

"What you gonna do? Well next time, when you design a small unit, take into account the ambient heat losses."

"Okay, Andy. I will. But what am I going to do now, about the Hi-Fi Unit?"

"Pray, Lieberman! That's what I'd try. Prayer."

Manway Insulation

Actually, the heat loss through the manways was largely avoidable. Not by insulating externally, but by internal insulation. Insulating bricks should have been stacked inside the manway before the manway cover was bolted into place. These insulating bricks would likely have reduced the heat loss through the bare manways by a factor of five or six. As this could not be done on-stream, we did the next best thing. An insulating blanket was wired onto the end of the manway cover, but without covering the bolt circle. This cut the heat loss from the bare manways by 30% to 40% and allowed the reactor to reach an adequate aromatic saturation temperature.

Incidentally, almost the entire supply of Johnson & Johnson's Baby Oil was produced in the 1970s on the unit I had designed. We called it "Eleven triple six." It had no odor and was water white.

Calculating Ambient Heat Losses

On several process vessels, I have been able to make a direct field measurement of ambient heat losses based on the observed enthalpy lost inside the vessel by the process fluid. These vessels were all well insulated with about 3 to 4 inches of magnesium block or fiberglass insulation and no large exposed metal surfaces. The observed heat loss corresponded then to the following relationship:

$$Q = U \times A \times (\text{Delta T})$$

where Q = Ambient heat loss, in Btu/hr/ft^2/°F

U = 0.25 to 0.30 Btu/hr/ft^2/°F

A = The area of the vessel surface below the layer of insulation, including the top and bottom heads, but excluding attachments (i.e., nozzles, manways, ladders, etc.), ft^2.

Delta T = Temperature of the fluid inside vessel, minus the ambient temperature, °F.

The coefficient of 0.25 to 0.30 is the value that I've back-calculated based upon the average observed heat loss rate as a function of the vessel heat input rate. There is no theoretical basis for this number. Nor have I checked this against any literature reference. I use this 0.25 to 0.30 value not only for process vessels, but also for process piping and heat exchanger shells. As an example, I recall an assignment I had as a subcontractor at a Unocal refinery delayed coker. The problem with this coker was that the coke was too soft. The reason for the soft coke was low (i.e., 780°F) coke drum vapor outlet temperature. To produce hard coke, a coke drum vapor outlet temperature of 810°F to 820°F is required. The drum inlet temperature of 910°F was normal. The feed rate and coke drum geometry were:

- Coke drum feed = 100,000 lb/hr
- Drum = 20 feet. ID × 80 feet. T-T

Referring to Figure 17-1, I calculated the expected ambient heat loss based on the observed coke drum skin temperatures. By skin temperature, I mean the average temperature of the external weatherproofing metal jacket around the vessel. This I measured with my infrared temperature gun. The average skin temperature was about 190°F. Essentially, the entire coke drum was insulated, but not very evenly, as there were a number of relatively high-temperature areas, especially toward the top of the vessel.

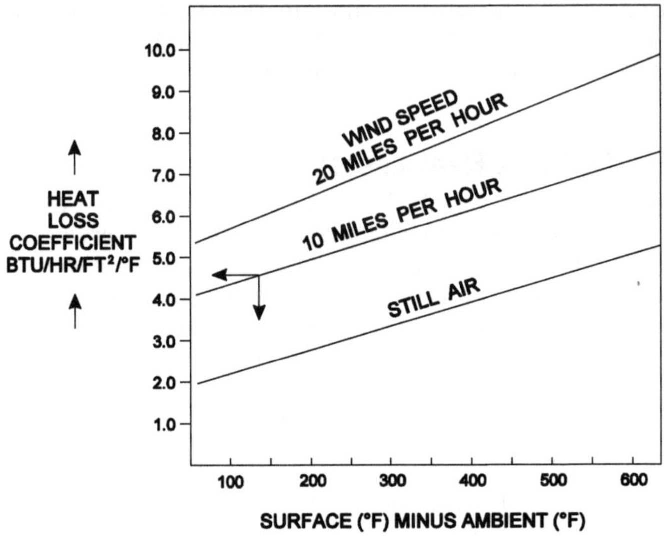

FIGURE 17-1 Ambient heat loss heat transfer coefficient varies with the wind and vessel surface temperature.

The air was quite still at about 5 miles per hour. Ambient temperature was 50°F. From Figure 17-1, the *U* value (heat loss coefficient), is about 3.3 Btu/hr/°F/ft². The area of the vessel (including the heads) was about 5,400 ft². Therefore, the ambient heat loss from the skin was:

$$(3.3) \times (5{,}400) \times (190°\text{F} - 50°\text{F}) = 3{,}500{,}000 \text{ Btu/hr}$$

The heat lost per pound of feed was:

$$(3{,}500{,}000) \div (100{,}000) = 35 \text{ Btu/lb}$$

If the coke drum had been properly insulated, the expected heat lost would have been:

$$(0.28) \times (5{,}400) \times (850°\text{F} - 50°\text{F}) = 1{,}200{,}000 \text{ Btu/hr}$$

The 0.28 factor is the heat loss coefficient or rate through a well-insulated vessel wall that I explained above (i.e., 0.25 to 0.30). The 850°F is the average process temperature inside the coke drum.

The extra heat loss above that which was normal in this service was:

$$3{,}500{,}000 - 1{,}200{,}000 = 2{,}300{,}000 \text{ Btu/hr}$$

The specific heat of the coker feed and/or products is about 0.7 Btu/lb/°F, at 800°F. Thus, the excess cooling of the coke drums, due to excessive ambient heat loss through the vessel walls, was:

$$(2{,}300{,}000) \div (0.70) = 33°F$$

The problem I found was that the weatherproofing metal jacket around the top head had fallen off. On coke drums, coke cutting water is spilled on the top deck every day. Thus, water soaked the insulation around most of the drum daily. This degraded the insulation integrity. Also, the insulation was 20 years old, and only half the current design standard thickness, for coke drums. As a result, the coke drum operating temperature was 30°F lower than it should have been (780°F, rather than 810°F) for the heater outlet temperature of 910°F.

It required 40 minutes to make my field observation: 20 minutes to calculate the heat loss, and 20 minutes to submit my report to Unocal. Then I drove back to the La Quinta Inn and went to sleep. I still recall how angry Roger Phelam was, who had subcontracted this job to me, when he found me sleeping. But when the coke drums were later both reinsulated, and the coke hardness was restored, he forgot about my napping on the job.

I employed a dual calculation method in this project. The skin temperature heat loss method is based on direct field measurements. It reflects the actual situation at the time the measurements were made.

The heat loss calculation based on the 0.25 to 0.30 coefficient and the process temperature assumes a well-insulated vessel, without bare spots and with as-new insulation and weatherproofing integrity.

These two methods should, and usually do, give the same (+ or – 20%) ambient heat loss values. But, if the skin heat loss method is double or triple the overall heat transfer coefficient method (i.e., the 0.25 – 0.30 method), then the insulation integrity is deficient.

In conclusion, note that the smaller the process flows, the more critical insulation is. In that sense, a 4-foot ID tower requires more insulation than a 22-foot-diameter tower. See my book, *Process Engineering for a Small Planet*.

Steam-Jacketed Pipe

There are some services where essentially no ambient heat loss is permissible. Sulfur plant product drain lines are such a case. The sulfur flow from the last reactor is extremely small. If the sulfur became too cold (240°F), it would solidify and plug the line. To avoid this problem, we must employ **steam-jacketed pipe**. For example, the process line size for the sulfur drain is 2 inches. This 2-inch pipe is inside a 4-inch pipe. The 4-inch pipe is sealed at either end of the

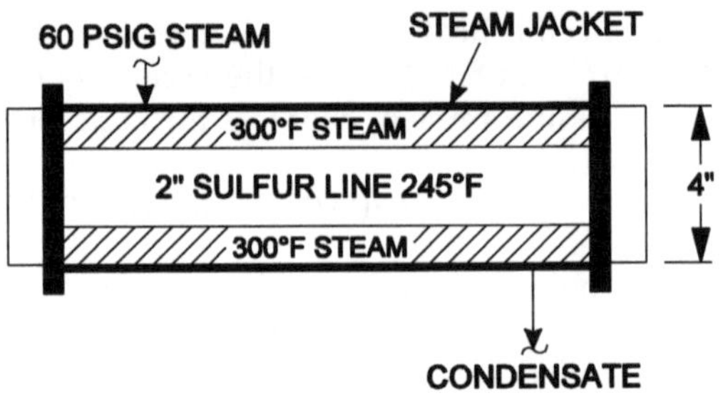

FIGURE 17-2 Steam jacketed sulfur piping eliminates ambient heat losses from a small flowing process stream.

spool piece of the 2-inch pipe. I've shown a sketch in Figure 17-2 of a steam-jacketed pipe. The steam supply pressure to the jacketed pipe is typically 60 psig. Steam at this pressure condenses at around 300°F. The common malfunction is poor condensate drainage from the steam jacketed pipes' annular space. Likely, the steam trap has plugged. Or, there is excessive backpressure from the condensate collection header. If bypassing the steam trap to the sewer increases the exterior temperature of the 4-inch steam jacket, then your problem is clearly condensate backup.

The problem with jacketed piping is that it's expensive. But you can pretty much obtain the same benefit by steam wrapping. This is not the same as steam tracing. A steam-wrapped line uses ½-inch stainless steel tubing and steam traps, just like ordinary steam tracing. However, the tubing is not run along the length of the line, but is coiled around the line. You will probably need a steam trap on every four or six coils. Meaning, you will probably require 10 times as much tubing and 10 times as many traps as would be used just for steam tracing. Insulation is then placed over the steam-wrapped pipe in the usual manner.

I have sometimes used steam-wrapped or steam-jacketed pipe in lieu of a heat exchanger in services where I had a small process flow to be heated by a few degrees. Preventing condensation at the suction of a compressor is one such example, especially for recip's.

Transfer Line Flanges

At a delayed coker unit, a new heater was constructed to replace an older unit. The feed rate to the heater was 8,000 BSD or 110,000 lb/hr. Because of plot plan constraints, the new heater was sited quite far from the existing coke drums. The new heater, while designed to all modern standards, had a pronounced tendency to coke rapidly and thus to suffer from short heater run lengths.

When I investigated the problem in the field, I found that:

- The new heater outlet temperature was 935°F. That is 15°F higher than the old heater outlet, which used to run at 920°F.

- The new heater dual transfer lines tied into the outlet of the two old idled heater transfer lines. The new connecting 6-inch piping was about 600 feet long.
- While the piping itself was well insulated, there were twenty 6-inch bare flanges in the convoluted 600 feet of piping for each of the two lines. Each bare flange had about 2 square feet of uninsulated surface area. Assuming a breeze of 10 mph, I would expect a radiant and convective heat transfer coefficient (see Figure 17-1) of about 10 Btu/hr/ft^2/°F. For high-velocity and low-viscosity fluids (viscosity of coker feed at 900°F is about one centistoke), the skin temperature of a line will be close to its flowing temperature. Therefore, I calculated that the heat loss through the bare flanges was:

$$Q_1 = (10)\ (850 - 50)\ (20)\ (2)\ (2) = 640{,}000 \text{ Btu/hr}$$

The 10 is the coefficient for the bare piping at 850°F skin temperature. The 850 is the skin temperature of the bare piping adjacent to the flanges that I measured. The 50 is ambient °F. The 20 and 2 are the number and bare area of the flanges, respectively. The last 2 factor is for the dual transfer lines.

The 600 feet of dual 6-inch pipe had an area of around 1,900 ft^2. So:

$$Q_2 = (0.25)\ (1900)\ (920 - 50) = 420{,}000 \text{ Btu/hr}$$

The 0.25 is the coefficient for a well-insulated pipe or vessel. The 920 is the average bulk oil temperature in the transfer line for the old heater and the 50°F is ambient. Therefore, the total heat loss is:

$$Q_1 + Q_2 = 640{,}000 + 420{,}000 = 1{,}060{,}000 \text{ Btu/hr}$$

To calculate the temperature loss, I'll note that the specific heat of the feed was about 0.70 Btu/lb/°F. Therefore:

$$\text{Ambient °F loss} = (1{,}060{,}000) \div (0.70) \div (110{,}000) = 14\text{°F}$$

This extra temperature of 14°F, required by the new heater outlet, increased the rate of coke formation in the new heater's tubes. My client corrected the malfunction by simplifying the piping configuration. Most of the piping flanges were eliminated. Also, the piping length was shortened. The higher the process temperature and the smaller the flow, the more important ambient heat losses from piping and vessels become. For an 85,000 BSD delayed coker I revamped, I ignored ambient heat losses. For large process equipment, running at greatly reduced rates, ambient heat loss can be an almost insurmountable malfunction, as I discussed in Chapter 7, "Reboiled and Steam Side Strippers."

Effect of Wet Insulation

Rain on bare piping doubles or triples heat losses. At the Tenneco refinery in New Orleans, steam consumption would jump by 200,000 lb/hr during a sudden rainstorm. Even well-insulated piping and vessels may exhibit a very large increase in ambient heat loss when they become wet. The problem is that wet insulation temporarily loses most of its insulating property.

In 1974 I was punished by Amoco Oil by being promoted to the position of operating superintendent of the Texas City Spent Sulfuric Acid Regeneration Plant. I have described the many malfunctions of this plant in my book, *Process Engineering for a Small Planet* (Wiley, 2010). One of these malfunctions was heat loss from the SO_2 catalytic converter:

$$SO_2 + \text{Air}\ (O_2) = SO_3 \tag{17.1}$$

This is an exothermic reaction where the heat of reaction ought to, and normally did, sustain the required converter temperature. My reactor or converter was a very large vessel, with few external manways or connections, and completely insulated. The hot converter effluent was used to preheat the converter feed (Figure 17-3). As I had a problem cooling the converter effluent, ambient heat loss from the converter was helpful during normal operating conditions. Until it rained.

While the insulation was of a type not permanently damaged by rain, during a heavy rainstorm of an hour or more, the ambient heat loss from the converter would exceed the exothermic heat of reaction. Then, the converter feed versus effluent preheat exchanger would start to function as a feed versus effluent precooler exchanger. The

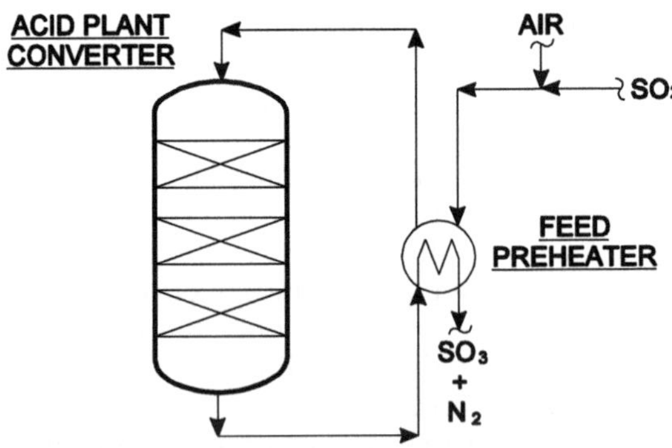

FIGURE 17-3 Bad weather-proofing on a sulfuric acid plant converter insulation, resulted in a plant shut-down due to rain.

reduced converter inlet temperature would reduce the reaction in Equation (17.1) and suppress the exothermic heat of reaction. A cooling wave would sweep through the converter's catalyst beds. I could monitor the gradual but irreversible loss of exothermic heat of reaction on the interbed TIs.

It required about 8 to 12 hours after the rain had passed for the converter catalyst beds to cool sufficiently, so that:

- The conversion of SO_2 to SO_3 would largely have died off.
- The strength of my product, sulfuric acid, fell far below its required 98 wt%.
- The concentration of SO_2 in the vapor plume from my unit increased exponentially.
- "Fuzzy" Griffin, my hostile shift foreman, screamed, "Mr. Lieberman! Are you crazy, or stupid, or both? We can't run the acid plant with a cold converter."
- Fuzzy would shut the plant down and relight the startup heater. Which would rarely light off until the 20th try.

The story did have a happy ending. I hired a contractor who renewed the weatherproofing. That's the thin steel metal cladding that covers the insulation and protects it from rain. This contractor sealed off all the seams between the cladding and all the gaps between the converter's connections and manways. He used roofing cement. Finally, the contractor covered the top hemispherical head with roofing asphalt. The converter was now totally weatherproofed. When it rained, I could now watch TV in peace at home, without Fuzzy phoning to ask, "Well, Mr. Lieberman, you all gonna get out here now or should I just shut her down?"

I estimated that this one insulation upgrading job increased my production of sulfuric acid in Texas City by about 10% per year by avoiding periodic unit outages due to rain.

Insulation in Refrigeration System

As I become older, I'm also getting smarter. For example, I once decided to prepare a vessel for entry by steaming-out the vessel to purge residual butane vapors. This was a standard method at Texas City to gas-free distillation towers. However, this particular vessel was a refrigeration flash drum on my sulfuric acid alkylation unit. It was insulated not to prevent ambient heat loss, but to prevent ambient heat gain. Heating the vessel shell with steam melted this insulation and destroyed its resistance to ambient heat transfer. The correct medium to purge out or gas-free such refrigerated vessels is to use not steam, but nitrogen.

Uninsulated Minimum Flow Spill-Back

Some of the malfunctions presented in my text are simple. This is not one of them. It occurred at the Tenneco Oil refinery. The problem was low overhead naphtha production from a diesel oil hydrodesulferizer H_2S stripper. The naphtha production rate was designed at 1,200 BSD, but it was very erratic and averaging around 600 BSD. I've presented a sketch of the stripper tower's overhead system in Figure 17-4. The small compressor shown pressurized the off-gas from 1 to 10 psig, into the refinery low-pressure 10 psig gas recovery system. The H_2S stripper operated at a 4 psig tower-top pressure. The compressor was a constant-speed machine, driven by a motor. The tower pressure was controlled by the 8-inch spill-back valve on the line from the discharge of the compressor, back into the top of the stripper tower.

As the tower set point pressure was 5 psig, but the actual tower operating pressure was only 4 psig, the pressure control valve on the 8-inch spill-back line was wide open.

The spill-back line was an 8-inch-diameter carbon steel line. No insulation and no steam tracing. Considering its service as a spill-back

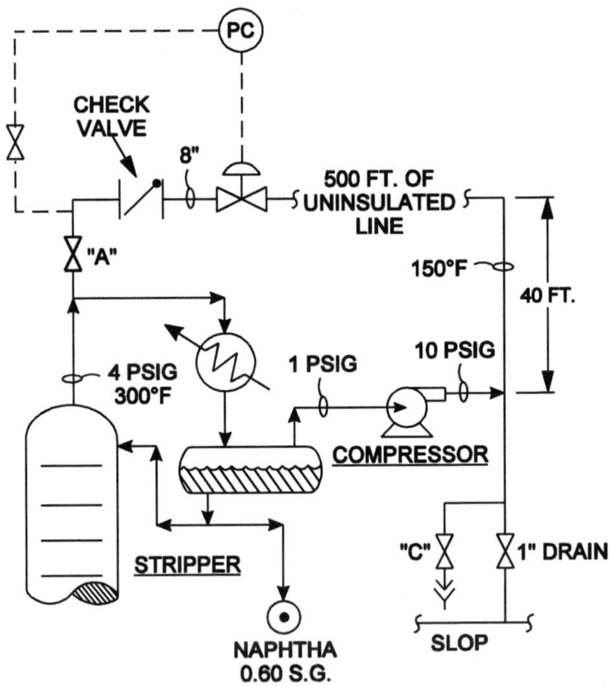

FIGURE 17-4 The uninsulated compressor spill-back line functioned as an air-cooled condenser, resulting in 500 BSD of naphtha lost to slop.

from the off-gas compressor, there does not appear to be any obvious need for heat conservation for this line. Condensation due to ambient heat loss is likely. That's why the designer routed the spill-back line into the upstream side of the stripper reflux drum. Of course, we would never want to permit the stripper overhead vapor, which is rich in naphtha, to flow into the refinery gas recovery system. But this ought never to happen for two reasons:

- There is a check valve in the spill-back line to retard the flow of stripper overhead vapor into the compressor discharge line.
- The pressure at the discharge of the compressor (10 psig) must always be higher than the tower-top pressure of 4 psig.

But plant process malfunctions are often unexpected. In this case, I noted that the 1-inch drain line flowing to slop on the compressor discharge was warm to the touch. This indicated flow. So I closed off the 1-inch drain valve. Instantly, the compressor discharge pressure started to creep up from 10 to 11 psig.

Then I opened the 1-inch drain line to the sewer. Liquid naphtha came out. It poured out in a solid 1-inch stream.

"Yeah, Norm! There's always a bunch of liquid in the compressor discharge," explained Franco, the outside operator. "That's why we've got to leave the 1-inch drain lined up to slop."

"But Franco, that's a lot of naphtha. I bet that's 10 or 20 GPM we're looking at," I answered.

"Could be, Mr. Norm. But that's normal. It's okay. Maybe it'll stop after a while."

But the flow didn't stop. It continued on and on.

"You all know, Mr. Norm, we're puttin' an awful lot of hydrocarbons down the sewer," said Franco.

"Yes. Close it. But I can't understand where all that naphtha liquid is coming from," I answered.

"From the compressor, I guess. Can't be comin' from anyplace else," concluded Franco.

But let's take a closer look at the 8-inch spill-back line. Certainly there's a possibility that the check valve could be leaking. But then, could there be hydrocarbon flow from the 4 psig tower into the 10 psig compressor discharge line? The answer is yes. Here's how:

- The compressor spill-back line, which was bare, had a surface area of 1,200 ft^2. The average skin temperature of this line was about 220°F. Using the chart in Figure 17-1, we would have a heat transfer coefficient of 4 Btu/hr/ft^2/°F. Hence:

$$Q = (4)\ (1200)\ (220 - 50) = 800{,}000 \text{ Btu/hr}$$

- The 50 is the air temperature, °F.

- One pound of 300°F naphtha vapor condensing to 1 pound of naphtha liquid at 160°F requires the removal of 202 Btu/lb:

$$120 + [(300°F - 160°F) \times 0.59] = 202 \text{ Btu/lb}$$

where 120 = latent heat of condensation of naphtha in Btu/lb
0.59 = specific heat of naphtha vapor in Btu/lb/°F

So, the heat loss from the bare 8-inch spill-back line of 800,000 Btu/hr will condense:

$$(800{,}000) \div (202) = 4{,}000 \text{ lb/hr of naphtha}$$

The 4,000 lb/hr of light naphtha is 18 barrels per hour or 440 BSD.

Note that in Figure 17-4, the spill-back line is running through an elevated pipe rack 40 feet above the compressor discharge. The head pressure of 40 feet of liquid naphtha is:

$$(40) \times (0.65) \div (2.31) = 11.3 \text{ psi}$$

where 0.65 is the SG of naphtha.
2.31 feet of water is equivalent to 1 psig of head pressure.

The sum of 11.3 psi liquid head pressure, plus the 4 psig of tower pressure, overcame the compressor discharge pressure of 10 psig. To prove my point, I then closed valve A shown in Figure 17-4. As a result:

- The stripper tower pressure went up to 5 psig.
- The liquid draining out of the compressor discharge line dried out.
- The production of naphtha from the stripper increased by about 500 BSD

The solution to this problem was not to insulate and steam trace the 8-inch spill-back line. The engineering solution was to control the stripper pressure by throttling at the suction of the compressor. Or better yet, by installing a frequency speed control, variable-speed electric motor driver. I've dealt in detail with these sorts of control problems in my book, *Troubleshooting Process Plant Control.*

Fuel Saving for Replacing a Channel Head Insulating Blanket

During a unit turnaround, the tube bundles are cleaned on crude distillation units. This requires removal of the channel head insulating blanket. This is not a problem as this insulation is wired in place and designed for easy removal. But often, I see the channel head insulation months after the turnaround, still lying on the deck. What does this cost in terms of energy? Let's make several assumptions:

- The exchanger diameter is 4 feet.
- The crude is on the tube side.
- The crude temperature is 440°F. Since the channel head is bare, we can assume its skin temperature is 10% lower—i.e., about 400°F.
- Ambient conditions are 40°F and wind speed is 5 to 10 mph.

The exposed area of the channel head and channel head cover is about 50 ft^2. The coefficient from Figure 17-1 is 5. The delta T driving force is 360°F (i.e., 400°F minus 40°F). Therefore, total ambient heat loss from crude charge is:

$$(50\ ft^2) \times (5\ Btu/hr/ft^2/^\circ F) \times (360^\circ F) = 90{,}000\ Btu/hr$$

If the crude heater efficiency is 90%, then the net heating value lost is close to 100,000 Btu/hr. If fuel is valued at $10/MM Btus, that's around $8,400 a year worth of insulation lying in a muddy puddle, to say nothing of our "carbon footprint." I bet if someone offered a plant operator $8,000, he would be glad to replace the channel head blanket. If not, give me a call and I'll be right over.

CHAPTER 18

Process Plant Corrosion

Hope for the best, but plan for the worst.

—My life's plan

As operating supervisor for American Oil in Texas City, I had the privilege of observing many types of corrosion failures in piping, heat exchangers, distillation trays, and vessels. Most of the leaks were associated with the attack of weak acid. The most common hydrocarbon leaks were due to pit-type corrosion inside carbon steel pipes. The interior of the failed piping looked rather like it had been struck repeatedly with a sharp spike. The leaks were always associated with the deeper pits. Rather disturbing was the overall thickness of the piping. It was still almost the original thickness. Thus, an external measurement by the Inspection Department failed to predict any corrosive failure due to thinning. The origin of the pits was known to me. Sulfuric acid had mixed with the wet liquid butane to form 1 pH water, which aggressively attacked tiny grains in the structure of the steel. Why some of the grains were singled out for destruction, I cannot say.

I saw the same pitting pattern inside my carbon steel heat exchanger tubes. However, in the thinner-walled exchanger tubes, failures would occur in months. In the piping, years would be required before they began leaking. I had replaced some of my **admiralty** (a copper alloy, a type of brass used on ships) tube bundles with the carbon steel tube bundles for several reasons:

- To save time on tube delivery.
- To save money.
- Because I didn't know any better.

The admiralty copper alloy tubes had also failed, but not due to wall pitting. They just became uniformly thinner until they began leaking. Such uniform thinning, while creating an overall greater rate of metal loss as compared to pitting, did not cause leaks after just a few months. My copper admiralty tube bundle would last for 1 or 2 years.

Weld Leaks

After 20 years of service, I found it necessary to replace some leaking 8-inch carbon steel piping on my alkylation unit reactor effluent system. Twenty years of service did not seem to represent to me an unreasonably short service life for carbon steel piping. However, two days after the new piping was installed, the welds were blowing out thin streams of isobutane. And two days of service certainly did not represent a reasonable service life.

It's true my chief operator, John Hunter, had allowed weak acid to carry over into the reactor effluent piping. But Mr. Hunter had been periodically allowing this to happen for years without leaks. Why now?

- One suggestion was that the new piping was not **stress relieved**. That means it was not post-weld heat-treated, to remove stresses in the metal's crystal structure created during welding.
- Another suggestion was that the workmanship of the welds was second rate. This was the most likely explanation.
- A third possibility was that the wrong sort of metallurgy of the welding rods were used.

I've done a little, very poor quality, welding myself since my time in Texas City. Welding is a craft that requires expert workmanship. And a poor weld, especially when exposed to corrosive conditions, will be subject to rapid rates of failure.

Erosion-Corrosion

I, your famous author, have created the largest liquid isobutane leak in the history of North America (see Chapter 10, "Shell-and-Tube Heat Exchangers in Sensible Heat Transfer Service"). It was in the floating head of a carbon steel water-cooled propane condenser. Here's how I did it.

First, as previously noted, I cleverly replaced an admiralty tube bundle with a carbon steel tube bundle. Next, chief operator John Hunter allowed acidic wet propane to slip into our depropanizer feed. This caused a single pinhole leak, due to pit-type corrosion, in only one of the new carbon steel tubes. The 300 psig condensed acidic propane liquid leaked into the cooling water flowing through the leaking tube. The liquid propane flashed to vapor as it joined with the warm, low-pressure (20 psig) cooling water in the tube. The water in the tube then became acidic. The propane vapor–water mixture blew out, with tremendous force and velocity, against the inside of the floating head (which covered the tube bundle back-end tube sheet). The combined effects of erosion, due to the high velocity striking the floating head, and corrosion, due to the acidic propane, chewed out a hole in the floating head larger than my fist. The 300 psig liquid propane from the shell side of the condenser rushed through this hole into the cooling water return line.

Based on my alkylation unit overall isobutane material balance, I estimated an isobutane loss of around 6,000 BSD. I could see butane gushing out, as a dense hydrocarbon vapor cloud, from the top of my cooling tower. John Brundrett, the maintenance superintendent, who didn't like me anyway, told the plant manager that he had never seen a leak of this size in his 30 years of refinery maintenance supervision.

When I cut the leaking tube lengthwise for inspection, 99% of its original thickness was still intact. Just a few dozen pits were apparent that had caused this giant failure inside the tube.

Some Common Corrosion Problems

Likely a 10,000 gigabyte computer memory chip could be filled to overflowing with books on corrosion in process plants—or even just petroleum refineries. My brain, which contains 10×10^{13} synapse connections (100,000 gigabytes), is also packed with corrosion failure malfunction experiences, alongside of ex-girlfriend malfunction experiences. So I'll just pick out a few examples.

My most recent example occurred yesterday. I was on the phone to Priam, an engineer from the Suncor plant in Ft. McMurray, Alberta. This is a giant plant that recovers synthetic petroleum from tar sands. About 3% of the world's crude oil supply comes from these Canadian tar sands.

Priam's problem is that they have just opened their diluent naphtha recovery fractionator. To their surprise, the 304 stainless steel naphtha pumparound spray distribution pipe grid was cracked. The 304 s.s. has nickel as well as chrome. The common 300 series stainless steels are all subject to chloride stress corrosion cracking. The feed to the fractionator contained ancient seawater salts. In particular, the $Mg(Cl)_2$ is subject to hydrolysis:

$$MgCl_2 + H_2O \rightarrow MgOH + HCl \tag{18.1}$$

The HCl vapors will condense at the 240°F to 280°F spray distributor operating temperature. This will promote corrosion and cracking of 304 s.s. pipes and fittings.

Priam called to obtain my opinion about temporarily using a carbon steel spray pipe distributor grid. He thought if he used 3x pipe (triple thick pipe), it might last a year until the next unit turnaround. I told him that although I did not know the exact corrosion rate for carbon steel in this service, it was sure to be high. But the real issues are:

- Spray nozzles themselves are almost always 304 or 316 stainless. Screwing a stainless spray nozzle into a carbon steel pipe will certainly create a flow of electrons between the stainless nozzle and the carbon steel pipe. This is called **galvanic corrosion**. This means that very high rates of metal loss will occur where the two dissimilar metals come into direct physical contact. Likely the nozzles would become loose and fall out of the carbon steel pipes as the threads in the piping failed.
- The corrosion products formed in the carbon steel spray piping would almost surely plug the openings in the spray nozzles. This would destroy the function of the spray nozzles.
- Using carbon steel spray nozzles would avoid the galvanic corrosion. But these nozzles would plug even faster due to wet HCl corrosion.

The general problem is that moisture would almost surely form at the temperature and pressure of operation. Then wet HCl, in the presence of H_2S, would rapidly form $Fe(HS)_2$, which is a water-insoluble corrosion product. My recommendation was to replace both the spray piping and nozzles with 410 stainless steel. This is a chrome-type steel, but without nickel. It is not subject to chloride stress corrosion cracking. But Priam's final comment was that while he agreed with me, they could not quickly locate 410 s.s. piping, and would most likely be forced to use carbon steel. Perhaps God, in his infinite wisdom and mercy, will intervene and help my friend Priam.

Naphthenic Acid Corrosion and Sulfur

Normally one would think increasing the sulfur content of hydrocarbons increases corrosion. But this is not necessarily correct. If the corrosion is due to napthenic acid, which is an organic acid, sulfur retards rates of corrosion. That means low-sulfur, high-acid-number crude oils are quite corrosive. When faced with naphthenic acids, I specify 316 s.s., or even better, 317 s.s. My experience with 304 s.s. in naphthenic acid service is bad. Using 410 chrome steel, at least for vacuum tower internals, is also really bad in naphthenic acid service. Areas of high velocity exposed to naphthenic acid are especially vulnerable to erosion–corrosion.

In a practical frame of reference, it's not worth a lot of time and effort to optimize metallurgy. The best alloy for a service is usually the cheapest. Cheaper if one takes into account replacement costs, safety considerations, and the detrimental effects of downstream equipment fouling with corrosion products.

If I suspect naphthenic acids may be present, I will typically specify 316 s.s. tower internal components without inquiring as to exact rates of corrosion. My clients have lots of money, and investments in superior metallurgy are better than many of their other investments.

On the other hand, for cracked hydrocarbon distillation service, that is coker, fluid cracking units, hydrocracker or visbreaker effluents, the use of 410 stainless steel is adequate as naphthenic acids are destroyed. Also, even in virgin crude service, I have specified 410 stainless steel stripping section trays for both the atmospheric and vacuum tower with no adverse effects.

Chrome Embrittlement—Piping and Vessels

At the Amoco refinery in Texas City, a high chrome (nine plus) steel alloy vessel was to be hydrotreated. It was a heavy-walled vessel in high-pressure hydrotreating service. The hydrotesting (i.e., the pressurization of the reactor with water) proceeded at ambient temperatures. The vessel shattered. It rather looked like a giant gray glass bottle had been dropped onto concrete. Apparently, the vessel would have passed its pressure test if it had been tested at its normal operating temperature (about 700°F), at which temperature it possessed greater ductility.

At the Amoco refinery in Whiting, Indiana, a delayed coker fired heater had experienced a rapid emergency shut-down. It was a cold winter day. An inspector was assigned to check the nine percent chrome furnace tubes for obvious visual damage. He tapped on a furnace tube with a hammer and it shattered. After ruining several more tubes, the inspector began to wonder if he was doing something wrong. The problem with the 9% chrome tubes was that they had become embrittled. They had lost their ductility by being cooled too rapidly. Simply by slowly reheating the tubes to their normal operating temperature, the operators would have re-annealed the tubes and restored their ductility.

An overhead coke drum high chrome steel line was also broken by a hard blow. It was cooled rapidly from 820°F down to ambient temperatures. Thus, it also lost its ductility and became subject to brittle fracture failure.

Surface Condenser Seal Strip Failure

The seal strips used in steam turbine exhaust surface condensers are critical to the development of a good vacuum in the condenser. These strips prevent steam at the inlet from short-circuiting the tube bundle. That is, they force the steam to flow around the air or vapor baffle, and through the tubes (see Chapter 27, "Vacuum Surface Condensers and Precondensers").

For vacuum systems serving steam turbines, the seal strips are constructed from a copper alloy material, which is totally suitable for steam service. However, in a refinery and certain chemical plant operations, the surface condensers are used to service the discharge of vacuum jet systems. Also, they are employed to partly condense the overheads of process vacuum towers. And this too often creates a seal strip corrosion problem.

A refinery vacuum tower overhead vapor stream will contain various corrosive elements:

- HCl
- NH_3
- H_2S
- O_2
- HCN
- CO_2

Exposing copper alloy seal strips to these components leads to rapid corrosion and failure. I've found badly corroded pieces of seal strips in the pump suction strainer serving the condensers in vacuum tower overhead service. The results of such seal strip failures are high condenser vapor outlet temperature and the loss of vacuum due to overloading the vacuum jets.

I don't actually know the optimum alloy for seal strip construction in various applications. However, unless one specifies something else, it appears as if the vendors supply copper alloy, which is unsuitable for most chemical plants and refinery vacuum tower services.

Carbon Steel versus Chrome Steel Piping

The most common cause of fires in process plants in my experience is piping failures. And the most frequent cause of piping failures is the accidental substitution of a carbon steel piping **spool piece** (i.e., a flanged section of piping) for a chrome steel spool piece. The problem is that both carbon steel and chrome piping visually appear very similar. The 300 series (304, 316, 317) alloy steels are shiny, and thus easily identified, and are hardly to be confused with ordinary carbon steel piping. Also, 300 series steels are not magnetic. But both chrome and carbon steels are magnetic.

Piping flanges ordinarily are marked with a symbol indicating their metallurgy. But that's an indication often missed in the field during construction (see *Troubleshooting Process Operations*).

The composition of metals can easily be checked using a handheld portable analyzer. And this should be, and often is, done after construction. But once a process line is insulated and weatherproofed, this metallurgy check becomes more difficult.

Both in theory and in practice, the rate of metal loss of a carbon steel process line in hot (400°F to 500°F), high-sulfur (1% to 3%) hydrocarbon liquid service is 10 to 20 times greater than for a 5% chrome steel process line. I say "in practice" because I have repeatedly noted failures of carbon steel piping after 3 or 4 years of service, while sections of chrome piping have only lost a few mils (a thousandth of an inch) of thickness in the same interval.

At the Good Hope refinery in Norco, Louisiana, a new delayed coker was built. There were several thousand feet of hot (450°F to 690°F), heavy, gas oil piping. The piping was inspected by **Sonaray** for thickness on an annual basis. Unfortunately there was a single short section of carbon steel piping in this large and complex piping network. Actually, I had commissioned a portion (i.e., the gas oil filtration system) of this system myself during the initial unit startup. During the startup, the piping had not yet been insulated and thus was visible. Neither I, nor anyone else, noted the short carbon steel spool piece in the extensive chrome steel piping network, because chrome and carbon steel look pretty much the same.

Of course, the odds of checking the thickness of a short spool piece in such an extensive piping network are small. After about 5 years of service, the carbon steel piping failed. The heavy coker gas oil, which was above its **autoignition temperature**, ignited. The resulting fire led to a long outage of the delayed coker, and an even longer lawsuit between the refinery owners and the construction company that built the delayed coker unit.

A complete check of chrome piping lines, with a portable meter that indicates the metallurgy composition, is the only practical method to avoid this sort of malfunction and failure.

Quantifying Corrosion Rates in Piping

A typical 6-inch piping section of ordinary process piping will have a thickness of ¼ inch (i.e., 6 to 7 millimeters). Part of this thickness is that needed to contain the internal process fluid pressure (perhaps 0.15 inch). The other 0.10 inch might be the corrosion allowance of 100 mils:

$$1 \text{ inch} = 1{,}000 \text{ mils}$$

If the predicted rate of corrosion is 10 mils per year, and the corrosion allowance is 100 mils (0.10 inch), then the process line will be predicted to have a service life of 10 years.

Vessels, distillation towers, furnace tube, and heat exchanger shells all have this sort of predicted service life. Some process equipment can outlast us. The wax sweating tanks at the Amoco Oil refinery in Whiting, Indiana, that I worked with in the early 1970s were 80 years old. A cast iron pump at my great-grandfather's winery in Argentina dated from the 1920s when I saw it in 2008. It still looked operable. Cast iron, unlike carbon steel, does not corrode at any noticeable rate.

A corrosion rate of 2 or 3 mils per year is generally considered to be acceptable. A corrosion rate of 20 or more mils per year is normally considered to be dangerous. At the Aramco refinery in Saudi Arabia, they had a stabilizer with the reboiler circulation lines exposed to wet chlorides. Corrosion rates were 40 to 60 mils per year. The reboiler inlet line blew out and killed two people.

Corrosion Effect on Process Efficiency

Structured packing is an open mesh material widely used to bring vapor and liquid into close contact in distillation towers for two reasons:

- Fractionation
- Heat transfer

When used for heat transfer purposes, we say that structured packing is being employed in a **pumparound** service (Chapter 6). The typical materials used for structured packings are 410, 304, 316, and 317 stainless steels. 410 has little nickel or moly. 316 and 317 have several percent moly and a lot of nickel as well. 317 has about 40% more moly than does 316.

Neither 410 nor 304 (lacking moly) are very resistant to naphthenic acids. In particular, I have observed especially severe corrosion to structured packing in the light vacuum gas oil pumparounds of refinery vacuum distillation towers. It really depends on the crude type. Some vacuum towers, such as those processing Russian Ural crude, use 410 s.s. in the light vacuum gas oil pumparound with zero corrosion. Other vacuum towers, such as those processing heavy California or Venezuelan crudes, have terrible rates of corrosion in the light vacuum gas oil pumparound, even when using 316 s.s. packed beds of structured packing.

Damage to the structured packing becomes obvious during normal operations. Bits and pieces of the structured packing perforated sheets appear daily in the suction of the gas oil circulating pump. The pump's suction screen becomes clogged with the corroded structured packing.

The temperature of the vapor leaving the corroded packed bed begins to increase. The normal operating response to this temperature increase is to increase the gas oil circulation rate. But as the packing deteriorates, this response becomes progressively less effective. Eventually, I have observed at the Coastal (now Valero) refinery in

Corpus Christi, Texas, that more gas oil circulation increases, rather than decreases, the structured packing pumparound vapor effluent temperature.

Hydrogen Ion Penetration in Steel

Regardless of the corrosion mechanism, there are always two products of corrosion:

- A salt like $Fe(HS)_2$
- A hydrogen molecule

For example:

$$Fe + 2(H_2S) = Fe(HS)_2 + H_2 \tag{18.2}$$

$$\frac{1}{2}Fe + H_2S = \frac{1}{2}Fe(HS)^- + H^+ \tag{18.3}$$

Incidentally, the preceding equations account for 90% of metal loss in petroleum refineries. When any acid, like H_2S, reacts with a metal, a hydrogen ion, H^+, is initially created. This ion is supposed to bond with another hydrogen ion. The resulting molecule of H_2 then joins up with the process fluid and leaves the vessel.

But, under certain circumstances, the hydrogen ion does not perform in the expected manner. It chooses to migrate through the metal lattice structure. Then it recombines, as molecular hydrogen, with a like-minded hydrogen ion fugitive, on the outside surface of the vessel. And that's okay.

But sometimes the migrating hydrogen ions get lost inside the metal lattice structure. This happens because of an imperfection in the metal plate itself, or an imperfection in the weld.

If the imperfection is in the steel plate, then a **hydrogen blister** is formed, rather like a metal bulge inside the vessel wall. If the ion combines with a like-minded fugitive to form molecular H_2 at a weld, then the weld begins to crack due to the pressure exerted by the hydrogen gas. Usually this happens because the weld is not stress relieved or post-weld heat-treated.

Sometimes the crack in the weld is propagated at a fantastic rate, like inches per minute. This happened at the Unocal refinery in Chicago in the mid-1980s. A propane treater vessel parted and ignited. The service was H_2S extraction with mono-ethanol-amine (MEA).

The source of ignition was the refinery fire truck. Sixteen men and one woman were killed. The fundamental cause of death was found to be hydrogen cyanide. The HCN in the propane feed retarded the recombination of H^+ inside the vessel, and thus promoted H^+ (i.e., hydrogen ion) penetration of the 1-inch-thick, carbon steel vessel wall.

The fact that the vessel was not stress relieved didn't help either. The UOP engineer who filled out the vessel data sheet circled "n.r." (not required) next to the line item, "Stress Relieve." I was hired by a smart Chicago lawyer to investigate the entire sorry incident.

Life is strange. In two weeks, I'll be at the same plant (now Citgo) to teach a seminar. I bet when I ask my students about the explosion, they probably will not have even heard about how it all happened.

Carbonic Acid Corrosion

Chapter 14, "Steam Condensate Collection Systems," largely dealt with the problem of carbonic acid corrosion of the steam side of reboilers. The CO_2 in steam will form H_2CO_3, which is quite corrosive, even at a pH of 6. If corrosion is observed on the steam side of exchangers, this is almost certainly a consequence of carbonic acid attack due to the CO_2 in the steam. For horizontal exchangers, with steam of the tube side, vent off about one-half of 1% of the supply steam from below the bottom pass partition baffle in the channel head.

For vertical exchangers, with steam of the shell side (i.e., the common vertical thermosyphon reboiler), the optimum place to vent accumulated CO_2 is far less obvious. Venting from the top portion of the shell will vent a lot of steam with the noncondensible CO_2. I've given this matter a lot of thought, but have failed to reach a good conclusion.

Shell Norco Explosion

In 1989, not far from my home in New Orleans, in Norco (New Orleans Refining Co.), Louisiana, a free air detonation blew a giant fractionator off its foundation. It landed on the control room and killed the entire operating crew. I was hired by greedy lawyers suing Shell for property damage to the local community.

A 6- or 8-inch vapor line, upstream of the depropanizer overhead condenser, failed at an elbow. I saw the blown-out elbow after it became a legal exhibit. It was obvious what had happened. The elbow was carbon steel, even though a long history of corrosion had forced Shell Norco to retube the condenser tube bundle with alloy steel.

To help protect the bundle from acidic attack, a weak alkaline (NH_3) water wash was injected through a small-diameter piece of tubing upstream of the elbow. The nature of the acidic component in the propane vapors was not investigated by Shell, and I do not know, either. However, my investigation concluded that the acidic components dissolved in the periphery of the water wash stream. The outside of the water stream turned acidic and cut a groove on the elbow and in about 18 inches of the downstream piping. Both inside and outside the elliptical groove cut by the acidic water, the piping was original thickness.

Shell could have avoided the failure by using a Bete 60°, full-cone spray nozzle, in a horizontal section of piping, directed 180° from the direction of flow. Also, carbon steel is not the best choice in an acidic environment when moisture is present.

One of the pleasant points in my career occurred after I had testified. Shell's young process engineering representative came over to me and said, "Thanks, Mr. Lieberman. I learned a lot." Needless to say, he waited for the dozen or so lawyers to leave first.

H_2S Promotes HCl Corrosion

Any time HCl and high concentrations of H_2S get together in an aqueous environment, trouble is sure to follow:

$$2\,HCl + Fe = FeCl_2 + H_2 \tag{18.4}$$

$$FeCl_2 + 2\,H_2S = Fe(HS)_2 + 2HCl \tag{18.5}$$

$$2HCl + Fe = FeCl_2 + H_2 \tag{18.4}$$

I think you can appreciate my point. Small amounts of HCl in the presence of 1,000 times greater concentration of H_2S can be very corrosive, especially to carbon steel. Ordinarily a weak acid (H_2S) cannot displace from its salt ($FeCl_2$) a stronger acid (HCl). But if the H_2S concentration is great enough, this does happen in an aqueous environment.

The only metallurgy that I know for sure that resists this sort of corrosion is titanium or gold—titanium being very expensive, but the cheaper of the two choices.

Internal Boiler Tube Corrosion

The corrosive agent here is certain to be oxygen. Air should have been stripped out of the boiler feed water in the deaerator. The deaerator is basically a simple steam stripper.

If residual oxygen is left in the boiler feed water, which is hard to avoid, then an oxygen-scavenging chemical (like pyridine) must be added to the boiler feed water. To minimize the use of expensive oxygen-scavenging chemicals, you have three choices:

- Lower the deaerator pressure.
- Increase the venting of steam from the top of the deaerator.
- Increase steam condensate recovery in the plant.

Oxygen is an extremely reactive and corrosive element. It will attack most metal surfaces, and certainly boiler tubes, very aggressively.

Carbon dioxide, while generated from carbonates in boiler feed water, is present in too small a concentration to harm the boiler tubes through corrosion.

Effect of High Velocity

In many services, products of corrosion adhere to the metal surface attacked and retard further corrosions. Aluminum oxide is the common example. But in the presence of high velocities, such protection may be eroded off the metal surface. In general, a low velocity is 2 ft/sec for liquids. A high velocity is 12 ft/sec for liquids. That's a pretty big range to be of much use to someone evaluating erosive velocities. The problem is that the protective layer of corrosion products is actually not removed by the high velocity of the liquid, but by the particulates in the liquid.

Adding catalyst fines to a stream of heavy oil makes the oil quite erosive at 8 ft/sec in the presence of a corrosive H_2S environment. Without the fines, there would be no particular problem.

In summary, as a former Amoco Oil Texas City operating supervisor, I had three main problems that occupied most of my time:

- Corrosion.
- Pump mechanical seal malfunctions (see Chapter 29).
- Shift foremen who signed fraudulent time cards authorizing overtime pay without reading them first.

Author's Comment

Is environmental change a real problem? Perhaps the measures now being implemented will correct the imbalance in CO_2 emissions. Possibly yes, but likely no. Here are the facts:

- CO_2 concentration in 2010 = 390 ppm (ambient air)
- CO_2 concentration in 1980 = 290 ppm
- Rate of CO_2 concentration increase = 2.1 ppm (or 0.51% cumulative since 1980)

The data for CO_2 increase show no sign of moderating. It's been the same 0.5% ever since the danger was recognized in the 1980s. The technical progress to reduce fossil fuel oxidation has been swamped by per capita increase in energy use and by population growth.

Incidentally, we have made fine progress on methane suppression in the atmosphere, which accounts for 10% of global warming. Also, ozone-destructive chemical emissions have been greatly reduced. That is encouraging.

But from my perspective, I see my clients doing everything possible to accelerate hydrocarbon production rates:

- Exxon—Natural gas from shale oil deposits.
- Venezuela—Heavy crude oil production.
- Qatar—Liquefied natural gas.
- Canada—Tar sands synthetic oil.
- South Africa—Gas to liquids, where the cheap gas comes from Mozambique.
- BP—Drilling in water way over their heads in the Gulf of Mexico.

All hydrocarbons produced will be oxidized to CO_2. Then why, one might inquire, am I writing this comment, on a plane to Australia, to work for BP? Because I have calculated that the effect of CO_2 on the pH and temperature in the upper layers of the oceans will not become critical until 2038. Then the world's oceans will become a net emitter, rather than a sink for CO_2. At which point a positive feedback loop of CO_2 emissions and global temperature will be created.

Then, 28 years from now, I'll be dead and won't care.

I showed my calculations to my pal Gerry. "Norm," Gerry said, "Your calculations are wrong. They are based on pessimistic assumptions. I've redone your calculations on a realistic basis. The timing for the positive feedback loop to be created is not 28 years, but 51 years."

"So Gerry, what's your point?" I asked.

"Norm, by 2061, I'll be dead, too, and I won't care, either."

"Yeah, Gerry, and how about that baby boy Margaret had last month?"

"Gee, Norm, I hadn't thought about him."

And the scary part of this story is that Gerry's the smartest engineer I know!

"Gerry," I suggested, "Why not read my book, *Process Engineering for a Small Planet*?"

But from my perspective, I see my clients doing everything possible to accelerate hydrocarbon production rates:

[illegible]

CHAPTER 19

Vapor–Liquid Separation Vessels

I wouldn't want to work in a refinery that would employ engineers like me.

—Jack Benny

Early process operations were mainly carried out in a series of vertical drums called "stills." A fire was lit underneath a large copper pot with a top vent. The off-gas was next condensed and partially revaporized in a secondary still. The process was repeated to make a progressively purer light component. The "flux" from the bottom of the pot of each still was recycled back to the previous still. That is, the flux flowed countercurrently to the vapor flow. This collection of stills and interconnecting pipes was called a "pipe still." Texaco was still operating such a unit in the early 1980s, down by the Neches River in Texas. I happened to visit the plant during its final month of operations.

Entrainment of heavier components as droplets of liquid into the vapor phase will partially defeat the purpose of the stills. Thus, very early in the process industry, criteria were established to determine a proper diameter of the stills to minimize entrainment to an "acceptable level." Of course the term "acceptable level of entrainment" is like beauty. It lies in the eye of the beholder.

To calculate an acceptable or allowable vertical vapor velocity (Va), in feet per second:

$$\mathrm{Va} = (0.227) \times \left(\frac{\mathrm{DL}}{\mathrm{DV}}\right)^{\frac{1}{2}} \tag{19.1}$$

where DL; DV = Density Liquid; Density Vapor.

This equation neglects droplet size and viscosity. The liquid viscosity doesn't matter, as it is not the **continuous phase**. The vapor viscosities are always low, like 0.01 ± 50% centipoise (c.p.), and thus they also do not matter. Droplet size does matter, even though it's usually ignored. But I'll discuss this more later in Chapter 20, "Knock-Out Drums; Demisters and Impingement Plates."

Kellogg Allowable Velocity

The engineering firm of M.W. Kellogg was a process design contractor staffed by incompetent engineers in New York City. It's true! They actually rejected me for employment in 1966, thus betraying their incompetence.

The Va value calculated above is called, "100% of Kellogg allowable velocity." What this means is that at this vertical vapor velocity, if all other design and operating parameters are okay, the vapor might contain a few weight percent of entrained liquid. By a "few," I mean maybe about 1%, but not as much as 5% liquid in the flowing vapor phase. Entrainment rates above 5% are unacceptable for most process applications.

As I describe in Chapter 20, a properly designed wire demister pad may reduce liquid entrainment rates by an order of magnitude. However, as explained in that chapter, partially plugged or improperly designed demisters can increase liquid droplet entrainment rates.

Most vertical separators are designed and operated at 50% or less than their Kellogg allowable velocity as calculated from Equation (19.1). If 5-plus weight percent entrainment in the vapor phase is not important, then operating at 160% to 180% of the Kellogg allowable velocity is okay. Above 180%, entrainment would be expected to increase exponentially. Neither I nor any other sane engineer would design a vapor–liquid separator to operate above 180% of Kellogg allowable velocity without a demister, inlet impingement plate, or feed distributor.

Inadequate Tower Diameter

Too often, unit engineers try to troubleshoot process equipment malfunctions when no such malfunctions actually exist. For instance, sometimes the diameter of a knock-out drum (i.e., a vertical vapor–liquid disengagement vessel) is simply too small. To make this determination, calculate the volume of vapor flow using Equation (19.2):

$$\mathrm{VOL} = \frac{(\mathrm{LB})}{(\mathrm{MW})} \times \frac{(379)}{(3600)} \times \frac{(^\circ\mathrm{R})}{(520)} \times \frac{(14.7)}{(\mathrm{P})} \tag{19.2}$$

where VOL = Vapor flow, $\mathrm{ft^3/sec}$
LB = Pounds per hour of vapor flow
MW = Molecular weight of vapor, pounds per mole
°R = Degrees Rankine (°F plus 460)
P = Absolute pressure, psia (psig plus 14.7)

Next calculate the actual flowing velocity by dividing VOL by the cross-sectional open area of the vertical drum in square feet to obtain the actual vertical velocity. Divide this by the Kellogg allowable velocity from Equation (19.1). This yields the percent of the Kellogg allowable velocity. If you're 50% over the allowable velocity and experience high rates of liquid entrainment, then just be thankful that it's not even worse. Your vessel diameter is just too small.

Inadequate Vertical Height

The space between the top of the inlet nozzle and the tangent line of the vessel's top head ought to be a matter of feet, not inches. In my book, *Process Design for Reliable Operations*, 3rd ed., I have detailed some of the knock-out drum's design dimensions. In general, this space should be a minimum of one vessel diameter, and better yet, more like two vessel diameters. Bigger being better.

Having the bottoms liquid level too close to the bottom of the inlet nozzle can create re-entrainment problems. Very often there can be a very high layer of foam in the bottom of a separator vessel. Vapor impinging on such a low-density layer of foam may cause the frothy foam to be blown out the top of the vessel.

I am not hypothesizing about this foam. I've measured it using neutron back-scatter technology. I have measured foam densities of only 10% of the clear liquid density. I have measured foam heights of 40 feet! No wonder that a high-velocity vapor stream can cause such a light, frothy foam to be re-entrained.

Not for any particular reason, I keep the bottom of the feed nozzle about 3 feet minimum above the top-level tap of the bottom LRC connections. Again, bigger is better.

Note that I consistently refer to the outer dimensions of the feed nozzle. Too often designers use the center line of a nozzle as a reference point. That's acceptable for 4-inch or 6-inch nozzles. But not really acceptable for a 24-inch or 30-inch feed nozzle.

Fog Formation

The finer the droplets, the lower the percent of Kellogg allowable velocity should be, to retard entrainment. A finely divided mist or fog can be lifted out of a vessel with very little vapor velocity. There are two theories of fog formation, neither of which I can support from my own experience:

- Mixing of cold and hot vapors. When a cold air cooler effluent of relatively light, dry gas was mixed with a hotter gas, heavier gas droplets condensed in a widely dispersed manner into the cold, dry gas. This makes sense, and I imagine it could happen.

- Accelerating a vapor–liquid mixture above sonic velocity causes the combined phases to exceed the speed of sound. This creates a sonic boom–type phenomenon that shatters the liquid droplets and thus creates a fog that cannot be recoalesced or settled. This idea is widely accepted in the industry.

I actually had a vacuum heater feed line that supposedly suffered from this malfunction. Hence, the downstream tower suffered from severe entrainment. But when we opened this tower at the CVR refinery in Coffeyville, Kansas, we found the feed distributor was busted. So we fixed the distributor and the entrainment problem vanished. I've described this whole sad story in my book, *Process Engineering for a Small Planet*. It's just one small data point that stands in lonely contradiction to a generally accepted industry theory of fog formation in high-velocity lines.

Incidentally, sonic velocity in air at ambient conditions is about 1,000 ft/sec. Sonic velocity under vacuum conditions (about 100 mm Hg) is about 500 ft/sec. The lower the density of the continuous phase, the lower the sonic velocity. Sonic velocity is also called critical or choke flow. Or better yet, it's just the speed of sound.

Excessive Inlet Velocities

High inlet vapor velocity will promote liquid droplet carryover. For an ordinary vapor density of about 0.1 to 0.8 lb/ft^3, I consider an inlet velocity of less than 20 or 30 ft/sec to be low. An inlet velocity about 120 to 150 ft/sec is high in my experience.

If the vapor outlet nozzle is several vessel diameters, or perhaps 20 feet above the inlet nozzle, quite a high inlet velocity is not particularly critical. But if you have an excessive inlet velocity and minimal vertical space available above the inlet nozzle, then consider the new baffle shown in Figure 19-1.

I first used this baffle design on a vacuum fractionator at a Texaco Bakersfield, California, refinery and it worked very well. The idea was to reduce black residue entrainment by halving the vapor inlet velocity. The intent of the design change is rather obvious from my sketch. One critical word of caution: Baffles like these must never restrict the flow of vapor as the vapor enters a vessel. There's plenty of evidence that such restrictions cause the vapor to develop localized high vertical velocities, which then promote entrainment.

Also, there is a great temptation to place a roof over the top of the baffle to suppress entrainment. That's a good idea, if the liquid level in the bottom of the vessel is 10 or 20 feet below the bottom edge of the baffle. But if the top tap of the level controller is just a few feet below the bottom of the inlet nozzle, this is not such a good practice. The vapor will impinge on the liquid layer below and promote re-entrainment.

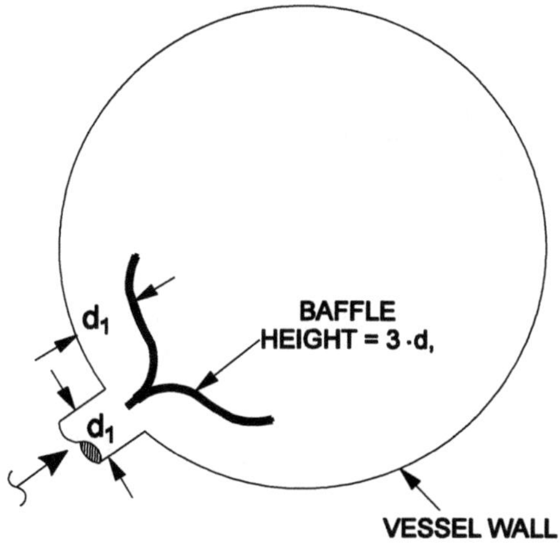

FIGURE 19-1 New baffle effectively halves inlet nozzle vapor velocity.

Foam Formation in Bottoms

A lot of my early success in troubleshooting came with the realization that the external indicated level in many, if not the majority of process vessels, was lower than the actual level in the vessel. Whenever vapor and liquid begin to separate, time is involved. It takes a finite amount of time for liquid droplet de-entrainment to proceed from the continuous vapor phase.

It also takes a finite amount of time for vapor bubbles to escape from the continuous liquid phase. This intermediate phase is called foam or froth. Thus, there is always foam in the upper liquid portion of all knock-out drums. I remember an incident at the ARCO refinery in Cherry Point, Washington. A vessel was carrying over vapor with a light, black froth from a 60-feet-high tower and at the same time the vessel's bottoms pump was cavitating due to heavy foam in its suction line! Amazing but true.

If you want to see such foam with your own eyes, then refer to Figure 19-2. This sketch assumes the foam is above the top gauge glass connection, C. Then proceed as follows:

- Close B 100%.
- Open C fully.
- Open A just a little bit.

The foam should now drain through the gauge glass. I've only done this in the bottom of low-pressure amine absorber towers. It worked great. Except that I almost killed myself once with hydrogen sulfide gas as I neglected to don my BA (breathing air equipment).

The problem with foam inside a vessel is its density. If the SG of foam is half the SG of the clear liquid in the external level-trol or

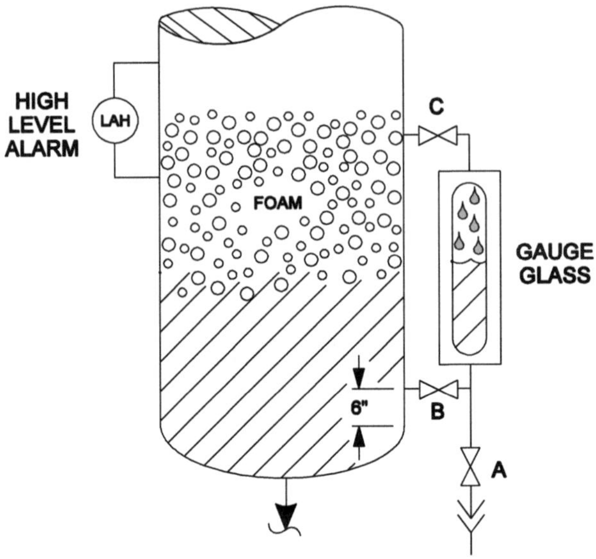

FIGURE 19-2 Liquid dripping through the gauge glass is an indication of foam above the top gauge glass connection.

gauge glass, then the indicated level will be 50% of the range of the level indication device. As the foam rises above the top-level indication tap from 4 inches to 4 feet, the indicated level of 50% will not greatly increase (see Chapter 16, "Level Control Problems."). Also see my book, *Troubleshooting Process Plant Control.*

Factors Promoting Foam Formation

A high foam level is not truly a malfunction with your knock-out (KO) drum. It's just a nasty type of high liquid level. If you have a properly calibrated and installed **high level alarm**, as detailed in Figure 19-2, it should alarm even though the indicated level is only 50%. If your high-level alarm is stupidly connected to the same set of level taps as your gauge glass, then you're in big trouble. But what factors promote the formation of aerated liquid or foam, which is the fundamental problem?:

- Particulates—Corrosion product; catalyst fines; precipitated salts.
- Surfactants—Heavy aromatics in aqueous system, sodium naphthanates in crude oil.
- High-viscosity liquids.
- Aeration—Rapid boiling.

- Turbulence—Rapid rates of circulation.
- Lack of residence time.
- Small surface area for foam to break.

Regarding this last factor, the need for surface area to accelerate the breaking of foam, a horizontal vessel will be a far better choice for a vapor–liquid separator in a foaming service than a vertical separator. For example, a 5-foot ID by 20-foot T-T vertical vessel has 20 ft^2 of surface area for foam dissipation. The same vessel in a horizontal position has 100 ft^2. Of course, when half full of liquid, the vertical vessel's liquid level will be 10 feet below the vapor outlet. For the horizontal configuration, the liquid level will only be 2½ feet below the vapor outlet. And that's a big problem, too.

Still, at the Amoco Refinery in Yorktown, Virginia, in 1978, we substituted a horizontal KO drum for a vertical vessel, to stop entrainment in a heavy gas oil-hydrogen flash drum with excellent results. (See my book, *Process Engineering for a Small Planet*, Chapter 16, for a complete description of this nasty incident.)

Defoaming Agents

The best way to stop separator carryover problems due to foaming is to remove the agent that causes the foam. In one unit that suffered foaming due to iron sulfide particulates (i.e., $Fe[HS]_2$), I stopped the upstream corrosion that was creating the particulates. The corrosion mechanism was:

$$H_2 + Cl_2 = 2HCl$$

$$2HCl + Fe = FeCl_2 + H_2$$

$$FeCl_2 + 2H_2S = Fe(HS)_2 + 2HCl$$

I did this by reactivating an idle NaOH scrubber that was supposed to remove the chlorides from a naphtha reformer hydrogen effluent stream. Why the caustic scrubber was bypassed at the ARCO refinery in Cherry Point is a subject best forgotten as the guilty parties are now dead, retired, or BP employees.

On closed-loop systems such as a refrigeration or an amine system, filtration is a good method to remove particulates. Most of my clients will use a cartridge-type filter, which is inexpensive and easy to operate. But then, what does one do with all the spent cartridges?

A rotary precoat filter is expensive and complex. But it's really efficient, and there are no old, dirty cartridges to contend with. Plate-and-frame–type paper filters are very efficient, and are also environmentally friendly.

However, after 40 years of filter selection, I find that our industry will use cartridge filters because they have the least-cost up-front investment and require a minimum of operator involvement.

Some surfactants, such as heavy aromatics in aqueous streams, can also be removed by filtration. In this case, the filter is typically an activated charcoal–type filter. Such filter media are only surface active and have very little capacity to absorb the heavy aromatics. Also, if the charcoal escapes into the circulating aqueous phase, foaming and carryover will result due to the particles of carbon.

Antifoam agents are in widespread use to suppress carryover of liquids from separation vessels. Five or 10 ppm of a silicon defoamer is used in many hydrocarbon processes. For example, at the Suncor tar sands operation in Alberta, diluent naphtha and hot water are used to extract the tar from the sand. The naphtha is then flashed out of the heated tar. But the sand promotes foaming and carryover. So the silicon defoamer is employed to suppress carryover. But any silicon that is not recycled with the naphtha eventually deactivates the downstream hydrotreater catalyst. More seriously, silicon defoaming agents are used in:

- Crude preflash drums
- Delayed coker drums

Both have serious consequences for downstream hydroprocessing units. My point is that when injecting defoaming additive agents to suppress carryover, consider what will be the final resting place of the additive.

A vanadium-based antifoam additive is often used to suppress foaming and carryover in circulating amine system flash drums. But the additive degrades with time to a foaming agent. Then, to combat the resulting foaming, more of the additive is employed. This is okay, if your charcoal or carbon filter is up to par. But its absorptive capacity will quickly be exhausted by the continuously degrading Va additive. So you'll then add more defoaming agent, and then. . .

Plugged Level Taps

False liquid level indication is the most common cause of knock-out drums malfunctioning. As the level rises to the vapor inlet, liquid will be re-entrained. Plugged level connections are the usual cause of this malfunction. Often, the plugged connections are only ½-inch ID In potentially fouling services, I use 1-inch level taps. Some of my clients use 2 inches. But perhaps that's a little too cautious. Also, the lower tap should be 6 inches above the bottom vessel tangent line, and not on the bottom head itself (see Figure 19-2).

The only sure way of preventing level connections from plugging is by continuous purging. The top tap should be purged with natural gas, nitrogen, or clean refinery fuel gas. The bottom tap should be purged with clean process fluid, or water in aqueous systems. The purge fluid rate is best adjusted by a permanent **restriction orifice**. The diameter of the orifice should be roughly 25% of the diameter of the level tap connection. Too big a restriction orifice will distort the level measurement. Too small an orifice will not protect the taps from plugging.

Effect of Pressure and Temperature

One of the surest ways to suppress excessive entrainment is to increase the drum pressure. A 10% increase in the absolute operating pressure (psia) will decrease the vapor vertical velocity by 10%. However, vapor density will increase by 5%. Thus the effective entrainment velocity:

$$(\text{Vapor Flow}) \times (\text{Vapor Density})^{\frac{1}{2}}$$

will decrease by about 5%.

Of greater significance is that a gradual increase in pressure suppresses foam formation, and thus entrainment.

On the other hand, a rapid drop in operating pressure will promote a drum foam-over in many services. I have in mind refinery crude preflash drums. These drums often produce a high foam front due to the flow improvers added to the crude. These flow improver chemicals are added to crude oil to reduce frictional losses and pressure drop in pipelines. At 300°F to 400°F, they will promote foaming and entrainment, as the naphtha fractionation flashes out of the crude charge.

To stop the entrainment, an online colorimeter analyzer is used to continuously monitor the color of the naphtha. Entrainment will turn the clear naphtha brown. As the naphtha color degrades, the online colorimeter will automatically raise the flash drum pressure by partially closing a pressure control valve on the flash drum off-gas line. Later, the console operator slowly, and carefully, using the naphtha colorimeter as a guide, manually lowers the flash drum pressure. The usual cause of the foam-over is free water entering the flash drum, or a small reduction in flash drum pressure.

I've used the expression "small change in pressure" in the context of about 2% pressure change in 5 minutes. This is, in a 35 psig drum, a 1 psi change in 5 minutes is small. A 1 psi change in 1 minute would be large.

Increasing a drum temperature increases the volume of vapor and hence the flash drum's vertical velocity. The higher temperature also increases the tendency of many systems to form a higher foam level and thus also promotes entrainment. In Chapter 21, I describe an effective way to mechanically suppress carryover from a crude preflash tower.

Effect of Surface Tension

High surface tension suppresses the formation of foam. Many very light hydrocarbon systems (i.e., methane, ethane), which have low surface tensions, tend to foam. That's also why particulates promote entrainment. Particulates such as corrosion products reduce surface tension. So in general, vapor–liquid separators in a low surface tension service would tend to require a lower vertical vapor velocity than a higher surface tension system.

On the other hand, droplets of entrained liquid can more easily coalesce into larger, heavier droplets if they have a lower surface tension. The heavier the droplet, the more readily the gravity can force the droplet to fall back into the bottom of the knock-out drum.

In order to gauge the effect of different process variables on entrainment, we ought to have a method to quantify liquid entrainment rates.

Measuring Entrainment Rates

Let's assume I have a fuel gas KO drum operating at 50 psig and 120°F. The vapor phase is mostly methane. The liquid phase is mostly pentane. Your boss wants you to justify the purchase of a larger separator to remove the entrained pentanes from the fuel gas. So you'll have to measure the rate of entrained liquid. You already know the gas rate, but what is the weight percent of liquid hydrocarbon in the flowing gas?

Step 1—Vent a small amount of the gas through a 1-inch bleeder to the atmosphere. However, install about a ⅛-inch restriction orifice in the vent line. The restriction orifice will prevent the gas from converting a substantial amount of enthalpy into kinetic energy in the downstream 1-inch line.

Step 2—Measure the flowing temperature of the gas as it exhausts from the 1-inch line. Let's assume it's 90°F.

Step 3—The specific heat of the gas is 0.50 Btu/lb/°F. Thus each pound of gas has lost 0.5 (120°F – 90°F) = 15 Btu/lb.

Step 4—The latent heat of pentane liquid flashing to pentane vapor is 150 Btu/lb.

Step 5—One should correct the delta T of the gas noted in Step 3 for cooling due to the gas's Joule–Thompson expansion. I will neglect this factor.

Step 6—Divide the sensible heat reduction of the gas (i.e., 15 Btu/lb) by the latent heat of vaporization of the gas (i.e., 150 Btu/lb), to obtain 10 wt% liquid entrainment.

This is pretty similar to the technique I've described in Chapter 15, "Steam Quality Problems," to control BFW levels in kettle-type waste heat boilers. It's just an application of the standard laboratory instrument described in front of your steam tables, the **throttling calorimeter**.

The main caution about this measurement I've just described is to make sure that the velocity of the expanded fuel gas does not increase by more than 30 or 40 ft/sec as it's expanded from 50 psig down to atmospheric pressure. Otherwise, a large part of the observed temperature reduction may be due to the conversion of enthalpy into kinetic energy, rather than the conversion of sensible heat (temperature) into latent heat of vaporization of the entrained butane liquid.

Vapor Separation from a High-Viscosity Liquid

I've been working on a solvent de-asphalting unit this week. The solvent is pentane. The pentane extracts heavy gas oil from a crude unit vacuum tower residue stream. The asphalt from the bottom of this extractor has about 20% dissolved pentanes. When the extractor bottoms are heated and flashed into a lower-pressure vessel, the pentane is vaporized to leave the heavy asphaltines behind as a highly viscous liquid. The viscosity of the asphaltines is about 3,000 centistokes. At such an extremely high viscosity, the bubbles of vapor are trapped as a foam or froth in the liquid phase. We know this is the case because it's very difficult to pump the flash drum bottoms. There are three ways that we have found to be effective to assist in driving out the pentane bubbles trapped in the asphaltine liquid:

- Time—Raising the liquid level to a maximum has proved marginally beneficial. However, even at a liquid retention time of 20 minutes, the foam is not totally eliminated from the flash drum bottoms.
- Antifoam—Injection of a silicon-based antifoam chemical at a concentration of 5 parts per million (ppm) has helped to a somewhat greater degree than increasing the liquid level. This is expensive and far from totally effective. The silicon works by reducing the surface tension of the liquid phase.
- Temperature—Raising the drum operating temperature by 45°F has proven to be, by far, the most important and effective way, as indicated by the bottoms pump performance, of reducing the aeration of the viscous liquid asphaltines. I've calculated, based on published viscosity versus temperature curves for paving asphalt, that an increase of 45°F, from 510°F to 555°F, will cut the flash drum viscosity in half to about 1,500 centistokes (i.e., 6,000 SSU).

Unfortunately, the feed heater to the flash drum is also subject to rapid rates of fouling, and it cannot be cleaned onstream. So while the higher temperature has certainly proven to be an effective technical solution to this malfunction, from the operational perspective it's not an answer to this difficult phase separation problem.

CHAPTER 20

Knock-Out Drums; Demisters and Impingement Plates

Walk more; see more.

—Senior South African shift foreman

A demister is a mesh pad placed above a vapor–liquid inlet in a knock-out drum, or at the outlet of a horizontal separator. It looks rather like a giant Brillo pad with coarse wires. The typical demister is 4 to 8 inches thick. Its purpose is to promote small droplets of liquid in the entrained gas to coalesce on the surface of the demister's wires. The large droplets of liquid will then drop down, out of the vapor phase into the bottom of the drum.

Demisters plug with both fouling and corrosion deposits. This chapter's main theme is that a partially plugged demister is far worse than no demister at all. Demisters should never be used unless they are actually needed to aid in de-entrainment of high-velocity vapors. Use demisters with care.

Demister Malfunctions

I had been working for Texaco in Convent, Louisiana, on a visbreaker unit malfunction. Entrained black tar components were contaminating the previously clean gas oil product. My entrainment velocity calculation indicated a reasonably low vapor velocity. My client had been shutting down this tower every 6 to 9 months to replace the fouled demister for 20 years. As the visbreaker was not needed to run the rest of the refinery, and was barely profitable, this was not a problem that greatly troubled plant management. The reason that the visbreaker was not particularly profitable was that normally its main

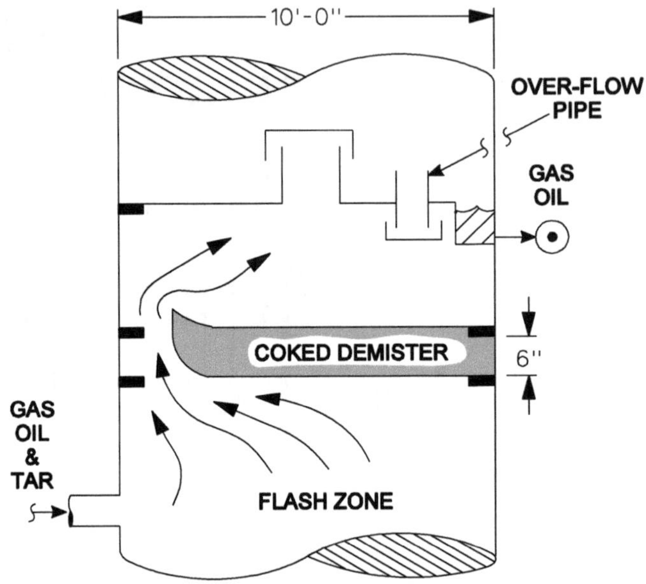

FIGURE 20-1 Completely fouled demister has torn away from vessel wall and promoted entrainment of tar into the gas oil product.

product, gas oil, was contaminated with black tar, thus degrading its product value.

Immediately after the demister was replaced (see Figure 20-1), gas oil quality was excellent. But within a week or two, the gas oil would gradually darken and turn black. Within 6 to 9 months, the gas oil product was so badly contaminated with entrained tar that a shutdown was taken. It had happened so often that the renewal of the demister had assumed the status of routine maintenance.

I recall inspecting the demister with my wife Liz one afternoon. "Norm! The demister is plugged solid with hard coke. I can't imagine how even one molecule of gas oil vapor could escape from the fractionator's flash zone."

"Look Liz, at the upper, left-hand side of the demister. See how the demister is turned away from its support ring. That's the path the vapor followed."

"But that's just 1% of the demister's area," Liz said. "That would cause super-high localized vapor velocity. But how did that hole develop?"

"Well Liz, as the demister coked-off and plugged, the pressure drop of the vapor through the demister became extremely high. Eventually, the differential pressure caused the demister to pull away from its support ring. The vapor then rushed up through the small hole. Then, the rest of the demister just gradually plugged solid with coke."

"Yes. And that localized high velocity promoted entrainment of the tar droplets into the gas oil product. That's all clear enough. But how are we going to stop the demister from coking and plugging?" asked Liz.

"Well, dear, we're not," I answered. "Not in visbreaker service. At Texaco's operating flash zone temperature, the demister is going to coke no matter what's done. But I've calculated that the C factor, or entrainment velocity, is only 0.10 ft/sec:

$$C = V \times (PV \div PL)^{\frac{1}{2}} \tag{20.1}$$

where PV, PL = Density of vapor and liquid
V = Superficial vapor velocity, ft/sec

Liz, not only is the C factor low, but the vertical distance between the flash zone inlet and the gas oil draw-off chimney tray exceeds the tower diameter. Anytime the C factor is less than about 0.12 ft/sec, and nozzle entry velocities are low (less than 50 or 60 ft/sec) and there's a large vertical separation between the inlet and draw-off nozzles, gravity settling is usually adequate to suppress entrainment."

"And so, Norm, what you're planning to do is leave the demister out entirely and rely only on gravity settling. Is that right?"

"Just so, Liz."

And that's what we did. Texaco produced clean gas oil when they started back up. Happily, the clean gas oil production persisted for many months of visbreaker operation.

From this incident, we can draw several conclusions:

- A partially plugged demister is worse than no demister.
- Gravity settling is adequate for reasonable de-entrainment if the entrainment velocity, as calculated above, is low. Also, a reasonable vertical height must be available for the gravity settling to take place.
- Longstanding process malfunctions can be eliminated by simple field observations, combined with elementary analysis.

I just taught a seminar in South Africa last month. An older shift foreman became quite angry when I even suggested that demisters could be useful in certain services.

"Demisters are the Devil's work. God would never have created such an evil instrument."

Catacarb CO_2 Absorber Carryover

Half the pure hydrogen produced in the world relies on the absorption of CO_2 from the product hydrogen. Residual CO_2 is typically converted back into methane, in the methanator reactor, which wastefully consumes the hydrogen product. The most common absorption solution is a potassium carbonate salt, and the most common technology employed is the circulation of a "catacarb" solution.

In Aruba, the Coastal refinery had a large hydrogen plant limited by carryover of catacarb solution. Meaning, as the unit throughput increased, the catacarb solution losses increased exponentially. As hydrogen availability was the limiting refinery operating factor, this was a critical matter for Mr. English, the refinery manager.

The extreme nature of the entrainment problem seemed strange to me. The problem was that the C factor, or entrainment velocity (see Equation [20.1]), as defined previously, was only 0.08 ft/sec. I would not have expected very much entrainment at a calculated entrainment velocity much below 0.15 to 0.20 ft/sec. However, to add to my confusion, this tower had a demister pad above the top packed bed. With a demister, I would expect no significant entrainment below 0.24 to 0.28 ft/sec entrainment velocity.

Something must be very wrong with the demister. Rather like the Texaco demister story in the previous section of this chapter. But the demister had been inspected many times with no obvious fouling or damage. The top packed bed of the absorber was clean and intact. During the previous plant turnaround, the demister had been renewed at Mr. English's instructions—all to no avail.

So I decided to inspect the demister myself. I climbed through the top vessel manway. Figure 20-2 shows pretty much to scale what I observed. The demister only occupied about 15% of the tower cross-sectional area. The cross-sectional area of the tower devoted to the demister was over 20%. But the rather wide structural supports underneath the demister did indeed reduce the demister open area to 15%. I have encountered this type of malfunction many times in my career as a process design engineer. I will size a certain area of a vessel for vapor or liquid flow. Then a mechanical engineer consumes 25%

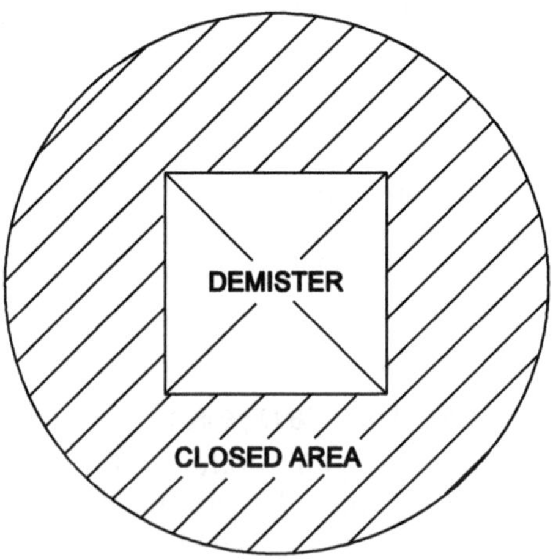

Figure 20-2 Too small a demister area increases rather than decreases entrainment.

of my open area with overly massive structural support members. The lesson is to sign off on the final fabrication drawings before fabrication begins. Sometimes, I'm too lazy for such efforts; the alternative is the final inspection of the installation prior to startup.

In the Aruba catacarb absorber tower, if the entrainment velocity based on the tower open area was 0.08 ft/sec, and the open area of the demister was 15%, then the entrainment velocity of the absorber off-gas through the demister was:

$$C = \frac{0.08}{0.15} = 0.53\text{ft/sec}$$

This is about double a maximum demister design velocity. In other services, I have observed that a C factor above 0.40 to 0.42 will result in severe entrainment, regardless of:

- The vertical space between the vapor inlet, and the collector tray above.
- The velocity of the vapor through the inlet nozzle.
- The type of de-entrainment device used: vapor horns, distribution tubes, demisters, etc.

My next step was to phone Mr. Lieu, the catacarb process expert in their Illinois headquarters.

"Mr. Lieu, this is Norm Lieberman calling from Aruba."

"Mr. Who?" asked Mr. Lieu.

"Mr. Lieu, why is your demister only 15% of the tower cross-sectional area? I've calculated too great an entrainment velocity."

"It's our design standard, Mr. Loberman."

"Lieberman. But why?" I asked.

"It's optimized. Very sorry. But I'm very, very busy. Call back next week, Mr. Lobsterman," said Mr. Lieu.

"It's Lieberman. But how is it optimized?" I asked again.

"I just told you. It's our standard." And then, *click*.

So I told Mr. English that I had discussed my proposed change to their demister with the Catacarb Engineering Division. And after a detailed review, the Catacarb engineer and I had agreed to increase the demister open area from 15% to 35%. This would reduce the entrainment velocity from 0.53 to 0.23 ft/sec. When the hydrogen plant was re-streamed, the limitation of catacarb solution losses was gone. Mr. English later told me this was one of the single best process changes I had ever initiated in Aruba.

Demister Suppresses Overspray

I always use Bete full-cone spray nozzles. But I've observed that when the delta P exceeds 40 or 50 psi through the nozzle (for water), the nozzle begins to generate an overspray or a mist. For a delta P of

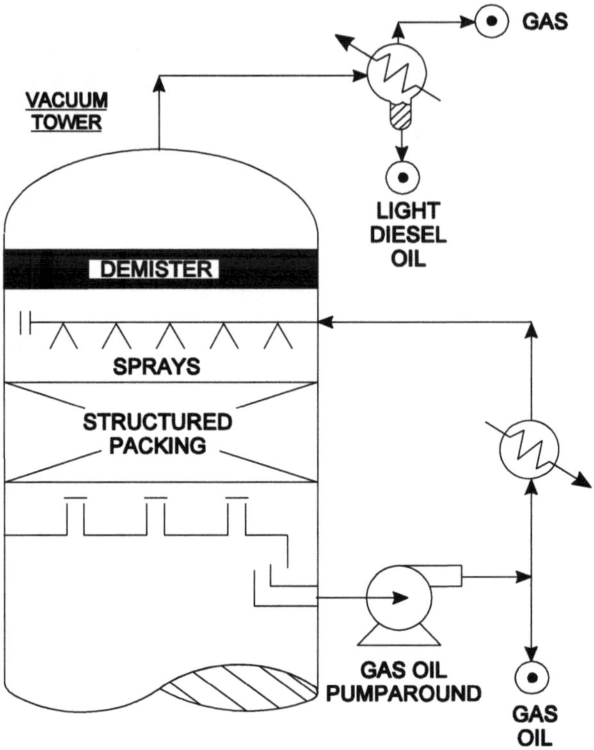

FIGURE 20-3 Addition of demister improves color of diesel product.

100 psi for water (or 70 psi for a hydrocarbon of 0.70 SG), the overspray represents a significant portion of the spray nozzle effluent flow rate.

At the now-defunct Pacific Refinery north of San Francisco, they produced a light diesel oil product from the overhead of their vacuum tower, as shown in Figure 20-3. At higher feed rates, the diesel product had a high end point because the tower-top temperature was excessive. So I increased the top gas oil pumparound to suppress the tower-top temperature, with the objective of condensing out the heavier components in the vapor flowing to the light diesel product condenser.

Before this increase in pumparound rate, the **virgin*** (i.e., without cracked components) light diesel was water white, and the gas oil was clear, but quite yellow. After I increased the pumparound rate, a sample of light diesel indicated:

- A reduction in end point—which was good.
- A definite yellow color—which was bad.

---

* *Words in bold are defined in the Glossary, at the end of this text.*

The measured delta P across the spray header had increased from about 30 to 50 psi, due to the 30% increase in gas oil pumparound circulation.

During the next unit turnaround, the demister shown in Figure 20-3 was added. The demister was 4 inches thick and constructed of 317 s.s. mesh to resist naphthenic acid corrosion and fouling due to corrosion. After the unit was restreamed, the gas oil pumparound rate could be varied without any effect on the color of the vacuum tower overhead diesel oil product.

On the other hand, I was working at the Texaco refinery in Convent, Louisiana, on a very similar tower. The problem with this vacuum tower had been high tower-top pressure. I write "had been" because a few days before my arrival, the problem had quite suddenly, without human intervention, disappeared. The operator who was conducting me through the unit explained his theory as to the unexpected improvement.

"Mr. Norm, this all happened before. The pressure slowly builds at the top of our vacuum tower below the demister. Builds up over a year or two. Then, it suddenly releases. Drops like 10 to 15 mm Hg in minutes. Same thing happened two years ago. Later, during the turnaround, we opened the top of the vacuum tower. The top demister was blown out. It was all eaten up with corrosion. Wrong type of material, I reckon."

The overheads of many vacuum towers are subject partly to naphthenic acid attack, and also to hydrochloric acid. The HCl originates from the hydrolysis, or the thermal decomposition of $MgCl_2$ and $CaCl_2$ (i.e., seawater salts) at about 700°F:

$$MgCl_2 + 2H_2O = Mg(OH)_2 + 2HCl$$

HCN, CO_2, NH_3, O_2, H_2O, and lots of H_2S are also commonly present. Selecting the proper metallurgy for your demister should be done with care, depending on the corrosive nature of the overhead vapor stream.

Demister in Sulfuric Acid Service

Demisters work by causing finely divided liquid droplets to coalesce into bigger droplets, which are too heavy to be lifted by the vertical vapor flow. The finer the droplets, the more dense the mist, and the more important the demister.

It seems to me that polar compounds like H_2O or H_2SO_4 would tend to form a finer, more opaque mist than gasoline or diesel. One thing for sure, the vent tail gas from my sulfuric acid regeneration plant in Texas City produced the most awful white plume. One day my first wife phoned me at the plant from the mall in Texas City.

"Norman! Are you trying to kill us? Your white plume is dropping right down on the parking lot."

The tail gas that created the white acid plume was vented from the top of our final absorber tower. I had been told by several acid plant experts that the demister at the top of the absorber should largely eliminate the plume. So I inspected the demister every time we shut down the plant. Looking down from the absorber's top manway, it seemed intact and clean. But the demister just didn't seem to work.

After one short outage of our plant, Joe Hensley, my inspector, dropped by my office. "Norm, you know that your absorber demister was loose."

"Joe, I just looked at it yesterday. It was right in place and looked fine."

"No, Norm. It sure was loose on one end. I got underneath it through the side manway and pushed up on it. One side's loose and the other side's tight. So, when you got that gas flow a-going, it pushes open like an old screen door in the wind. But Norm, I done fixed it. I wired up the loose side with No. 9 wire. That demister ain't going nowhere."

And Joe was right. A piece of No. 9 wire stopped the acid plume almost entirely. Demisters in the right service, made of the correct metallurgy, and sized for a proper entrainment velocity, can be extremely effective in suppressing droplet entrainment and mist carryover.

Improper Use of Demisters

In the last 50 years, there's been hardly any grassroots refineries built in North America. The Come-by-Chance Refinery in Newfoundland being one. They have suffered repeated bankruptcies. Maybe this story will explain why.

I was hired to revamp the tower shown in Figure 20-4. The problem was repeated coking of the demister. Every time the tower was opened, the demister was found to be badly coked and had to be replaced, which was quite expensive.

Note that both the black overflash product and the heavy gas oil product are going to the same fuel oil tank. What then was my client trying to accomplish with the demister?

They explained that when the refinery was built 20 years ago, the two products then went to different tanks. But there was never any market for the heavy gas oil in Newfoundland. Thus, the heavy gas oil product was always blended into the fuel oil tank.

So I, your clever author, had them discard the demister and instructed my client not to replace it. And for this, people pay me $2,400 per day. No wonder so many of my former students have become process consultants, and now head their own sizeable engineering companies and have now become too important to return my phone calls.

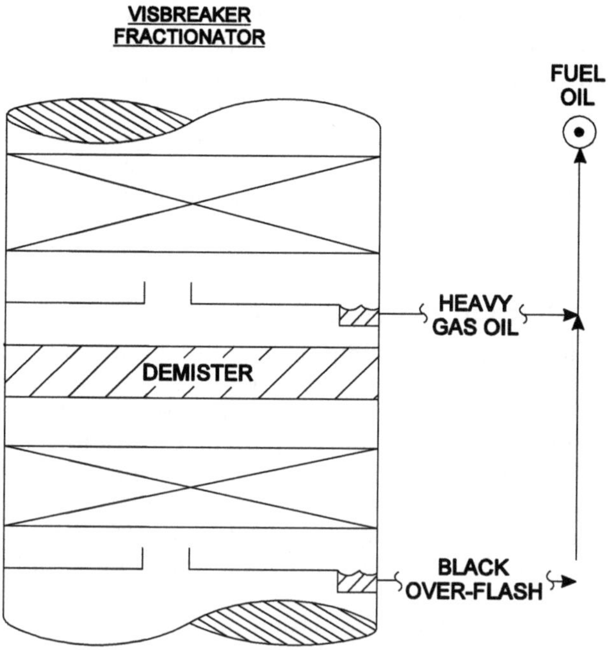

Figure 20-4 A useless demister service.

Demister Malfunction in Compressor Suction

One of the reasons I dislike demisters is because of an unpleasant incident I had at the Amoco refinery in Texas City on a refrigeration compressor. A demister in an intermediate knock-out drum came apart and was drawn piecemeal into the suction of the compressor, resulting in a premature shut-down.

All process equipment is naturally subject to failure. But in this case, I felt that the demister was not actually required in the first place. A moderate amount of entrainment into the suction of a centrifugal compressor does very little harm unless the entrainment is carrying salts or hardness deposits into the rotor. It may even prevent fouling deposits from accumulating on the rotor wheels.

This was not a closed-loop refrigeration system and was indeed a moderate fouling service. Any fouling service should be viewed as likely not to be suitable for demisters. A partially fouled demister increases, and does not decrease, entrainment. Also, it clearly runs the risk of creating downstream failures and outages.

Internal Vortex Breaker Tubes

I've been teaching a seminar this week at the Marathon refinery in Robinson, Illinois. Twenty years ago, I designed a very large crude flash drum (500°F, 120 psig) to flash off most of the naphtha in the desalted crude feed. The resulting naphtha vapors were contaminated with a few tenths of a percent of entrained crude, which slightly discolored the naphtha product. To remove this small amount of entrained crude, an "internal vortex breaker" six-tube cluster, manufactured by EGS Systems, Inc. (Houston), was installed at the feed inlet of this 15-foot ID horizontal flash drum. The results have been a naphtha product totally free of entrainment.

Impingement Plate in Knock-Out Drums

Placing a feed deflector or impingement plate in front of an inlet nozzle is a reasonable way to reduce localized high pressures. As I discussed in Chapter 1, a localized high pressure in the region of the bottom tray's downcomer seal pan can cause a tower to flood. However, as far as entrainment goes, inlet impingement plates promote, rather than reduce, entrainment. This is especially true if the edges of the impingement plate are not smooth. Rough edges promote fine droplet formation and thus liquid droplet entrainment.

Probably the best way to design an inlet for a minimum of entrainment is to use a tangential entry with a vapor horn. The vapor horn is a curved baffle that restricts the inlet flow to a curved path that hugs the vessel inside diameter for about 90°.

For typical vapor densities of between 0.2 and 1.0 lb/ft^3, a nozzle inlet velocity of 30 ft/sec (for the 1.0 vapor density) to 60 to 70 ft/sec (for the 0.2 vapor density) will not cause very much entrainment, and an ordinary radial feed entry is adequate. If entrainment is a problem, then a more practical solution to the problem than converting to a tangential entry is a demister or internal vane separator.

I have read in the literature (see the Robichaux reference that follows) that having an inlet velocity of about 15 ft/sec for a vapor density of about 1 lb/ft^3 will, by itself, minimize entrainment. While I'm sure this is true, no piping designer will be likely to oversize process lines to this extent, so we are rather committed, because of external piping sizing, to the velocities I have noted above.

I have a lot of experience with vapor horns and tangential entries, especially in vacuum separators. They work extremely well. But if the vapor horn inlet is placed so that it restricts the inlet flow, localized entrainment in the region of the inlet nozzle will create more carryover than without any vapor horn. I also have a lot of experience with demisters. They also work extremely well. But if the demister partially plugs, then vapor flow will be channeled and higher local velocities will occur with a consequent greater rate of overall liquid carryover.

Fog

I've extracted this section from a colleague's book, *Gas–Liquid Separation*, by R. Robichaux. Fog consists of droplets of liquid too tiny to settle. It is formed when the effluents (A & B) from two coolers mix:

- Effluent A is a warm vapor that is partially saturated with a condensable component.
- Effluent B is a colder vapor.

When A and B mix together, a fog is formed. If stream A had been cooled separately in a heat exchanger, larger droplets of liquid would have been formed at the cold metal heat exchanger surface.

Droplet Properties Affect Settling Rate

Droplets of very light liquid hydrocarbons, like ethane, have a tendency to entrain. But droplets of very heavy cold hydrocarbons also have a tendency to entrain. Why is that?

The lighter droplets entrain because of low surface tension. Low-surface-tension liquids produce tiny droplets, which have trouble making big droplets. When they do make big droplets, the velocity of the vapor easily shears the big droplets back into baby droplets. According to Stokes' Law, the settling rate of droplets is inversely proportional to their size.

The droplets of cold, heavy hydrocarbons have a tendency to entrain because of their higher viscosity. According to Stokes' Law, the settling rate of a droplet is inversely proportional to the droplet's viscosity.

Anything then that reduces the surface tension of a droplet, or increases its viscosity, will promote carryover. For light hydrocarbons, surface tension is reduced by particulates or dirt. For heavier hydrocarbons like heavy vacuum gas oil, viscosity is increased by a colder separator temperature. Vessels in such services should be designed for lower vertical velocities, and also for larger vertical spacing above the feed entry nozzle.

CHAPTER 21

Hydrocarbon–Water Separation—Electric Desalters and Gravity Settling

Just do it.

—Michael Jordan

The separation between hydrocarbons and an aqueous phase can be accelerated by applying an electric field across the two liquid phases. Let's assume the dispersed phase is brine and the continuous phase is crude oil. The brine or salt water is generated when wash water is injected into the crude oil with the objective of extracting salts from the crude supply. Applying an electric current across the crude oil–brine two-phase liquid will cause the more polar water droplets to coalesce. The bigger droplets will settle out far faster than the smaller droplets, all in accordance with **Stokes' law**. The type of device I just described is called a desalter.

Crude Oil Desalter Malfunctions

I just returned from a troubleshooting assignment at the Hunt refinery in Tuscaloosa, Alabama—the home of the Crimson Tide football team of the University of Alabama. Hunt's problem was desalter carryover. That means the water phase in the desalter would suddenly carry

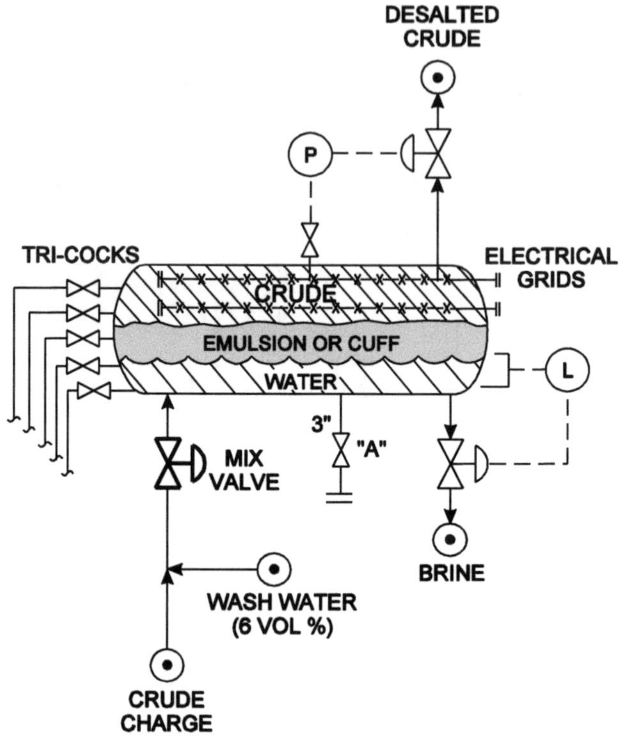

FIGURE 21-1 Refinery raw crude electric desalter.

over into the hydrocarbon phase. Figure 21-1 shows a sketch of their desalter. Note that I've shown three phases in the vessel:

- Water or brine
- Emulsion or cuff
- Crude or hydrocarbon

Most of the volume of the desalter ought to be devoted to the hydrocarbon phase. A reasonable residence time for the crude oil is 10 to 15 minutes. For design I've typically used 20 minutes.

The height of the water phase is typically about 2 feet. However, it's the height of the emulsion or cuff that is so critical. A small cuff height is 4 inches. A big cuff height is 20 inches. If the cuff height becomes excessive, some bad things will happen:

- As the emulsion expands up to the elevation of the lower grid, the grid voltage will decline, the grid amperage will increase, and eventually the grid will start to short out.
- The emulsion will start to carry over out of the desalter. Water will escape into the downstream process equipment. As the water flashes to steam, the evolved steam will upset the unit's stability. Hydrocarbon flow to a heater pass may be lost.

Measuring Emulsion Levels

Referring to the **tri-cocks** shown in Figure 21-1, we measure the height of the emulsion by opening the tri-cock valves in series. A clean strip of stainless steel is used to check for oil, emulsion, or water. With practice, dripping the effluent from the tri-cock valves onto the steel's surface will allow the operator to quickly determine the phase at any elevation. Typically there may be several tri-cocks, set 1 foot apart. Some desalters have a "swing-arm" inside the desalter that permits sampling of the liquid in the desalter at elevations an inch apart. We had this facility at the Hunt refinery, which made it far easier to track changes in the emulsion layer height.

Causes and Cures of Emulsion Layer Growth

At the Hunt refinery, I found that the cause of the increase in the height of their desalter cuff was related to a number of factors, which I tried to correct in a two-day period. Since I applied all the potential solutions pretty much simultaneously, I really cannot quantify their relative importance.

- *Temperature*—Their desalter had been operating at 230°F to 240°F. By optimizing the crude preheat exchanger train heat balance, I increased the desalter temperature to 260°F to 270°F. Increasing temperature by about 50°F cuts viscosity in half. According to Stokes' law, settling rates are inversely proportional to viscosity of the continuous phase. Thus, increasing temperature by 25°F to 30°F in the desalter should have speeded water settling rates by about 40% to 50%. Also, increasing the temperature by 25°F to 30°F reduced the hydrocarbon density from about 0.83 SG to about 0.82 SG Again, according to Stokes' law, settling rates are proportional to the square root of density difference. The density of water doesn't change much with temperature and is about 0.95 SG (at 250°F), and 1.00 at 60°F. Thus reduction of the crude oil density from 0.83 SG to 0.82 SG by the increased temperature, increases the density difference from:

$$(0.95\ \text{SG}) - (0.83\ \text{SG}) = 0.12\ \text{SG delta}$$

  to about

$$(0.95\ \text{SG}) - (0.82\ \text{SG}) = 0.13\ \text{SG delta}$$

- As settling rate is proportional to the square root of this difference:

$$(0.13 \div 0.12)^{\frac{1}{2}} = 1.04$$

Which means settling rates are enhanced by about 4%, due to the reduced hydrocarbon specific gravity. This smaller SG effect is added to the more important reduction in hydrocarbon viscosity.

Actually, I increased the desalter temperature at one point all the way from 231°F to 279°F. My client was initially concerned about the increasing amp load on the desalter grids. However, the higher amperage flow was not caused by any growth in the emulsion. Rather, it was a result of the increased solubility of water in the crude oil at a higher temperature. Increasing the temperature of the crude by 50°F can increase its water content by 1,000 ppm (0.1 wt%). This does not hurt the efficiency of the desalter, as dissolved water in crude does not increase the amount of salts in the crude. So even though the desalter voltage dropped by 20% and desalter amps increased by 20%, no arcing or shorting of the electrical grids was encountered.

- *Kerosene*—A second method I tried to improve desalter efficiency was to recycle part of the kerosene product back to the crude charge tank. This method also works based on reducing the crude specific gravity and viscosity. It also helps increase the desalter temperature. This happens because the heat of condensation of the recycled kerosene in the crude tower pumparound is used to preheat the crude charge upstream of the desalter. When I left the Hunt refinery, they were recycling, as a percent of crude charge, 5% to 7% of the product kerosene. The disadvantage of recycling kerosene to crude charge is that it reduces unit capacity by 5% to 7% and obviously wastes energy.
- *Sludge-busting steam*—One of the refinery's problems was that its tendency to carry over emulsion from its desalter was getting worse with time. It is likely that sludge, corrosion products, mud, or solids were accumulating in the bottom of the desalter vessel. Most desalters are equipped with "sludge-busting steam." That's a steam pipe that runs 2 feet above the bottom of the desalter. Holes at a 45° angle (pointing down) emit high-pressure steam. The idea is to stir up the mud that accumulates in the bottom of the vessel, and thus reduce liquid residence time. There was only one problem with my client's sludge-busting facilities. That problem was that they did not exist! So I did the next best thing. Referring to Figure 21-1, we connected a steam hose to nozzle A. Using high-pressure steam, we blew back through nozzle A. When we drained down the brine from the nozzle, a significant amount of sludgy water drained out. Probably, nozzle A was never intended for this purpose. But we can do some good by working with what we have.
- *Demulsifier chemical addition*—To accelerate the separation between crude oil and brine, 10 ppm of an oil-soluble

demulsifier chemical was added to the desalter upstream of the mix valve shown in Figure 21-1. Unlike many process additives used in refineries, chemical demulsifiers really do work. They noticeably speed oil–water separation, even without an imposed electrical field, and at low temperatures and relatively high viscosities. But, since the demulsifier is soluble in hydrocarbon and not in water, why not add the 10 ppm of chemical further upstream?

At the Hunt refinery in Alabama, half the crude was received from barges and the remainder via pipeline. The barge receipts had quite high (several percent) water content mixed in with the heavy (0.90 plus SG) crude. The barges offloaded to a dedicated tank. The crude was cold and viscous, and the water content of this tank settled very slowly before it was drained manually. Thus, as shown in Figure 21-2, the unsettled water was charged to the crude unit via a centrifugal pump.

One might think that such water would be drained out of the desalter shown in Figure 21-1. But this is largely not true. The problem is that once water and liquid hydrocarbons pass through a high head centrifugal pump, the resulting shear forces will form an extremely tight emulsion. This emulsion is sure to increase the height of the cuff in the desalter vessel and promote emulsion carryover from the desalter.

When the demulsifier chemical was added at a rate of about 10 ppm to the tank as the barges were unloaded, far more water settled out in the tank, prior to the crude being charged to the crude unit, via the centrifugal pump. As the

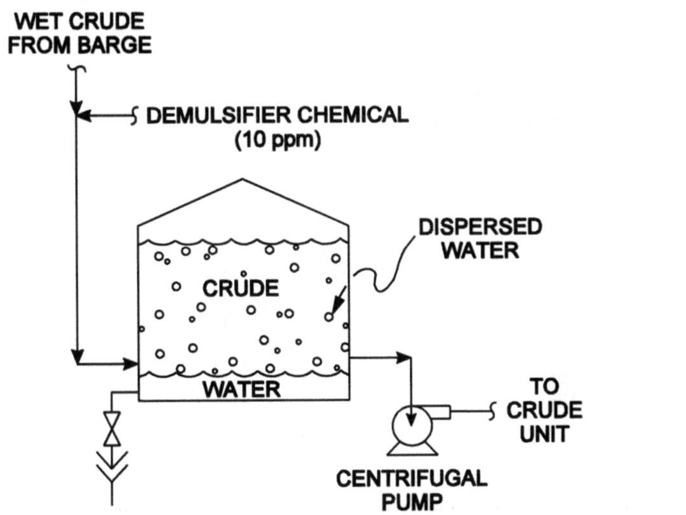

FIGURE 21-2 Passing water and hydrocarbons through a centrifugal pump creates an extremely tight emulsion.

demulsifier chemical is oil soluble, it was carried into the desalter, where it provided the same benefits as when it was added directly to the desalter crude charge.

The tank field operators did not particularly appreciate this innovation. They complained that it greatly increased the time required to manually drain the water heel from the bottom of the tank.

Incidentally, if the pump shown in Figure 21-2 had been a positive displacement pump, such as a reciprocating pump, a tight emulsion would not have been formed. At the Amoco refinery in Whiting, I had to pump a mixture of kerosene and water from a vacuum tower hot well or seal drum. If I used the centrifugal pump, the discharge flow looked like milk and would take an hour or more to separate. If I used the recip spare, the discharge flow would separate cleanly in 2 minutes or less.

The general lesson to be gleaned from this story is: do not put a two-phase liquid mixture through a centrifugal pump, unless downstream emulsion separation time is not considered to be a limitation.

- *Tank mixers*—I noted previously that some of the crude charge to the Hunt refinery came by pipeline. This crude also had an appreciable water content. I've shown the pipeline charge tank in Figure 21-3. I was puzzled when I reviewed the operation of this tank:
    - Why was the tank mixer running? Certainly, this would preclude water drainage from the tank.
    - Why was the oil-soluble demulsifier chemical not added to this tank? This would maximize water

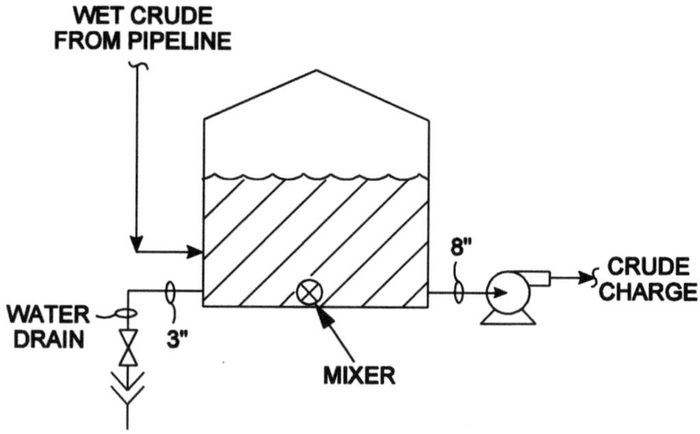

FIGURE 21-3 Having the water drain nozzle and the crude outlet nozzle at the same elevation forced the operators to use the tank mixer and precluded the use of demulsifier chemical in the tank.

extraction from the tank upstream of the centrifugal crude charge pump.

The design malfunction was that the 3-inch water draw-off nozzle and the 8-inch crude charge pump suction nozzle were both located at the same elevation. If a separate water phase did indeed accumulate at the bottom of the tank, a slug of water would periodically be charged to the crude unit. This would certainly destabilize the crude unit's operation. To prevent this from happening, the tank mixer was run continuously. Adding demulsifier chemical with the mixer running would then have served no purpose.

Obviously, the draw-off nozzle for water must be a few feet below the draw-off nozzle for crude. Which is, of course, what I recommended to my client. Okay, this is super-obvious! Okay, you probably feel that you're wasting your time reading this simplistic stuff. Yet, kindly note the following:

- My client looked at this mis-design for decades and never fixed it.
- Hunt Oil paid me $300 per hour to advise them to elevate the 8-inch nozzle 2½ feet above the 3-inch nozzle.
- They possibly failed to appreciate that once the crude was mixed with the water in the heel of the tank as it passed through the centrifugal pump, separation difficulties in the desalter would be magnified.

- *Effect of solids*—Particulates tend to stabilize emulsions. To enhance particulate settling rates, a water-soluble wetting agent is added directly to the desalter. This chemical helps the particulates to be extracted into the water or brine that is drained out of the bottom of the desalter. Why not add this chemical further upstream to aid solid extraction in the crude oil charge tanks? Well, because the water-soluble wetting agent would never make it into the desalter vessel. It would be lost to the sewer when the water was drained down from the tanks.

  However, by aiding water extraction in the crude charge tanks as follows:

  - Kerosene recycle
  - Demulsifier chemical addition
  - Shutting off tank mixers
  - Higher tank temperature

  We also increased the solids removed rate from crude charge, as many of the undesirable solids had already settled out in the water phase.

- *Tank temperature*—You will note that I listed higher tank temperature as one of the variables I employed to enhance settling rates. Hunt Oil had a very new and extremely large external steam heating circulating system on their barge receipt crude charge tank. Yet, this multi-million-dollar facility was not in service, even though everyone understood that warmer, less dense, less viscous crude would enhance water settling rates in crude oil tankage. I discussed this problem (see Figure 21-4) with the tank field operators.

  "See, Mr. Lieberman, the circulating pump will gas-up and slip. It's not operable. Bad design by dumb engineers," said Buckey, the tank field operator.

  "Buckey, could you show me?"

  "Okay, Mr. Lieberman. But that's a lot of work to line up the steam supply and the pump. It's gonna take me a while."

  "Don't worry, Buckey. I'll personally guarantee you that Hunt Oil will pay you for your time."

  Buckey opened the pump discharge valve a few turns, started steam flow into the circulating heater, and lined up the steam condensate drain to the sewer. And the pump ran just fine.

  "I don't see that the pump's cavitating. Seems like it's running just fine," I said.

  "Well, yeah. But that's 'cause I didn't open the discharge valve all the way," explained Buckey.

  "So," I asked, "Why not open it all the way and show me?"

  Buckey opened the discharge valve a few more turns and the pump still did not cavitate.

  "Well, Mr. Lieberman, that's 'cause the tank level ain't low. We got 20 feet of crude in the tank. See, the pump's suction pressure [at P1, Figure 21-4] is 7 psig. If we only got 3 psig on that there gauge, then the pump will slip."

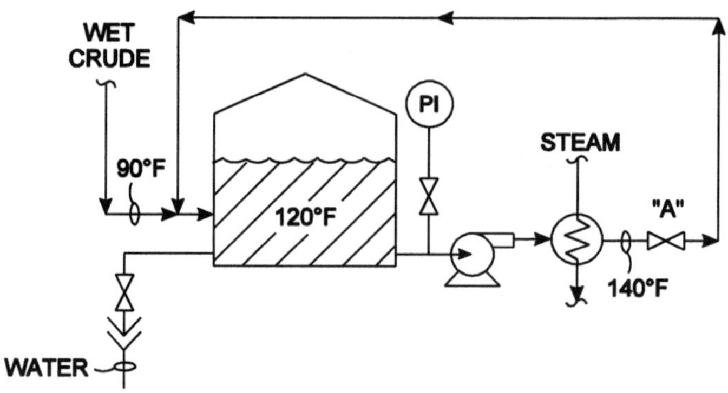

FIGURE 21-4 Too low a level in the charge tank causes the circulating pump to cavitate.

"But Buckey, don't you always try to keep at least 20 feet of crude charge in this tank?"

"Sure, Mr. Lieberman, but sometimes we don't."

"So then, when you only have 10 feet in the tank, throttle back on the pump discharge flow at valve A. Pumps need less suction head to suppress cavitation at a lower flow rate [see Chapter 29, "Centrifugal Pump NPSH Limitations"]."

"Yeah. I guess we could do all that stuff. But that's an awful lot of work," complained Buckey.

"But that's what you're paid for."

Buckey eyed me strangely, "Paid for what?"

"For work!"

So that circulating tank field heater was used to increase the tank temperature from 90°F to 120°F. Of course, at some tank level and some higher tank temperature, the pump would have cavitated due to lack of available NPSH. Likely, the pump was not purchased for a low enough required NPSH. But still, it was usable at most operating conditions.

In truth, the tank field operators (and everyone else) at Hunt Oil in Alabama were dedicated, intelligent, and serious people. The story about the tank field heater is true. The difficulty was that the tank field operators did not realize the true function of the circulating heater. The crude unit operators did not realize that their crude charge tanks (located a half mile away) were equipped with steam heaters. So often I see that apparent process malfunctions are really poor communications between different operating areas in the plant. My real contribution at the Hunt refinery was to get everyone in the same pick-up truck and drive around the plant together to look the problem over in the field.

- *Mix valve delta P*—Maintaining a mix valve delta P (see Figure 21-1) of about 15 psi is normal for crude unit desalters. The greater the mix valve delta P, the more efficiently the salt is extracted from the crude oil into the brine. But a larger mix valve delta P also causes a tighter emulsion between the oil and water phases. This increases the height of the cuff, and eventually a carryover of the emulsion layer from the top of the desalter. Of course, emulsion carryover will cause a step increase in the salt content of the effluent crude.

  Thus, as a final fall-back procedure, I instructed the operators to begin opening the mix valve as required, to keep the height of the emulsion cuff to less than 20 inches, with a minimum mix valve delta P of 2 psi. "There's no sense," I explained, "in holding an arbitrary mix valve pressure drop if it promotes brine carryover."

Many process malfunctions can be prevented, if we take the time to explain to the plant operators what it is that we are trying to accomplish, and how to do it. Try to avoid giving the operators arbitrary targets that they may not fundamentally understand. It's more difficult to explain to people what it is that you're trying to achieve and how to do it. It's easier to just issue instructions. But in a process plant, the road to death and disaster is paved with such arbitrary instructions.

Excessive Vapor Pressure

While working at the Coastal refinery in Aruba, I noted that they had neatly solved the tendency of their desalter to carry over water. They did this by completely eliminating the use of wash water. Of course, this also eliminated the whole purpose of the desalter. The problem in Aruba was that the desalter pressure was too low for the desalter operating temperature. If the vapor pressure of the wet crude oil exceeds the operating pressure of the desalter vessel, the contents of the desalter will start to boil. This stirs up the emulsion phase or cuff, and leads to water carryover with the effluent crude. I solved this malfunction quite easily by increasing the desalter set point pressure 20 psi above my calculated wet crude vapor pressure. I emphasize the term "wet," as one needs to take into account the moisture content of the crude (i.e., the partial pressure of water), in addition to the hydrocarbon composition of the crude itself, at the desalter inlet temperature.

Water–Oil Separation by Gravity

Let's assume I have steam and propane condensing together. At what rate have I observed the droplets of condensed water falling through the liquid propane phase? If both phases are clean, about 80 feet per hour. At 90 feet per hour, entrainment of water in the propane is observed. Actually, that's the only system for which I have any direct field observations. So the rest is just based on extrapolation using Stokes' law as follows:

Settling Rate is Proportional to the Square
Root of Density Difference

So for naphtha, which has a 0.70 SG, compared to propane (0.50 SG), I would extrapolate a maximum water (1.00 SG) settling rate of:

$$\left(\frac{80\ \text{ft}}{\text{hr}}\right) \times \left(\frac{1.00 - 0.70}{1.00 - 0.50}\right)^{\frac{1}{2}} = \frac{62\ \text{ft}}{\text{hr}}$$

That's a maximum sort of settling rate. At American Oil, our process engineering design settling rate for new water–oil separators (for low-viscosity service) was 25 feet per hour. But we often ran at much higher settling rates without incident.

Low viscosity means about 5 centistokes or less. Particulates, poor liquid feed distribution, vortexing, aggressive phase mixing, and surfactants will all reduce the settling rates between phases.

Therefore, before you start looking for a malfunction to explain water carryover from a settler, calculate the water settling rate first. It could be that there is no malfunction at all. It's just that you may have a situation where:

1. The vertical height through which the dispersed phase fluid has to travel is too tall, or
2. The residence time of the continuous phase is too short, or
3. The specific gravity difference between the two phases is too small.

Liquid Phase Interface Level Indication

Often our problem with water–hydrocarbon separation is one of measurement rather than control. Of course having a thick cuff, rag, or emulsion height between the two phases complicates the problem. When I design propane-amine liquid–liquid H_2S extractors, I'll have a dedicated draw-off nozzle at the expected interface level so that the operators may periodically draw off the rag layer.

One acceptable method to follow the interface level is with a conductivity probe. Water being far more conductive than hydrocarbon, the long probe measures electric conductivity every few inches. But for this text, we have to learn to work with the equipment at hand. So here's how to find an interface level in a gauge glass:

Let's refer to Figure 21-5. First, note that the true interface level may be between the gauge glasses. Second, you have to blow out the taps. This will clear any plugged taps and clean out the sludge in the gauge glasses themselves:

- Open A and C together to clean the top tap. B is closed.
- Open B and C together to clean the bottom-level tap. A is closed.

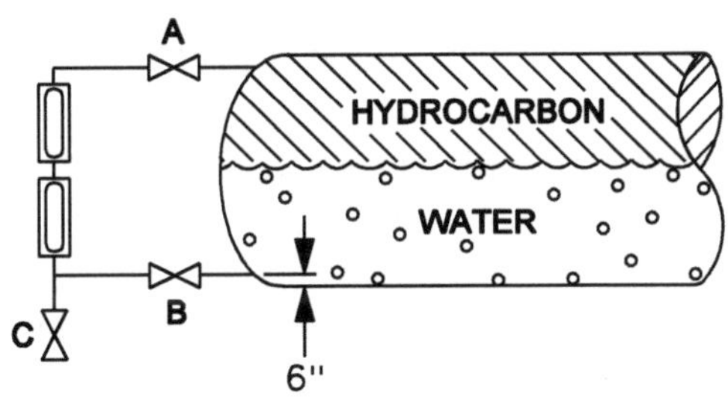

Figure 21-5 It's difficult to find an interface in a gauge glass when both phases are water white.

Next, with B shut and A open, open C the smallest possible amount. Hold your flashlight behind the gauge glass. If the water is blackish and/or the hydrocarbon is yellow, the flashlight is not required. But if you have clean virgin butane and clean water, observing the meniscus between the two clear phases is going to be a challenge even with a flashlight to illuminate their interface. In practice, the only way I can find the interface is to make it move past a beam of light. It's really easy to mistake a scratch or streak on the gauge glass surface for the interface, unless the interface is moving.

Of course, this moving interface in the glass is not the real vessel interface. So next open valve B a bit and close off C. How one does this while still moving your flashlight so as to continuously observe the interface is easier said than done. It's really a two-person job on a tall gauge glass assembly. Eventually, you should wind up with A and B fully open and C shut, with the interface either still visible or between the two gauge glasses.

For many services where we wish to control the hydrocarbon–water interface level automatically, the big problem is that the level taps plug with dirt. This is especially true for the lower-level tap in the water phase. Only too often, the bottom-level connection is located on the very bottom of the vessel. This is always wrong. Dirt will plug this tap. As shown in Figure 21-5, the bottom-level tap ought to be 6 inches above the bottom of the vessel.

Also, the bottom-level tap should not be ½ inch. Some of my clients always use 2-inch level taps to retard plugging. Maybe that's a bit of overkill. I believe that a 1-inch connection is adequate for most fouling services.

But this is not a design text. We have what we have. So you have a ½-inch bottom-level tap located off the bottom of a vessel, and iron sulfide scale is forever plugging the level connection. What's to be done?

- Connect a source of clean flush water to the bottom tap.
- Install a ⅛-inch restriction orifice in the flush connection.
- Leave the flush water running continuously. If the level connection is ½ inch, the delta P of the flushing water flowing back into the vessel will only be:

$$\text{Delta } P_2 = [(D1)^4 \div (D2)^4] \times \text{Delta } P_1$$

where D1 = Diameter of restriction orifice

D2 = Diameter of level tap

Delta P_1 = Pressure difference between the flushing water supply and the pressure in the vessel

Delta P_2 = Pressure difference between the gauge glass and the pressure in the vessel

Delta P_2 expressed in inches will represent the amount of extra height in the vessel above the interface level in the glass. The correct sizing of the restriction orifice, using the preceding equation, will minimize this potential error.

Safety Note—Manual Water Drains

What exactly is a "dead man's valve?" This is a valve intended to prevent you from becoming a dead man (or woman). At the Good Hope refinery in 1981, an operator was manually draining water from a debutanizer reflux drum to the sewer. While so engaged, he was called away for a moment and forgot about the open drain. After a few more minutes, light naphtha began to pour into the sewer. A welder's spark above a remote sewer opening ignited the naphtha. Fire spread through the sewer system.

To prevent this sort of accident, the **dead man's valve** is used. The one I have is a ¾-inch spring-loaded globe valve. If I exert several pounds of force on the handle, the valve's globe will rotate by 90 degrees, and the valve will be wide open. If I let go of the handle, the valve will be pulled shut by the spring tension. The best place to obtain such a valve is to steal one from the Exxon refinery in Baton Rouge. At one time, Exxon had a lot of these valves. Several months earlier, at another location, an operator had been killed while draining water from a wet H_2S knock-out drum. She was found lying next to the open drain valve, which she couldn't close—because she was dead. The trick is to use this technology before, not after, an operator is killed from H_2S gas inhalation.

Coalescers

I try to avoid the use of mechanical coalescers because of their tendency to foul and plug. In 1974 I operated a coalescer to remove dilute caustic from a liquid butane stream. Every few days, I had to have the coalescer washed to remove caustic salts that accumulated on the coalescing medium and reduced the coalescer efficiency. This was bad, because I had to bypass the coalescer while it was being water washed. Hence, I asked an old operator the nature of the coalescing medium.

"Norm, it's just straw. But if it ain't broke, don't fix it."

So I changed the old straw pad to a modern plastic-type coalescing medium. And when I returned the coalescer to service, it barely worked at all. Then I put new straw back into the coalescer, and it worked just fine.

"See, Norm," the old operator reminded me, "If it ain't broke, don't fix it."

Probably the best sort of coalescer works without any sort of coalescer fibers or packing. A stack of sloped, parallel plates, as shown

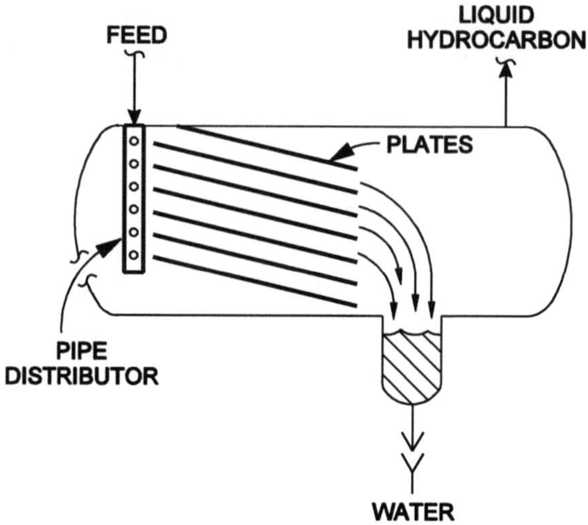

FIGURE 21-6 A coalescer using solid, sloped, parallel plates, is preferred for dirty services.

in Figure 21-6, is used to reduce the vertical height through which the droplets of water have to settle. The plates may be 1 or 2 inches apart. I don't actually know their vertical separation or angle of slope, as the coalescer is purchased as a package. The droplets of water coalesce and run down the sloped plates into the water draw-off sump. This type of plate coalescer does not have much of a tendency to foul and plug.

I have shown a pipe distributor in the vessel. Good distribution is a compromise between:

- Too low a hole velocity, which allows most of the flow to escape from just a few of the holes.
- Too high a hole velocity, which causes the effluent liquid to agitate the two liquid phases and retard settling.

Of course, if the distributor breaks off at the inlet nozzle, phase separation will also be retarded. Failure of inlet distributors is always a common malfunction in water–hydrocarbon separation drums. I've told an interesting story about such an incident at a jet fuel naphthenic acid treater in my book, *Process Engineering for a Small Planet.* The incident happened at the Mobil refinery in Coryton, England, in 1987.

CHAPTER 22

Natural Draft-Fired Heaters

Mr. Lieberman. Can you tell our management that fire comes out of our stack when we run 3% excess O_2.

—Operators at Mobil Beaumont Refinery, 1985

When I encounter a malfunction in fired heaters, I first classify the problem into one of five categories:

- Lack of draft
- Lack of combustion air
- Heat release limitation
- Excessive radiant heat density
- Flame impingement

As I discuss in Chapter 24, "Fired Heaters: Air Preheaters," leaking and fouled air preheaters are the single largest cause of draft limitations and lack of combustion air. First, however, let's discuss the meaning of draft.

Calculating Available Draft

In order to perform an engineering calculation correctly, it is necessary to understand the underlying physical principles. Draft is not a pressure. It is a pressure difference between points. These two points are always at the same elevation. And one of the two points is atmospheric pressure. The pressure difference is caused by the density difference between two columns of gas. One column of gas I'll call air. The other column of gas I'll call flue gas. When I calculate draft, I ignore the change in the molecular weight as the combustion air is converted into flue gas. The only significant parameter that affects density of air versus flue gas is absolute temperature. For example, let's calculate the draft produced by a stack:

- Stack height = 100 feet
- Stack temperature = 600°F (or 1060°R)
- Ambient temperature = 70°F (or 530°R)
  (°R = Degrees Rankine = °F + 460)

I'll assume that the density of air at the location that I'm working at is 0.068 lb/ft^3. For those who have forgotten how to calculate gas density:

$$\text{Density of air} = \frac{29}{379} \times \frac{520}{°R} \times \frac{P}{14.7}$$

where 29 = Molecular weight of air
379 = Volume of a mole at 60°F and standard atmosphere, ft^3
°R = Ambient temperature, degrees Rankine
P = Ambient pressure, psia
14.7 = standard atmosphere, psia
Density of air = pounds per cubic foot

Returning to my stack problem, if the stack is twice as hot as the air (in °R), then the flue gas density is one-half the density of the ambient air. Thus, their density difference is $\frac{0.068}{2} = 0.034$ lb/ft^3. Then to calculate draft:

$$(0.034) \times (100) \times (0.2) = \text{inches of water draft}$$

where 100 = stack height, in feet
$0.2 = \frac{28}{144}$, a factor that converts draft in lb/ft^2 to inches of H_2O

The draft that the 100-foot stack is developing is 0.68 inches of H_2O. But maybe not. I'm working at the Coastal refinery in Eagle Point, New Jersey, and the crude heater is draft limited. Draft limited in the sense that the heater has a positive pressure (or negative draft) just below the bottom row of convective tubes at point A, shown in Figure 22-1. This is contrary to Coastal's safety operating guidelines. So, using a bottle of water and a piece of plastic tubing (see Figure 24-3 in Chapter 24 for details), I measure the draft at the base of the stack, above the convective section tubes. It's 0.4 inches of water less than my calculated draft. Was something then wrong with my calculations?

I look up at the stack damper, and it appears to be in a fully open position. But maybe the indication is wrong? So I pull a little on the chain connected to the damper. The arrow indicating the stack

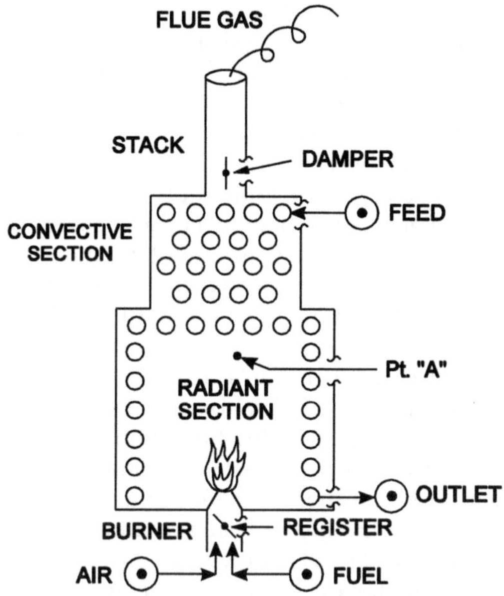

FIGURE 22-1 Components of a natural draft-fired heater.

damper's position moves quite easily. Really, too easily. As I "closed" it halfway, the draft that I observed in my bottle did not change. So I pulled the chain in the other direction and still the draft as indicated in my bottle did not change. Finally, I began to wonder if I was actually moving the damper, or just the indicator arrow which was supposed to be attached to the damper.

I dragged over a 16-foot ladder and propped it up on the side of the stack. Using my 12-inch pipe wrench, I tried to turn the shaft connected to the damper directly, without pulling on the chain. But at this point, the unit maintenance supervisor, Pete Livingston, walked up to my ladder.

"Hey, you," he called, "What the (expletive deleted by the publisher) are you doing up there?"

"I'm trying to get the stack damper to move."

"Who the (expletive deleted) are you?"

I climbed down from my ladder. "Mr. Livingston, I presume? I'm trying to get the stack damper open. I'm Norm Lieberman. I think the gear between the chain operator and stack damper shaft is broken. The indicator arrow's moving, but not the damper itself. I'm a consultant from New Orleans. Coastal hired me to help increase the crude rate to this heater. It's out of draft. You all got a positive pressure up in the convective section inlet."

"Mr. Lieberman, don't you know that you should get an operator to move the stack dampers? You aren't allowed to wander around our refinery, moving valves and dampers yourself. Do you even have a permit to be out here?"

Now I've been listening to this, "You're not allowed to," and "Permit stuff," my whole life. But suddenly, contrary to my expectations, something amazing happened.

"You say the damper's stuck," said Pete Livingston. "You're sure not gonna move that damper shaft with that little wrench. You engineers got no common sense. Probably you'll just fall off that ladder and crack open your empty head."

Then Pete walked over to a nearby **cherry picker** and picked up a 36-inch pipe wrench. Ten minutes later the water in the plastic tubing in my bottle had been sucked up another third of an inch. And the draft below the bottom row of convective tubes in the heater increased from 0.05 to 0.40 inches of water.

I can't remember Pete's real name, but other than that, it's true that a maintenance supervisor actually helped. You just never know.

The other aspect of the heater limitation that opening the stack damper relieved was lack of combustion air. The extra draft permitted the burner's secondary air registers to be opened more fully, without the heater going to a positive pressure. The extra air reduced the CO levels in the flue gas below the legal limit set by the state of New Jersey. I'll discuss air limitations later in this chapter. But what other sorts of malfunctions, other than broken stack dampers, create draft limitations?

Fouled and Distorted Convective Section Tubes

In the USA, we don't fire fuel oil very often. Most refineries in America are self-sufficient in gas from their cracking and coking operations. But when I worked in Aruba we fired heavy, high-sulfur fuel oil all the time. With lots of vanadium. The resulting vanadium salts deposited on the radiant and convective section tubes. This cannot really be considered a malfunction. Ash deposits on convective tubes are just a consequence of firing fuel oil rather than gas. Anyway, the ash deposits are easily removed by the use of soot blowers. These are steam pipes inserted between the convective tubes. The high-velocity steam blows off the vanadium ash deposits.

A more serious and totally avoidable loss of draft occurs due to afterburn or secondary combustion. Here's what happens:

- Carl attends a seminar where he learns that 3% O_2 in the flue gas is the optimum.
- Carl then goes to his crude unit (3 B Pipe Still in Texas City) to instruct Pedro, the lead operator, to throttle back on the burner air registers to reduce excess oxygen from 6% to 3%, as indicated by the stack flue gas excess oxygen analyzer.
- Pedro closes his burner air registers to drive the O_2 down to the 3% analyzer reading that Carl wants.

- Because of poor air–fuel mixing efficiency in the burners, partially oxidized hydrocarbons escape into the convective box.
- Tramp air leaks in the convective box cause the radiant section effluent flue gas to reignite.
- The carbon steel convective section tubes, now exposed to direct radiant heat transfer, overheat to above 1,000°F and consequently begin to sag.
- The sagging convective tubes restrict the flow of flue gas and cause a loss of draft. A positive pressure develops at the top of the radiant section firebox.
- Combustion airflow must then be further reduced to relieve the positive pressure. A reduction in crude charge must also follow the reduction in the combustion air rate.
- Pedro is demoted from lead operator to the labor pool, where he can do less damage.
- Carl is promoted to the computer science division, where he can also do less damage.

Sometimes draft is lost because of restrictions in the ducts connecting the convective section to the stack. In Chapter 24 on air preheaters, I've shown how to calculate such draft losses.

Often, draft is lost due to cold air being drawn into the flue gas due to leaks in the stack ducts or the convective box. The problem arises not from the extra volumetric flow of flue gas, but from the reduction in the stack temperature. If the stack temperature were to drop from 600°F to 500°F due to cold air in-leakage, then the available draft would decline by 20% due to the increase in flue gas density. Finding and fixing tramp air leaks is easy. Use some powdery chalk to locate the air leaks. And then cover the leaks with aluminum duct tape or silicon sealer. I've heard of plants that inject steam (with a molecular weight of 18) into the stack flue gas (with a molecular weight of 30) to increase draft. This hardly seems an energy-efficient way to improve draft, as compared to stopping cold tramp air leaks into the stack.

Combustion Air Limitations

When a heater is limited by air, both the stack damper and secondary burner air registers are wide open (see Figure 22-1). Then, when the fuel gas rate is increased, the heater outlet temperature goes down rather than up. Basically, in an air-deficient environment, the heater firebox progressively becomes a reducing zone rather than an oxidizing zone. As the ratio of air to fuel diminishes, the incremental gas does not burn, but undergoes **thermal cracking**. Whereas oxidation is an exothermic reaction, thermal cracking is an **endothermic reaction**. Meaning, heat is absorbed, not liberated.

I'm not speaking about the theoretical operation of fired heaters, or even about process plant heaters. I'm referring to my electric kiln I use to make ceramics in my garage in New Orleans. The kiln has six heating coils, two of which are broken. So rather than replace the broken coils, which is expensive, I stuck a natural gas supplementary burner into the kiln's bottom sight port. The top sight port I converted into a chimney. I'll describe the results of my efforts:

1. If I open the gas a little, my kiln temperature will rise about 50°F. The flue gas coming out of my ½-inch ID by 15-inch chimney tube is clear.
2. If I open the gas a lot, my kiln temperature drops by about 150°F. The flue gas exhausting from my tiny chimney tube turns gray-black.
3. If I open the gas halfway between a little and a lot, my kiln temperature still drops by about 50°F, but the flue gas is perfectly clear.

Note that in step 3 above, the flue gas is clear even though I have certainly created a reducing zone and an endothermic reaction in my kiln. Before I figured all this out, I would get impatient to heat my kiln to finish off my ceramic productions. So I'd open the fuel gas a lot, which not only cooled off my kiln, but would actually cause the burner to go out.

This really annoyed me. So I'd pull my homemade burner out and relight it. This would take a minute or two. Then, when I stuck the lit burner back into my electric kiln, it would go, "Bang!" And my wife Liz, would run into the garage.

"Norman! What happened? Now what are you up to? Are you stupid, or crazy, or both?"

But this is no joke. Two operators were killed at the Good Hope refinery in Norco, Louisiana, in 1982 doing the same thing. Not on a ceramic kiln, but on a 250 MM (million) Btu/hr vertical crude heater. They were trying to light off the heater without lighting the pilot lights first. The pilot lights were all plugged. So they turned on the gas to a main burner, then looked around for a pole and a rag. They next lit the rag with a cigarette lighter, and blew the bottom of the heater down on themselves. The dual malfunctions were:

1. Pilot lights need to be connected to clean natural gas, not inherently dirty refinery fuel gas.
2. Inherent human nature that understands and accepts risks on the basis that bad things can happen—but not to me.

Incidentally, the same thing happened at the Tenneco refinery in Chalmette, Louisiana, in 1984. I showed the operators that inside the pilot light burner, there is a small orifice plate with a pinhead-sized hole. If the pilot light is plugged, take the burner apart and poke a wire through the little hole. It's a 10-minute job. I thought since they had just blown up the hydrodesulferizer heater, they would be interested in this bit of technology. But I was quite wrong.

"Listen, you smartass engineer. Cleaning burners ain't our job. Who let you damn Yankees into our plant, anyway? You got a permit to be out here? It sure ain't our fault that heater blew itself up."

The Point of Absolute Combustion

If I am firing the heater at a fixed fuel rate, and varying the airflow, I will generate a curve as shown in Figure 22-2. The heater outlet temperature reaches a maximum, which I call the "**point of absolute combustion**." Below this point, lack of air promotes endothermic reactions. Above this point, increased flue gas flow increases the heat loss up the stack. This optimum point is a function of the burner's air–fuel mixing efficiency. The more efficient the mixing, the lower the point of absolute combustion.

What's typical? What's a good target for the operators? What O_2 level in the flue gas represents energy-efficient operations? What O_2 indicates a malfunction of the burner? What is the optimum age to marry? All rather foolish questions which I, not being a fool, will not answer.

At the Murphy Oil refinery, I noted recently that they had optimized their heater at 0.6% O_2. At a boiler at the Tenneco Oil refinery, they had found, by trial and error, the point of absolute

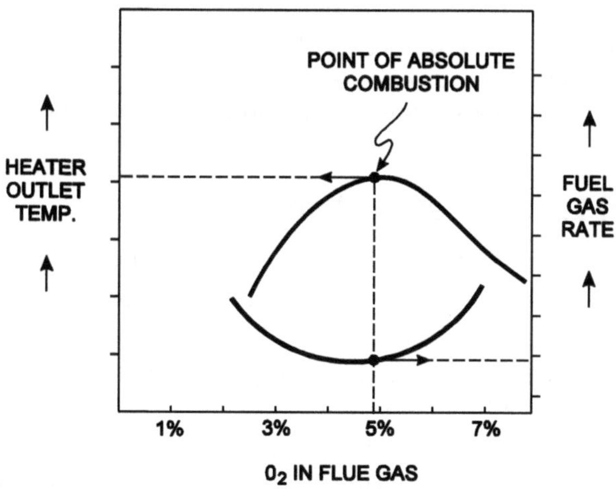

FIGURE 22-2 The point of absolute combustion is a variable. Varying airflow with a fixed fuel rate determines the maximum heater outlet temperature. The O_2 in flue gas at the peak temperature is the point of absolute combustion, which for this heater on this day was 5%.

combustion was 3% O_2. At an old crude heater at the Gulf refinery in Port Arthur, Texas, their optimum O_2 was 7%. At our gas-fired turbines in Laredo, the optimum was around 11%. I rather doubt this helps you set a target. The optimum O_2 is what it is! And anyone who gives you a target for your heater, without a plant test, is a fool.

What sort of malfunctions reduce the air–fuel mixing efficiency to the burner and thus increase the amount of extra O_2 needed to achieve the point of absolute combustion?

Cause of Poor Air–Fuel Mixing Efficiency

- For fuel oil burners, wet dispersion steam or too low a steam pressure causes poor atomization of the fuel oil.
- Any openings in the firebox that allow tramp air to enter the box. This source of air, which bypasses the burners, will not mix with the fuel. Secondary air registers that leak on burners that are out-of-service can be a big problem. Observation ports or pilot light covers that have fallen off are also common areas for tramp air leaks in the firebox.
- Plugged burner tips require more air to optimize fuel combustion than clean burner tips. Dirty gas is usually caused by air contamination of the fuel, which promotes gum and sulfur deposits in burner tips.
- Excessively low burner tip pressure is bad for mixing efficiency. It's easy to check this pressure. Place a pressure gauge downstream of the fuel gas TRC valve. Removing burners from service is the correct way to increase this pressure. Of course, if you can't tightly close the air register to the idled burner, this will do more harm than good.

Indication of Lack of Combustion Air

- Firebox is very hazy. Slight hazy appearance is okay.
- Flames are long and licking, with smoky tips. Big, bushy orange flames are also bad.
- Heater outlet temperature drops as fuel is increased. This is really a bad sign of trouble.
- Fire comes out of stack. I've seen this at the Mobil refinery in Beaumont. Yet the operating manager still wanted to use less air, because his O_2 analyzer showed 3%, and his target was 2½%.

Note that I have not included any reference to the use of an O_2 analyzer in the list of parameters indicating the lack of air. Again, some heaters may be air deficient at 4% O_2, while others may have a great deal of excess oxygen at 4% O_2 in the flue gas. Also, I haven't suggested that the flames should burn blue. The color of the flame is largely a function of the carbon-to-hydrogen ratio of the fuel. For example, heavy fuel oil always burns yellow. H_2S (no carbon) always burns blue.

I have not mentioned the concept of complete combustion. The only way to have complete combustion is to achieve perfect air–fuel mixing efficiency in the burner. Which is impossible, as is complete elimination of CO from the flue gas.

Let's say I'm controlling the fuel gas rate on TRC (temperature), and I throttle back on my airflow. If I was above the point of absolute combustion then, as shown on the right side vertical axis in Figure 22-2, the fuel gas rate would decline. If I was below the point of absolute combustion, the fuel gas rate would increase without stopping. This is called a **positive feedback loop** (see my book, *Troubleshooting Process Plant Control*).

This ought to make one wonder exactly what is the value of an oxygen analyzer? It's really quite a useful tool, providing that the air–fuel mixing efficiency of the burner does not change. But if this mixing efficiency is very variable (perhaps because of rapid swings in fuel gas composition), then the O_2 analyzer becomes progressively less useful.

Safety Note Regarding Adjusting Excess Air

If you believe that your heater has fallen below the point of absolute combustion, so that the fuel gas valve is opening but the process outlet temperature is declining, proceed as follows:

1. Do *not* open the air registers on the burners or open the stack damper. The resulting explosion in the fuel-rich firebox may kill you.
2. Switch the heater firing controls from auto to manual. You can't break the positive feedback loop in automatic control.
3. Manually back off on the fuel gas rate. You will see the heater outlet temperature increase.
4. When you are above the point of absolute combustion, increase airflow on the burner registers, or the stack damper, or both.
5. Switch back to auto (TRC) control.

Your brilliant author almost killed himself one night in 1980 in Texas City on a sulfur plant incinerator by violating these rules. I was blown right off my feet by too rapidly opening an air register to the combustion zone that was obviously short of air.

Air Limitations Due to Physical Restrictions

At the Coastal refinery in New Jersey, they had retrofitted a naphtha reformer heater with low-NO_x (nitric oxide) burners. Such burners do a great job of mixing the air and fuel. It's done in two stages. But mixing requires delta P. That is, two-stage, low-NO_x burners are typically designed with a higher delta P than regular, single-stage burners. But at the Coastal refinery, the older burners were operating in a wide-open position. Also, the stack damper was 100% open. Thus, the new, low–nitric oxide burners restricted the airflow, as the available draft loss, or pressure drop, through the burners could not be increased. With reduced airflow, the heater capacity was reduced, leading to a loss in unit capacity.

My client considered this to be a burner malfunction, which it was not. With the manufacturer's guidance, I modified the burner by enlarging the hole size in the orifice plate mixer. This reduced the burner delta P and restored the required combustion airflow.

There's a general lesson in this story. When purchasing process equipment, whether it's heat exchangers, filters, spray nozzles, or control valves, specify the max allowable delta P. Otherwise, the manufacturers will use whatever pressure drop they require to optimize the equipment efficiency.

Another example of an air restriction was on a crude heater. The heater was limited by air. Using my bottle of water and plastic tubing (see Chapter 24, Figure 24-3), I determined that the pressure drops, or observed draft losses, were distributed as follows:

- Across 100% open stack damper—nil
- Across 100% open air registers—0.6 inch H_2O
- Across internal intake screen—0.4 inch H_2O

The intake screen was inside a giant round box that was upstream of the burners. I guess the screen was intended to keep trash from getting sucked up into the burners. But, the screen itself was mostly plugged with trash. Apparently, it was fulfilling its intended function. But my client failed to grasp that the screen had to be cleaned. So the unit engineer and I cleaned the screen. Airflow then increased by 15%. However, the entire assembly—box and screen—has now been discarded by my client, as it really served no purpose.

A large percentage of airflow restrictions are associated with air preheaters, which I'll describe in Chapter 24. Restrictions to flue gas flow reduce airflow in two ways. First of all, draft that could be used to overcome the air register delta P is lost by overcoming the flue gas pressure losses. Secondly, a positive pressure might develop in the firebox. This forces the operators to partly close the air registers so as to reduce the firebox operating pressure.

Allowing air to flow into a combustion zone at any point may help increase the available O_2. But unless the incremental air enters through the burner air register, the point of absolute combustion will increase due to degraded air–fuel mixing efficiency. I suspect that often promoting such tramp airflow may actually do more harm than good in promoting fuel oxidation.

To summarize, do not run a heater on automatic temperature control with a marginal amount of air. You may kick off a positive feedback loop. When you then manage to restore the air-to-fuel ratio too quickly, the heater can explode and kill you. It happens all the time.

Heat Release Limitations

When we say a burner is limited by fuel gas pressure, this does not ordinarily mean that we cannot physically supply more gas pressure to a burner. Normally it means we have reached that maximum allowable burner tip pressure. The burner manufacturer has established this pressure limitation. The manufacturer has published a curve for the heat release versus the burner pressure.

If you exceed the pressure shown at the end of this curve by x psi, then the excessive burner tip velocity will blow out the flame. Then you're blowing unlit gas into the firebox, which is potentially explosive. As I do not know if x is 1 psi or 10 psi, I will always adhere to the maximum burner pressure set by the manufacturer. This pressure is a pound or so lower than the pressure gauge reading anywhere downstream of the fuel gas regulator control valve.

One of the most common malfunctions that reduce burner capacity is lean gas. This means that the heating valve of the fuel gas is lower than design. In a refinery, the main culprit is H_2. Most of the hydrogen we find in refinery fuel gas doesn't belong there. It's caused by excessive venting from our hydrotreating unit's recycle gas loop. I've made a careful study of this problem in Coastal's former refinery in Aruba and found:

- H_2 content of fuel gas less than 11% is excellent.
- H_2 content of fuel gas between 12 and 18% is normal, but could be reduced.
- H_2 content of fuel gas over 30% indicates careless operations and a waste of hydrogen. Whenever I brought this problem to the attention of my pal Ray Buckley, the operations manager, he would claim that nothing was wrong. But the next day, the H_2 in the fuel gas would drop to 11% anyway.

Projects that recover C_3s and C_4s from refinery fuel gas will also make the gas leaner and will lower its heating value. Leaner gas can be compensated for by drilling out the holes in the burner tip. But what I actually do is buy new burner tips (spuds) from the manufacturer for the specified new fuel gas composition.

Partly plugged burner tips will reduce heat release. This is due to dirty fuel gas. To me, dirty fuel gas means air contamination, which causes gums and solid sulfur to form in the gas header. Don't look at the oxygen concentration in the fuel gas. It's all reacted by the time you check the sample. Instead, check for nitrogen as an indication of an air contamination.

Steaming-out burner tips to increase the burner's heat release is a sure sign of sulfur deposits, which are fundamentally due not to H_2S, but to air in the fuel gas.

CAUTION! *We do not open bypasses around fuel gas regulators to increase the burner heat release. Suppose there is a unit upset or fire. And the panel operator needs to shut off the fuel to the heater quickly and the control valve bypass is open. Then what?*

Excessive Radiant Heat Density

From a design point of view, I might set the average radiant heat density or flux (Btu/hr/ft^2) for different services at:

- Delayed cokers: 9,000
- Crude: 12,000
- Hydrotreaters: 14,000
- Vacuum heaters: 11,000

The ft^2 term refers to the external surface area of the tubes in the firebox, including the bottom two rows of roof tubes. The concept of excessive radiant heat density is expressed from the operator's perspective, in that the firebox is too hot.

Excessive firebox temperature, or high radiant heat density, promotes coke accumulation inside the tubes. Localized tube wall overheating will result in tube wall bulging and failure.

Two common causes of increased radiant heat density are:

- Air preheat
- Optimizing excess air

Reducing excess O_2 to the point of absolute combustion (see Figure 22-2) or preheating air will both increase the firebox temperature. Both also reduce the flow of flue gas. This reduces the heat pickup by the process fluid in the convective section. The loss of convective duty is offset by an increased radiant section duty, as a result of the higher firebox temperature. The hotter firebox is the price we pay for saving energy with air preheat or reduced excess air.

Reduced feed preheat also requires higher radiant section duties. As most heaters have a convective section in feed preheat service, reducing the feed temperature is partially offset by more heat recovery from the flue gas.

If you're burning fuel oil with a high ash content (especially vanadium), the deposits coat the exterior of the radiant tubes and appear to reflect the heat away from the tubes. I can't say that I've ever seen that even ¼-inch-thick layer of vanadium salts coating the radiant tubes reduces heater capacity.

Flame Impingement

Localized overheating due to flame impinging on radiant section tubes will cause excessive heat flux rates, even though the average flux rate is quite acceptable. Flame impingement problems can sometimes be seen more easily at night. A positive pressure below the roof tubes can cause the flames to be pushed out against the upper few rows of radiant wall tubes. I had this problem in a crude heater at a refinery in Colombia. The operators had restricted the airflow to the registers to stop the flame impingement. But then they were short of combustion air. Their real problem was not a burner malfunction as they suspected, but a problem with the stack damper. The damper indication was wrong. Using my plastic tubing and water-filled bottle as a guide, I moved the damper to restore draft. This relieved the firebox positive pressure and stopped the flame impingement.

A shortage of air will typically cause flame impingement. Working in Holland on an oil-fired heater, I observed that most of the burners were producing giant, orange, bushy flames. These flames were touching the surrounding tubes. But a few of the burners had more compact, yellow, brighter flames. So I increased the airflow to those burners that had the orange flames. Also, I cut back on the fuel oil pressure to those burners where the air registers were already fully open.

Ordinarily, I would expect increasing airflow to shorten the length of the flame from a burner. But this is not always so, especially in vertical (i.e., can-type) heaters with floor-fired burners. Using a great quantity of excess air can result in higher flame lengths due to greater flue gas velocity. Sometimes, even though the flames appear to be air deficient, using more air in floor-fired burners increases the flame length. That's caused by excessive burner velocity, and it represents a burner malfunction.

Partly plugged burner tips will also produce an erratic flame pattern that may impinge on the lower wall tubes. Reducing fuel gas or increasing airflow will mitigate this problem.

Trimming burners to minimize flame impingement is an ordinary operator function that needs routine attention. I've read in a John Zink training manual that throwing handfuls of soda ash into a firebox helps reveal the flame pattern and is an aid to an operator adjusting the air and fuel flows to individual burners. I have not done this, but it sounds like a practical suggestion to help reduce flame impingement in the firebox.

Erratic Draft

Dear Mr. Lieberman:

I've just read in *Troubleshooting Process Operations* your dumb recommendation that draft in a natural draft heater should be controlled between 0.05 and 0.10 inch of water below the bottom row of convective tubes. This is bullshit! My new crude unit heater draft oscillates erratically between 0.03 and 0.18 inch of water over a period of a few seconds. The sort of tight control you suggest is insanely impossible.

Kevin Jones
Process Superintendent
Sinclair, Wyoming

Dear Kevin:

I've been in southern Wyoming, and it's really windy. When wind blows across the top of your stack, it creates a draft. Try opening your car window next time you're on I-80. You'll feel the reduced pressure in your ears, due to the air rushing past your car. Or blow across a straw in a glass of water. It will create a draft of a few tenths of an inch of water. The water will be drawn up in the straw.

Thus, your crude heater draft will increase due to the additional draft created by the highly variable winds in the high plains country of Wyoming. As the wind is erratic, especially 200 feet above the ground, so will be the heater's draft.

My primary recommendation then is to move your refinery to New Orleans, where the wind is very moderate, and usually absent. If this is not an acceptable alternative, then my secondary recommendation is to never contact me again with your inane problems.

Norman Lieberman
Important Person
New Orleans, Louisiana

P.S. *If you do not have any air in-leakage in your convective box, then draft control to avoid excessive draft is less critical, as long as the pressure below the bottom row of convective tubes does not go positive.*

CHAPTER 23

Tube Failure in Fired Heaters

In this unit, there are no such things as stupid questions, only stupid actions.

—Sign over FCU control room in
Sugar Creek, Missouri, 1968

I became an expert in heater tube failures while waiting for meetings with John Brooks. Mr. Brooks was the tech service manager of the Amoco Whiting refinery. His useless and frequent meetings would never start on time. So I would study the dusty Inspection Department's artifact cabinet in the dingy hallway, waiting for the meeting to start. The cabinet was full of longitudinal cross sections of failed furnace tubes. All the ruptured tubes exhibited had several characteristics in common:

- The thinning of the tube wall was extremely localized (see Figure 23-1, which is a radial cross-sectional view).
- There was no evidence of substantial corrosion inside the tube.
- The ruptured tube wall was thin at the point of failure, but uniformly thicker at other points.
- The rupture was associated with a bulge in the tube wall.
- The bulge coincided with an inch or two of dense coke inside the failed tube.
- There was no indication of external tube oxidation. That is, the tube had not "burned up." It had failed at the thin tube wall bulge associated with the coke.

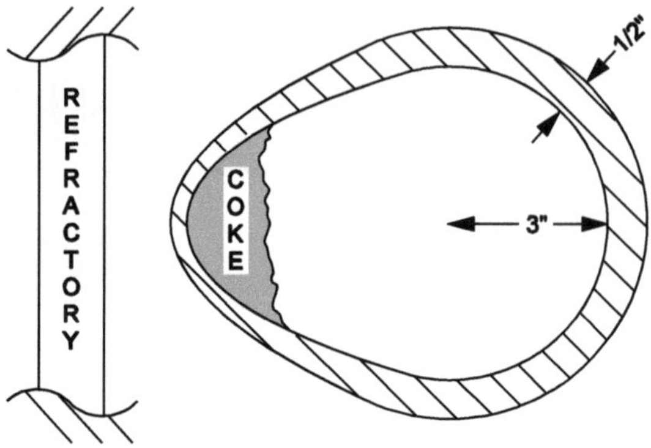

Figure 23-1 Tube bulging causes wall to thin and fail due to coke formation inside the tube.

High-Temperature Creep

Process heater tubes are typically constructed of 12% chrome steel with a small amount of silicon. The silicon (half of a weight percent) increases the tube's external oxidation resistance. At about 1,375°F, high chrome steel becomes subject to plastic deformation. This means that the internal pressure inside the tube causes the tube wall to slowly expand. This is called **high-temperature creep**. The tube wall is never uniformly heated for several reasons:

- Localized coke formation or inorganic deposits such as caustic and salts inside the tube
- Flame impingement
- High localized heat flux rates, especially on the side where the tube faces the furnace's refractory wall
- Poorly balanced fuel and air distribution to the burners

As the tube expands, its wall becomes thinner in the area of increased diameter. At some point, the wall thickness becomes inadequate to constrain the tube's internal pressure. Then the tube wall bursts open. Such was the story of the failed tubes displayed in the cabinet outside John Brooks's office.

Tube Hot Spot Identification

We can follow tube metal temperatures based on skin TIs. These are thermocouple wires attached to the exterior surfaces of the tubes. I dislike relying on such TIs because they only read a few points on the tubes. Also, they have a tendency to pull off the tube surface and then read the much hotter firebox temperature. A better way to gauge tube temperatures is to rely on a portable optical pyrometer. This device correlates the glowing color of the tubes with temperature. This technique can also be used to visually approximate tube external temperatures:

- White—2,800°F, 1,540°C
- Light yellow—2,450°F, 1,343°C
- Yellow—2,150°F, 1,177°C
- Orange—1,880°F, 1,027°C
- Bright cherry red—1,600°F, 871°C
- Cherry red—1,380°F, 750°C
- Dark red—1,200°F, 650°C

These colors do not depend on the metallurgy of the tube. My only problem with this method is that ash deposits on the tube surface create a false impression that a tube is overheated, when in reality it's only the ash that's glowing too bright a color. I've knocked the ash off a tube with a metal pole and noted that the exposed tube surface underneath the brightly glowing ash was a moderate dark red color.

Causes of Coke Formation in Tubes

At the American Oil refinery in Sugar Creek, Missouri, they had a delayed coker heater that coked up regularly every 4 months. It was a four-pass heater, meaning that there were four parallel tube pathways. The mass velocity of coker feed (tar) through the tubes was 65 lb/ft^2/sec. The ft^2 term refers to the tubes' cross-sectional area. I modified the pass configuration from four-pass to two-pass. Mass velocity then increased the 65 lb/ft^2/sec to 130 lb/ft^2/sec. Heater run length between decoking intervals increased to over a year. This was in 1967. It was my first big success in correcting refinery malfunctions. And I learned two important concepts:

- The best way to retard coke formation inside tubes is to maintain a high rate of fluid shear at the walls. This is best done with a high mass (not linear) velocity. I imagine the fluid shear keeps any solids formed inside the tubes from sticking to the interior walls of the tubes. A nice mass velocity is 200 lb/ft^2/sec. A great mass velocity is 400 lb/ft^2/sec. A really bad mass velocity is less than 100 lb/ft^2/sec.
- The other concept I learned is that solving process problems is not the path to promotion in a big oil company.

I mentioned that high linear velocity does not suppress coke formation in tubes like high mass velocity does. I only say this because a little velocity steam in heater coils does suppress coke formation. But as the steam flow increases past a point, it doesn't seem to help very much more. It may be that we're getting annular flow. Meaning, steam is rushing down the center of the heater tubes, but oil is dragging slowly along the ID of the coil.

High heat flux rates (Btu/hr/ft^2/°F) in the radiant firebox will also cause coke formation. The closer the tubes are to the wall and the closer the tubes are to adjacent tubes, the higher the heat flux rate will be on some portion of the tube surface. The malfunction that concerns me in this regard is sagging radiant tubes in the firebox. Sagging tubes are not necessarily damaged tubes, unless they result in a reduction of tube spacing or in the tubes moving closer to the radiant section refractory walls.

Feed Interruption

When fuel is burned in a heater, the heat generated is not mainly absorbed by the tubes directly. Most of the heat is absorbed by the refractory brick and then re-radiated to the tubes. Especially in older furnaces that have massively thick refractory brick walls, heat is stored and re-radiated to the process tubes. Thus, when the feed flow suddenly diminishes and the firing rates are accordingly reduced, heat input to the tubes from the radiating bricks continues at an excessive rate.

I became smart on this subject by a nasty experience in Texas City in the late 1970s. A delayed coker heater that I had designed was experiencing rapid rates of coke buildup and tube failure. When I investigated, I found the following sequence of events to be common:

- Flow of coker feed (tar) to the heater would trip off.
- The backup velocity steam, or sweep steam, to the radiant tubes would open to blow the tar out of the coils before the tar could thermally crack to coke. This was the purpose of the sweep steam.
- Sensing low feed flow, the same signal that activated the sweep steam would shut off all the fuel to the burners, except the pilot lights. Pilot lights are supplied from a natural gas line and not from the fuel gas line. They contribute about 1% of the furnace heat release, which we can ignore.
- The short-term re-radiation from the refractory walls however cannot be ignored. A relatively small weight of sweep stream (1 or 2 pounds of steam per barrel of feed) passes through the tubes. The heater outlet temperature rises rapidly above its normal outlet temperature before sinking back down to a low temperature, as shown in Figure 23-2.
- If feed is restored during the period of time when the heater outlet is close to its peak temperature, then coke will form in the coils. That is, coke accumulates in the tubes, not when flow is lost, but when the flow is prematurely restored.

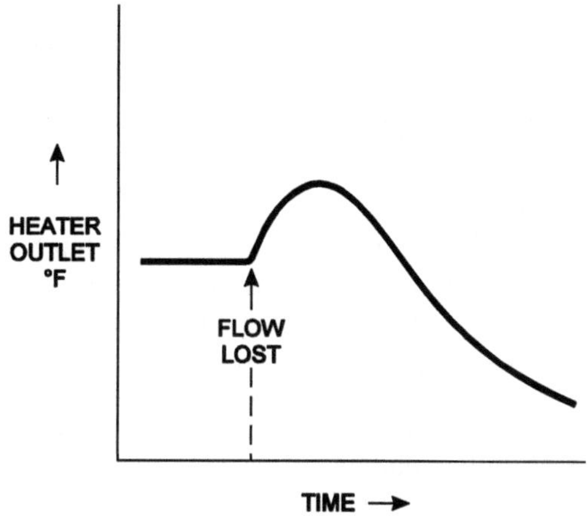

FIGURE 23-2 Effect of re-radiation of heat from refractory walls.

For new delayed coker heaters with lightweight reflective refractory tiles, it seems that the tubes have less of a tendency to accumulate coke. Likely this is due to the smaller amount of heat that can be stored in the less massive refractory tiles. But regardless of the type of refractory involved, the lesson is that we should not reintroduce the heater feed until the coil outlet temperature drops back close to its normal operating temperature range.

Flame Impingement

Partially plugged burner tips will often result in flame impingement. The flames can best be brought back closer to the burner tips by reducing the fuel gas rate to the plugged burner. Excluding tramp air from the fuel gas supply to a burner is generally the best way to avoid burner tip plugging. In refineries, burners also plug with entrained amine. Improved amine filtration is the best way to reduce carryover from the amine fuel gas H_2S scrubber. Degraded amine will form to a black, rock-like deposit in the burner tip.

Lack of combustion air will cause the flames to be long and licking. Try opening the air register associated with the burner, which has a yellow, smoking flame. This will shorten the flame. Fuel oil burners that lack optimum combustion air will have big bushy, orange flames.

If you have carbon deposits on top of the roof tubes in the firebox, that's a sign of localized lack of air. Even though you may have excess air leaving the radiant section, poor horizontal air distribution can cause a local lack of oxygen.

Too much air can cause high flue gas velocity above the burner. This may cause flame impingement at the radiant roof tubes. Reducing airflow in this case should reduce the flame length. But don't reduce the airflow to the point where the burner becomes air deficient and smoky.

A positive pressure below the roof tubes will also push the flames against the upper radiant firebox wall tubes. It's best to open the stack damper a small amount, so as to draw the flames back away from the wall tubes.

Sodium in Feed

When I worked at Amoco Oil, I learned that Na^+ ions (caustic, NaCl) would promote coke formation in furnace tubes. The concept was that sodium would accelerate thermal cracking. Later on, I read that sodium in a delayed coker feed may contribute to **shot coke** formation in the coke drums. Shot coke is caused by asphaltene precipitation due to the thermal cracking and the subsequent evaporation of oils and resins that had been maintaining the asphaltenes in solution. This suggests that residual seawater salts left in crude, or caustic that is injected into the heater feed, contributes to furnace tube failures by promoting coke formation. But I cannot say I have any personal evidence that would support this theory.

On the other hand, I have seen residual salt or caustic deposits inside heater tubes after the coke has been burned out of the tubes. This procedure of burning carbon deposits out of the inside of furnace tubes is called **steam-air decoking**. Basically it works like this:

- The heater is isolated from the process flow.
- A moderate amount of steam is introduced into the heater tube passes or coils.
- The temperature of the coils, as measured at the outlet, is slowly raised by firing harder over a period of 1 to 2 hours, from 1,250°F to 1,300°F, assuming 11% to 13% chrome tube metallurgy.
- A much larger amount of steam is quickly introduced. At the same time the firing rate is cut to a minimum. The sudden shrinking of the rapidly cooled tubes should break off the interior coke deposits due to the differential rate of thermal shrinking between the metal (which is large) and the coke (which is smaller). Then the velocity of the steam ought to blow the coke particles out of the tubes. This is called **spalling**.
- After the tubes have been spalled a few times, a small amount of air is introduced into the hot tubes. This ought to ignite any residual coke. The air has to be introduced carefully to avoid localized overheating of the tubes. This can easily result in tube damage and failure if the operator is not careful. By careful, I mean the operator must observe the progress of a ring of fire progressing through the length of the coil by adjusting:
  - Airflow
  - Steam flow
  - Firing rate

- The tube color should be kept no brighter than a dull red. If this red ring creeps slowly and continuously along the coil, it indicates the spalling step should be repeated, as too much coke has been left in the tubes.

Clearly, the preceding steam-air decoking procedure runs the risk of localized overheating and damage to a tube. On the other hand, an overly cautious approach to adding air to the coils is worse. The most detrimental outcome of steam-air decoking is to leave a layer of coke in an isolated portion of the tube. Should this happen, a hot spot will quickly form on the tube wall as soon as the heater is returned to service. This will all too often result in rupture of the tube at the hot spot.

Inspecting a tube after the steam-air decoking procedure will sometimes reveal the presence of a whitish deposit left in the tubes after the coke has been removed. That's either caustic, some other heat-stable salt, or an inorganic ash. This too will cause localized tube overheating and tube failure. It's best to water wash the tube to avoid this malfunction after decoking the tubes. If the tubes are **hydrotested** for leaks after decoking, this adds very little time to the decoking procedure.

Tube Pigging to Remove Coke

Forcing a **pig** through the coil with water pressure is just as common as steam-air decoking. The inorganic deposits are efficiently removed. But the problem of bypassed coke sticking to the tube walls still exists. To prevent tube failure after startup, I'll ask my clients to try to ignite any coke with air, to prove that all combustible organic material is truly gone. Without this "proof burn step," how can we be sure that the contractor who "pigged" our heater has actually completed the job?

Incidentally, I prefer this hydro-pigging to steam-air decoking to decoke a furnace tube. It's expensive compared to steam-air decoking. But it sometimes happens that hourly operators use far too much air, for far too long, when burning out coke. As described in my book, *Troubleshooting Process Operations,* the effluent quench water turns orange-red due to the evolved iron oxide from the tube when excessive air is used during the steam-air decoking procedure. This naturally is quite disheartening.

Tube Dry-Out Deposits

There are two types of problems that cause inorganic deposits to accumulate far into the heater passes. For naphtha reformer units, the liquid feed should be totally vaporized upstream of the reactor. This should take place after the naphtha exits from the furnace. That is, the naphtha ought to be 80% to 90% vaporized inside the final tube. Then, when the furnace effluent mixes with the recycle hydrogen, vaporization should be completed. If vaporization is completed inside the heater tubes, then dry-out will result. That is, any solids in the naphtha will drop out of the feed and accumulate inside the tubes. A local hot spot could then cause a tube to rupture.

For steam superheat coils, the story is similar. Entrained droplets of water from a waste heat boiler contain the same concentration of total dissolved solids as does the blowdown. The droplets are sure to dry and leave their salts inside the superheat tubes. Convective tube failure, due to local hot spots, will then result.

I've discussed how best to retard the carryover of moisture from a waste heat steam generator in my book, *Troubleshooting Process Plant Control.*

I've also seen sand in a heater feed cause localized inorganic deposits in the radiant section tubes. This happened in a tar sands delayed coker furnace in Alberta, Canada.

Convective Tube Failure in Steam Generation Service

Steam generation coils in convective sections are subject to failure, due to localized hardness deposits. The malfunction is due to low water circulation rates, excessive percent vaporization, or poor-quality boiler feed water (BFW). Phase separation in the tubes creating local dry spots is the underlying problem. The resulting deposits cause hot spots and accelerated rates of corrosion, leading to steam leaks in the convective tubes.

One proven method to mitigate this problem is to install a turbulator-type spring along the coil length. This technology is marketed by Total (France) as Petroval. The objective is to maintain wetting of the entire interior surface of the tubes. This offsets low mass velocity and helps avoid phase separation between steam and water, which is the underlying problem. High heat flux in the convective section magnifies the failure rate. The real difficulty arises when the steam coils have U bends. Then, to install the springs, the U bends have to be cut off and rewelded.

> The specific application for this addition of "scrolls" or "springs" in the convective section steam generation coils was at a Natref facility. The weight percent vaporization was about 16% and the linear velocity was 30 ft/sec. Mass velocity was approximately 200 lb/ft^2/sec.

Exterior Tube Failure in Oil-Fired Heaters

This is a particularly nasty malfunction that I encountered in Coastal's refinery in Aruba in 1995. It happened in a vacuum tower feed preheat furnace. The 6-inch tubes were in a vertical configuration. The fuel was very heavy (minus 4°API) fuel oil with a few hundred ppm vanadium plus a few ppm of sodium. The fuel oil sulfur content was approximately 5%.

On horizontal tubes, I had observed that the vanadium in the fuel oil had formed a uniformly distributed and smooth ash on the exterior of the radiant section tubes. This ash seemed to protect the tubes rather than harm them. However, for this vacuum heater, due to an unfortunate combination of circumstances, the story was quite different. The problem was due to a **eutectic mixture** of vanadium ash and sodium.

You may recall from your chemistry courses that a eutectic mixture combines two substances so that their combined melting point is lower than the melting point of each pure individual substance. For example, if I have vanadium ash deposits on the exterior of radiant section tubes, the ash deposits adhere as a dry solid to the tube wall, whether the tubes are horizontal or vertical. And this is true for a wide range of temperatures. However, if a few percent of sodium combines with the vanadium ash, then at around 1,300°F to 1,350°F tube skin temperature, the ash starts to melt.

If we have 11% to 13% chrome tubes, which is a common heater tube metallurgy, then the tube walls become plastic at 1,375°F. Therefore, most operators will keep even localized tube temperatures below 1,250°F to 1,300°F. However, due to flame impingement, some of the tubing sections will operate at 1,300°F to 1,350°F, even though 99% of the tube surface area is below 1,300°F.

I can't say exactly what will transpire with horizontal heater radiant tubes. The problem caused by the V–Na eutectic mixture is probably not so severe as I observed in Aruba on the vertical vacuum heater radiant section tubes:

- At the areas of localized flame impingement, a thick (about ½ to ¾ inch) viscous liquid began to form.
- The deposit then very slowly oozed or crept down the tube wall.
- This deposit became very corrosive as it slid down the outside of the tube. I rather assume (i.e., I'm guessing) that some SO_3 from the combustion gases was absorbed into the semiliquid eutectic of vanadium ash plus sodium.
- But what is not a guess is the corrosive nature of the eutectic. It ate a hole right through the ½-inch wall of the 6-inch vacuum heater high-chrome tubes. It was as if the tube had been a hollow wax cylinder, and I had held a lighted match too close to its wax wall.
- The resulting hole, about the size of a quarter, caused the residual crude oil vacuum tower feed to leak into the firebox, ignite, and thus shut down the vacuum distillation unit.

It may seem impossible to you that I could have made such detailed observations without entering the heater's firebox while it was in operation. Quite unlikely! But I did the next best thing. I had observed, shortly before the heater was removed from service, the areas of flame impingement. Then, when the heater was taken offline and cooled, I studied about a half-dozen areas on the vertical tubes that had the V–Na eutectic mixture frozen on the tubes' outside diameters. Actually, only two tubes were leaking. The other tubes were in various stages of tube wall thinning due to the eutectic deposits. It reminded me of the corrosion pattern when strong sulfuric acid leaks onto a carbon steel pipe.

In a refinery, the best way to handle this malfunction is to reduce the Na content of the refinery fuel oil. This is done by avoiding the injection of caustic into the crude unit desalter effluent. Improving salt extraction in the crude unit's **desalter** will also help.

I discussed the problem in Aruba with a corrosion control expert. He advised me that this malfunction could best be controlled, if I could not remove the sodium, by the addition of a commercially available chemical that somehow avoided the detrimental effects of the sodium. This seemed a rather costly way to control the tube failure pattern. So what we actually did in Aruba was to make a consistent and complete effort to carefully avoid flame impingement in the vacuum heater vertical tube radiant section. This didn't cost anything, and in practice, it did prevent further tube failures because it kept the localized tube wall temperatures below the melting point of the V–Na mixture.

Excessive Average Heat Flux Rates

Sometimes tubes fail simply because they are being overfired. Even though there are none of the following usual malfunctions:

- Tubes sagging or too close to the walls
- Flame impingement
- Low tube side mass velocity
- Solid contamination of feed
- Tubes too thin (too thin is ¼-inch walls. Normal design should be ½-inch wall for 4- to 8–inch-diameter tubes)
- Erratic flows
- Lack of velocity steam

Overfiring implies excessive heat flux rates in Btus per hour per square foot of the external radiant section wall and roof tubes. For liquid hydrocarbons that have not been previously thermally degraded, I would judge the following parameters can be used as an approximation as to how hard the heater is being fired:

- Under 8,000 flux rate: lightly loaded
- 10,000–12,000 flux rate: moderately loaded
- 14,000 flux rate: heavily loaded
- 16,000 flux rate: severely loaded

The design factors that permit increased flux rates are:

- Smaller-diameter tubes
- Larger tube spacing between tubes and refractory walls
- Larger combustion zone in firebox
- Larger mass velocity
- Use of moderate amount of velocity steam
- Higher-alloy tubes, with one-half percent silicon
- Thicker tube walls

At the American Oil refinery in Texas City, the tube failure rate on a delayed coker heater was a chronic problem. In 1962, they changed the tubes from ¼-inch walls to ½-inch walls, and tube failures disappeared.

Regardless of the preceding design features, I have never calculated any heater in a refinery operating above a heat flux rate of 17,500 Btu/hr/ft^2 on a continuous basis. For a new design, I would likely use 12,000 Btu/hr/ft^2. Heater tubes that fail because of excessive flux rates cannot be considered to be malfunctioning in a normal sense. Rather, the equipment is being overworked.

Effect of Air Preheat and Energy Savings

As I discuss in Chapter 24, "Fired Heaters: Air Preheaters," increasing the adiabatic flame temperature in a firebox will accelerate the rate of coke formation and promote tube failure. If the combustion air is preheated by 360°F to 400°F, the fuel gas consumption will decline by about 10%. However, the flames also become 360°F to 400°F hotter. The hotter flames can also be a consequence of the reduction in excess air, if the airflow before the reduction had been well above that amount required for optimum combustion conditions.

I recall at the Texaco refinery in El Dorado, Arkansas, that attempts to optimize excess air usage did save some fuel, but they also led to a threefold reduction in the run length of their delayed coker-fired heater, due to coke formation inside the lower radiant wall tubes. As I explained to the management of the refinery in 1987, more rapid coke formation inside the heater radiant tubes is the price they must pay for energy savings. Unfortunately, the energy conservation consultant who recommended the reduction in excess air was not around to profit from my comment.

Water in Feed to Coker Heater Radiant Tubes

The Citgo refinery in Corpus Christi had coked up and damaged one of the four passes to its crude charge furnace. A slug of water from the crude charge tank flashed to steam in the crude preheat exchanger train. The two-phase mixture then flowed to the four parallel heater pass inlets. Due to the nonsymmetrical nature of the piping, the evolved steam flowed preferentially to the first heater pass flow control valve. The valve went 100% open, but very little hydrocarbon liquid flow was charged through this valve.

Because of the low mass flow, the pass outlet temperature increased from 700°F to 900°F. The high temperature rapidly caused the coil to coke off.

There are two possible ways to avoid this problem:

- One method, as proposed by one of my former students who now has his own engineering company, is to install a preflash drum to flash off the evolved steam, before it can upset the heater pass inlet flow control valves. This will work.
- The second method, which is my proposed solution, is to drain the water heel from the crude tanks used to receive the crude before the water-contaminated crude is transferred to the crude unit charge tank. This method would also work.

My method is inherently inferior—inferior in the sense that no project would result that would then generate any engineering service fees.

Dear friends, let's all try to remember why we became engineers, rather than lawyers or stockbrokers or members of the Louisiana State Senate.

Carbon Steel Tube Failures

I have never specified carbon steel tubes for any heater. It seems rather pointless, as the overall installed heater cost will not change very much, compared to using 12% chrome steel tubes. But if you do use carbon steel tubes, avoiding phase separations in the tubes is critical. The top portion of a horizontal tube can operate at 200°F hotter than the lower half of the tube. That is because the liquid phase keeps the tube much cooler than the vapor phase. The flow regime at which phase separation occurs can easily be calculated.

If the feed to a heater with horizontal carbon steel tubes has a sulfur content of about 3%, and phase separation occurs, then corrosion rates in excess of 100 mils per year can be expected in the hot upper half of the tube. The lesson is the same. Always keep mass velocity high enough to avoid vapor–liquid phase separation (see the earlier section in this chapter, "Convective Tube Failure in Steam Generation Service," for an alternative approach to the problem of low mass velocity in the tubes).

CHAPTER 24

Fired Heaters: Air Preheaters

Cheap is cheap.

—My mother alluding to locating an O_2 analyzer in a heater stack

"Mr. Lieberman," she said, "Something is wrong with my boiler. Every time we line up the air preheater, steam production goes down. I would have thought that increasing the combustion air temperature to my fired boiler would have made more steam, rather than less steam. What's wrong? Bypassing the air preheater increases boiler steam flow by several percent."

Harriet produced the sketch shown in Figure 24-1 to illustrate the problem. It was not new to me. I had encountered the same apparent contradiction when revamping a delayed coker heater for the Amoco Oil refinery in Texas City. That is, air preheat reduces overall fired heater capacity when limited by excessive firebox temperatures, or excessive radiant heat density, or high bridge–wall temperatures, or high radiant tube skin temperatures. All of which means the same thing: the heater's firebox is too hot.

As I explained to Harriet, her boiler's steam production declined with the air preheater online because of the following:

- Combustion air is heated from 60°F to 420°F by heat exchange with hot (800°F) flue gas from the convective section of the boiler, as shown in Figure 24-1.
- The hotter combustion air raises the **adiabatic flame** temperature in the radiant section, which naturally makes the radiant section or firebox operate hotter.
- The operators then are forced to reduce the fuel fired in the radiant section to cool the firebox. For every increase of 360°F (200°C) of the combustion air, the operators will back off fuel flow by 10%.

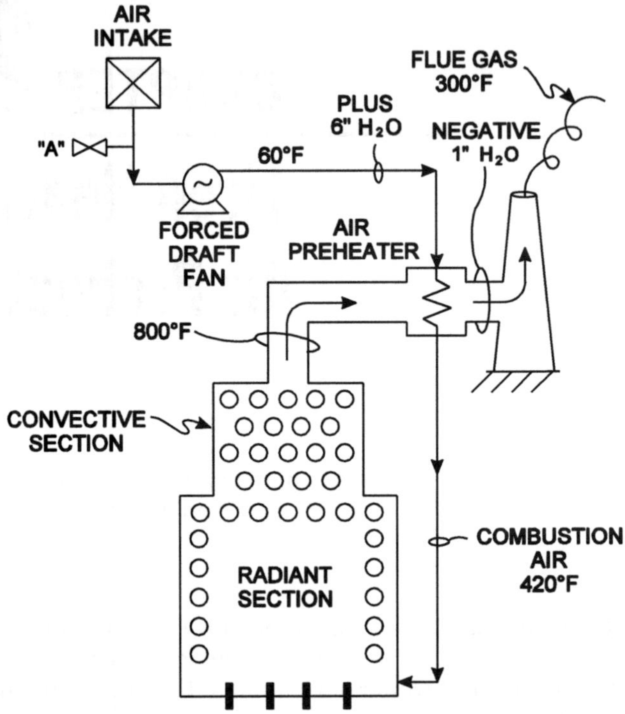

FIGURE 24-1 A fired heater of boiler equipped with a forced draft air preheater.

"But Norman," she interrupted, "That means that the heat duty in the radiant section will be constant. My question was, why does steam production go down, rather than remain the same?"

"It's like this. Your logic is correct as it pertains to the radiant section limitations. But since the operators are firing 10% less fuel, then they use 10% less air, because they are holding a constant 4% O_2 in the flue gas. With 10% less flue gas, the convective section heat recovery to your boiler feed water also drops by 10%. If the convective section contributes 30% of the boiler's duty, then you will lose:

$$(30\%) \times (10\%) = 3\% \text{ steam production}$$

"And that, Harriet, is the price we must pay for the improved energy efficiency associated with air preheat."

Note that while air preheaters do, in general, reduce fired heater capacity, this cannot be reasonably considered to be a malfunction of the air preheater. It is just not a well-advertised feature of air preheaters.

Preheating Air to a Vacuum Tower Heater

One of my first design projects as a consultant was the expansion of Chevron's #4 Vacuum Tower in El Segundo, California. Part of the Chevron expansion project was to increase the capacity of the vacuum heater by 15%. The heater was limited by high radiant heat density. I call this a heat flux limited situation:

$$12{,}000 - 15{,}000 \text{ Btu/hr/ft}^2$$

The ft^2 term refers to the exterior tube surface area in square feet in the radiant section. The 12,000 is my design number. The 15,000 is the sort of flux rate I see in well-designed, well-operated, and well-maintained vacuum heaters that are being pushed to their maximum operating limits. The Chevron unit was operating at a 15,000 to 16,000 Btu/hr/ft^2 heat flux rate.

Chevron had assigned a senior engineer, Mike Lowlaw, to handle this aspect of the expansion. My position on this portion of the project was that I did not know what to do, short of replacing the heater itself. Mike's proposal as to how to expand the heater's capacity was to add an air preheater.

My initial reaction to Mike's idea was quite negative. I explained that air preheat reduced heater capacity, if the radiant section is in a flux-limited situation. The reduced capacity was 5% to 10%, depending on the preheated air temperature.

Air Preheater Fouling and Delta P

When I first became involved with air preheaters in the 1970s, there were two prominent manufacturers:

- Decca—They produced an air preheat exchanger consisting of glass tubes. The tubes, I recall, were subject to excessive rates of breakage. They are currently not in widespread service, if at all.
- Lungstrom—A large, metal-mesh-type wheel rotated through the hot flue gas chamber and into the cold combustion air chamber. The air side had a propensity to leak massively into the flue gas side. This device is still in limited service. I've described it in greater detail in my book, *Troubleshooting Process Operations*, 4th ed. One of my first troubleshooting assignments as a consultant was on a Lungstrom air preheater for Tenneco Oil in Chalmette, Louisiana.

I believe that both Decca and Lungstrom are no longer manufactured. They have been replaced by a more ordinary heat exchanger device. Metal heat exchanger tubes are placed inside a large metal box, as shown in Figure 24-1. Often, there is also an induced-draft fan, as well as the forced-draft fan shown on my sketch. Most air preheaters of this type are absolute disasters. In the many refineries that I've visited, I cannot recall any plant where this sort of air preheater had not created major malfunctions of two sorts:

- Combustion air leaks in flue gas.
- Fouling and high flue gas side pressure drop.

Air Preheater Creates Draft Limitations

Unless you are burning natural gas (about 10 ppm of H_2S), there is going to be a substantial amount of sulfur in your fuel. For refinery fuel gas, the H_2S concentration will be 100 to 150 ppm H_2S. However, due to occasional problems in your amine plant, this value may increase to 1,000 to 10,000 ppm (1%) of H_2S. For refinery fuel oil, the sulfur content will typically be 5,000 ppm (0.5%) to the 5 wt% sulfur the Coastal Refinery used to burn at their refinery in Aruba. Some of the sulfur burned is oxidized to SO_2. A smaller quantity is oxidized to SO_3, which then reacts with the H_2O in the flue gas:

$$SO_3 + H_2O = H_2SO_4$$

The sulfuric acid will precipitate as a corrosive mist at:

- Around 300°F, if the sulfur content of the fuel is about 100 ppm.
- Around 450°F, if the sulfur content of the fuel is about 1 wt%.

The calculated precipitation temperature is rather meaningless. Let's say you measure a 390°F stack flue gas temperature. You now think, "Lieberman has said that if I'm burning 100 ppm H_2S gas, I'm 90°F above the H_2SO_4 precipitation temperature. So all is okay."

But all is not okay. Because you have to take several real-world factors into account:

- Sometimes your H_2S will not be 100, but 1,000 or 10,000 ppm.
- The 390°F is the average outlet flue gas temperature. Some portion of the flue gas will be a lot colder.
- Some portions of the tubes may be quite close to the 60°F air intake temperature.
- There will be cold air leaks into the hot flue gas side of the exchanger.

The problem is localized precipitation of H_2SO_4, which causes localized corrosion. Weak sulfuric acid (30%) is infinitely more corrosive than strong (98%) sulfuric acid. The product of the corrosive reaction is:

$$H_2SO_4 + FE = FeSO_4 + H_2$$

The iron sulfate is a gray, sticky deposit that absorbs moisture and SO_3 fumes from the flue gas. It's full of acid itself. I've had these deposits on my fingers and they burn my skin just like the sulfuric acid I made in the 1970s in Texas City. These deposits are themselves quite corrosive and promote more iron sulfate formation. And the resulting corrosion will cause cold air leaks, which create localized areas of cooling and hence localized H_2SO_4 precipitation in the lower pressure flue gas side of the air preheater.

The sludgy iron sulfide scale accumulates on the flue gas side of the air preheater. This causes high delta P on the flue gas side of the air preheater. To the operators, the problem manifests itself as loss of draft or high firebox pressure. The operators are then forced to open the flue gas bypass valve around the air preheater. The flue gas that's still flowing through the preheater gets even colder as its flow diminishes. This promotes more widespread H_2SO_4 precipitation, which increases $FeSO_4$ fouling. Which increases flue gas delta P. Which causes the operators to open the flue gas bypass further. Which then causes the . . . But I guess you understand my point.

Example of Flue Gas Bypassing

I had been hired by Texaco to provide a revamp process design for their crude unit. The stated limit was heater capacity. The operators observed that the heater was limited by draft. The following indicate a draft limit:

- The heater's outlet temperature is reduced as the fuel rate increases.
- The secondary burner air registers at the base of the heater are partly closed.
- The pressure below the bottom of the convective tubes is a slight positive draft (0.05 inches of water negative pressure). That is, opening the air registers to get more combustion air would cause a positive pressure to develop below the bottom row of convective tubes.

The Texaco air preheater was forced draft (i.e., from a forced-draft air blower) on the combustion air side and natural draft on the flue gas side. The hot flue gas was forced to flow through the air preheater by closing the stack damper, as shown in Figure 24-2, above the hot flue

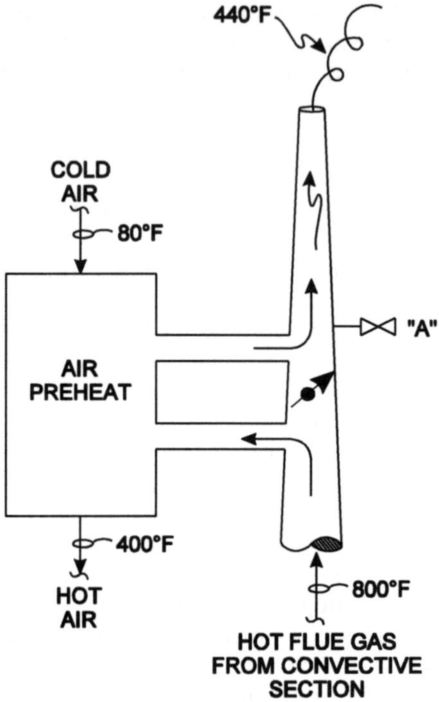

FIGURE 24-2 Opening stack damper promotes air preheater corrosion, but improves draft.

outlet duct to the preheater. The stack damper position was indicated by a large black arrow on the stack, quite visible from ground level.

I checked the pressure loss across the flue gas side of the air preheater with a portable manometer. It was over an inch of water. The total draft developed by the stack was also about 1 inch of water. The design pressure loss through the flue gas side of the preheater was 0.2 inches of water. Thus, the current delta P through the flue gas side of the air preheater was five times the design delta P. Discussion with the unit operators indicated that when the air preheater was installed 5 years before, they had run with their burner air registers wide open. But lately, they had to progressively restrict airflow with the registers to avoid the positive pressure at the top of the crude unit's firebox.

Clearly, the operators were describing progressive fouling due to $FeSO_4$ deposits in the flue gas side of the hot air preheater. Texaco's erratic H_2S extraction from refinery fuel gas contributed to the problem. Especially when a high fuel gas H_2S level corresponded to the occasional 20°F nights.

So I asked Rick Beabien, the Texaco refinery rep on our project, to open "the arrow" just a bit.

"No, Norm, our operation instructions state clearly that the arrow has to be kept in a horizontal direction."

"But Rick, let's try just one little degree. Just for a moment," I suggested. "Just to see what happens."

"We're not supposed to," insisted Rick.

"Yeah, Rick, but just for a minute. As an experiment. Just make a real small adjustment, then we'll go to lunch. It's already 11:30."

But when we came back from lunch, the operators had increased the crude charge rate to the heater by 2,000 BSD. And Rick could not very well ask the operators to reduce the refinery crude rate and reduce Texaco's profitability. So we opened the arrow just a bit more the next day and crude charge rate increased again. Within a week, the arrow was in a vertical position and the draft limitation on the heater was gone.

The results of opening the air preheater bypass valve (i.e., the arrow) were:

- Reduced air preheater delta P due to less flue gas flow, which directly increased the draft at the top of the heater's firebox or radiant section. Bypassing 30% of the flue gas reduced delta P by half.
- Increased stack draft. The heater flue gas is less dense and thus creates more draft. For a 150-foot-high stack, an extra 50°F of the stack temperature will create 0.1 inches more water draft in the radiant section.
- Opening the bypass arrow reduces the air preheater flue gas outlet temperature. This promotes sulfuric acid precipitation and fouling with $FeSO_4$ deposits. Which promotes cold air leaks and further fouling, which then causes... But, I guess you understand my point.

I suppose that I forgot to mention to Rick the longer-term detrimental effects of opening the arrow on the air preheater's mechanical integrity.

Air Preheat Using Low-Pressure Steam

I recently presented a seminar for the five refineries in South Africa. The attendees felt, as I do, that the flue gas versus combustion air preheater shown in Figure 24-1 will fail in a few years. Thus, the common practice in South Africa is to preheat air with low-pressure steam. I'm not too sure this makes much sense, either. Probably the air cannot be heated much above 220°F with 30 psig steam. Also, the energy value of the 30 psig steam has to be close to zero for air preheat with steam to be economical. On the other hand, air versus low-pressure steam preheat will definitely not result in the sorts of failures that are described in this chapter. And, if nothing else, they will at least result in the recapture of valuable hot, clean steam condensate for recycle to the boiler feed water deaerator.

Preheater Leaks Reduce Combustion Air

Figure 24-1 shows the pressure of the air side of the preheater is 7 inches of water greater:

$$+6 \text{ inches } H_2O - (-1 \text{ inch } H_2O) = +7 \text{ inches } H_2O$$

than the flue gas side. You may be sure that if corrosion failures have occurred, that combustion air will flow from the cold side (air) into the hot side (flue gas). To calculate the percent of the forced draft discharge combustion air leaking into the stack with the hot flue gas, proceed as follows:

Step 1—Obtain an O_2 analysis of the convective section outlet. Let's assume that is 2%.

Step 2—Obtain an O_2 analysis of the air preheater outlet. Let's assume this is 9%.

Step 3—(21% – 9%) ÷ (21% – 2%) = 63%

The 63% is approximately the percentage of combustion air flowing to the burners.

Step 4—The percentage of combustion air leaking through the air preheater directly up the stack is then 100% – 63% = 37%.

Incidentally, the preceding values were taken from the Petronor refinery in Spain on a relatively modern crude unit. This unit was so severely limited by combustion air that the operators had opened all of the sight doors on the radiant section box, in the hope of drawing in a few stray molecules of atmospheric oxygen. Yet they were quite unaware that around 37% of their combustion air supply was leaking through their air preheater.

The reader may also note that the flue gas stack temperature of 300°F shown in Figure 24-1 was quite cold compared to the 420°F preheated air temperature. Those readers familiar with the design of heat exchanger equipment such as air preheaters may justifiably claim that the indicated **temperature cross** of 120°F (i.e., 420°F to 300°F) is impossible, and that my data is flawed. It's not my data that's faulty. It's the air preheater. It's all the cold 60°F air leaking up the chimney and quenching the hotter flue gas.

I can summarize the detrimental effects of corrosion due to sulfuric acid attack in air preheaters as follows:

- Loss of draft due to a high delta P caused by accumulation of iron sulfate deposits on the flue gas side of the air preheater.
- Loss of combustion air directly up the stack and bypassing the furnace firebox.
- Reduced air preheater heat transfer efficiency due to quenching the hot flue gas with cold air in-leakage.

Loss of Draft Through an Empty Air Preheater Exchanger

At the Murphy refinery in New Orleans, many years before my visit, they had experienced all of the malfunctions with their crude unit air-preheat that I have just enumerated. So they bypassed 100% of the combustion air around the air preheater exchanger. All the tubes were removed from the air preheater, and the air inlet and outlet nozzles were both blinded-off. This was done a decade before my visit. So I do not know how much benefit resulted from this change. But what I do know is that the crude unit capacity was still limited by the available heater draft in 2008. What Murphy Oil had failed to understand was that a large draft loss associated with the defunct air preheater still existed. That loss was associated with the acceleration through the inlet nozzle of the hot flue gas through the preheater box. I measured the pressure drop, or draft loss, between the inlet and outlet of the empty air preheater box. But I always like to calculate a variable and then compare it to the measured parameter. To calculate the draft loss, I used the following equation:

$$\text{Delta P} = 0.34 \cdot \frac{\text{Dv}}{62} \cdot \frac{(\text{Vg})^2}{28}$$

where Delta P = Draft loss in inches of water.
Dv = Density of flue gas, in lb/ft^3.
Vg = Velocity of flue gas through the 36-inch outlet duct, ft/sec.

The pressure loss through the inlet duct was quite small, as it was much larger than the outlet duct. The inlet duct was designed for hot flue gas, while the outlet duct (36 inches) was only sized for the cooler flue gas.

Both my calculated and observed draft loss values were about 0.5 inches of water. This was about 40% of the total available draft. Installing a 36-inch bypass duct would have reduced the draft loss by about 0.3 inches H_2O, and thus permitted the firing rate to be increased by roughly 15%.

What happened? You may think Murphy Oil gave me a gold watch or at least a certificate of appreciation. Not quite. Somehow, I accidentally insulted the technical manager and Murphy canceled my service contract. So I don't know what happened. But I suspect nothing.

Overheating Combustion Air

If we reduce the combustion airflow to a fired heater too much, partially oxidized hydrocarbons in the form of aldyhydes, ketones, carbon monoxide, and light alcohols will flow into the convective section. For a natural or induced draft, and even for most forced-draft heaters, there will be a positive draft or a small negative pressure in the convective section. Then, tramp air leaks may reignite the combustible components in the flue gas. This is called **afterburn** or **secondary ignition**. Afterburn may be quite damaging to the convective section carbon steel, finned tube bank. Obviously, afterburn can also damage the air preheater.

At the new delayed coker unit at the Co-op refinery in Saskatchewan, Canada, the feed heater's firebox would periodically increase in temperature from 1,300°F to over 1,600°F over a day or two. The problem was afterburn. Combustion in the flue gas caused the air preheater hot air outlet temperature to rise to about 1,000°F. That's just a guess. There was no temperature indicator on the preheated air on the panel, and it was too cold to go outside to read the local dial thermometer. But what was not a guess was that the excessively preheated combustion air would cause the firebox temperature to greatly increase. In turn, the tubes in the heater would coke off due to the high radiant heat density.

Adding more combustion air had the effect of suppressing afterburn, reducing the air preheater outlet temperature and cooling the firebox. On this unit, the rate of coke formation inside the heater's tubes was reduced by a factor of three or four by simply adding more air to the heater. I've always felt that this was one of my most satisfying troubleshooting assignments.

Fire Damage to an Air Preheater

The following incident occurred at the American Oil refinery in Whiting, Indiana, in the early 1960s. I heard about it secondhand. A boiler in the power station was equipped with an older air preheater that had been in operation for several years. Apparently it leaked. But no one was particularly concerned.

One day a new oxygen analyzer was installed downstream of the air preheater. It read 6% O_2, but the reading was ignored by the operators. Then the unit engineer attended an energy conservation seminar. There she was told that excess O_2 targets should be 3%, not 6%. So she required the operators to reduce the combustion air rate, with the target of 3% to 4% O_2 in the air preheater flue gas effluent. But the air preheater was leaking. So the O_2 analyzer located downstream of the air preheater was reading several percent too high because of the leaks. Thus, when the operators reduced the O_2 analyzer reading from 6% to 3%, the firebox must have gone grossly air deficient. The indication of this was a fire in the air preheater. The air preheater was destroyed. The partially combusted hydrocarbons from the firebox, mixed with the air drawn into the flue gas side of the air preheater, reignited and burned up the air preheater.

If one is to adjust excess air based upon an online oxygen analyzer (not a good practice, as discussed in my book, *Troubleshooting Process*

Operations, 4th ed., or see Chapter 22, "Natural Draft-Fired Heaters"), the analyzer ought to be upstream of the air preheater. Then a portable analyzer should be used to monitor the oxygen content downstream of the air preheater, on the flue gas side, for air leaks. Checking once a month is a reasonable frequency.

As far as temperature indication is concerned, monitoring both the flue gas outlet and the air outlet from the air preheater will alert operators that they are either:

- Overheating the combustion air due to secondary ignition in the heater's convective tube section.
- Or, overheating the air preheater itself due to a combination of lack of combustion air to the firebox and air preheater leaks.

The correct operating response to either of these two malfunctions is to increase the combustion airflow and/or cut the fuel gas rate to the burners.

Restrictive Air Intake

I used to be the tech manager at the Good Hope refinery in Norco, Louisiana (now Valero's St. Charles Refinery). In 1990, after the plant had been idle for 7 years, it was restarted. One of the big problems encountered after restarting the vacuum tower was lack of combustion air to the vacuum heater. The owner of the refinery, Jack Stanley, asked me to see if the problem could be resolved without shutting down the vacuum tower.

The vacuum heater was equipped with a conventional air preheater, as shown in Figure 24-1. Naturally, I assumed that the air preheater was leaking. Or perhaps after years of sitting idle, residual sulfuric acid from the flue gas side had infiltrated into the air side and gradually caused iron sulfate deposits to form and restrict the airflow.

Note that I was not concerned with excessive fouling on the flue gas side of the air preheater. Had this been a serious problem, then there would have been a lack of draft (i.e., a positive pressure would have been observed below the bottom row of convective tubes).

As this was not really a combustion-type malfunction, but a fluid flow malfunction (the fluid being air), I decided to run a delta P survey on the air supply to the heater. To do this, I used the high-tech instrument shown in Figure 24-3. This patented device will accurately measure both positive and negative pressures. I produce this state-of-the-art instrument for sale. Contact Lieberman Industrial Instruments, Inc. Model MP-43. Cost: $398 plus S&H.

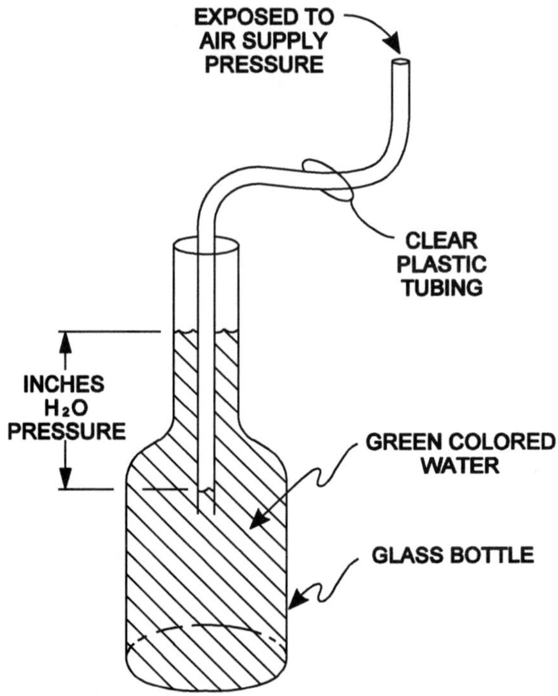

FIGURE 24-3 Measuring supply air pressure on air blower discharge.

The first point I checked was the suction of the air blower, shown as point A in Figure 24-1. Contrary to my expectations, point A was not atmospheric pressure, but a vacuum of 4 inches of water. Meaning the green colored water in my bottle had been sucked up in the tubing by 4 inches! This air blower was only designed to develop 8 inches of water differential pressure. As you can see, I was losing half of the available head across the air intake screen.

This screen was just a metal box made of expanded metal grating. I had walked past it 100 times when the plant was running in the early 1980s. Now it looked different. It used to be a nondescript rusty color. But now it was a bright blue. Pretty much the entire unit was a bright blue, as it had been painted during the 7-year shut-down. It rather looked as if the painters, with lots of time, had given the air intake screen several heavy coats of blue paint. In so doing, they had restricted the air intake screen open area by a noticeable amount.

So I thought, "I'll be a hero!" I'll unbolt a screen panel and increase the air blower suction pressure. Then we'll have lots of air to the heater. Next, I'll call up the owner, Jack Stanley. He'll likely reward me with a gold watch or at least a certificate of appreciation. So I took my 12-inch wrench and pulled all the bolts out of the rear end 4 × 4–foot screen. And it wouldn't budge. Why?

$$4 \text{ feet} \times 4 \text{ feet} \times 144 \text{ in/ft}^2 \times 4 \text{ inches} \div 28 = 320 \text{ pounds}$$

where The 4 feet × 4 feet × 144 = area of screen in square inches
4 inches = screen delta P, in inches of water
28 = inches of water per psi

Next, I called over the operations manager, Bill Wilguard, to help me. But Bill said, "Norm. Let's get the maintenance guys to do this. Come on, I'll buy you lunch. Just got to make me a quickie call first. The maintenance guys will get this screen panel off by the time we get back. You like Mexican? Maybe we should pull off a couple panels?"

But Bill's "quickie call" was to Jack Stanley. He needed to explain how he had solved the vacuum heater air limitation and that vacuum charge would be up to 80,000 BSD by 6:00 p.m. And how about my gold watch or letter of appreciation? I guess that's why we have Mom.

Correct Design Alternatives for Air Preheater

There are two ways to design air preheaters that recover waste heat from the hot flue gas, and that still avoid the dual problems of plugging and leaking, due to sulfuric acid cold-end precipitation. In practice, excursions in the H_2S content of refinery fuel gas, combined with occasional low ambient conditions, make ordinary air preheaters inappropriate for refinery use, or for process plants that burn fuel oil with sulfur levels as low as 0.1 wt%.

My preferred method for air preheat is the technology we used at the Good Hope refinery. A circulating hot oil stream was used to recover heat from the flue gas in the convective section. The circulating oil entered the convective section about 350°F and was heated to between 500°F and 600°F, in a conventional convective tube bank. Next, the hot oil would flow through the air preheater, where the air was heated to perhaps 300°F.

I call the heat transfer medium "oil," but this is not quite right. It was a specially purchased, sulfur-free liquid, designed to facilitate heat transfer and retard thermal degradation.

The whole process was 100% trouble free and never lost any efficiency. But it had several drawbacks:

1. It was the most expensive air preheat system available. Why Jack Stanley, the owner, who never cared much about energy efficiency purchased this facility, I could never grasp. But to me, it was worth the extra investment.
2. The hot oil circulation pump was a substantial consumer of electric power.

3. Rather than have direct heat exchange between the hot flue gas and the cold combustion air, heat exchange was indirect via the hot oil. This type of indirect heat transfer likely reduced the ultimate air preheat temperature by 50°F to 100°F.

These three debits are of a theoretical nature, because I'm comparing this indirect method of heat transfer (which works consistently) to the direct method of heat transfer between hot flue gas and combustion air (which fails consistently).

The other method that I've seen function reasonably well is practiced by Total-Fina-Elf in their European refineries. It's just an ordinary air preheater with one addition. Cold ambient combustion air is first heated from ambient conditions to about 110°F to 120°F with low-pressure steam. In effect, this eliminates the cold-end corrosion problem. However, at least 20% of the energy-saving potential of air preheat is lost because some of the air preheat comes not from flue gas, but from steam. Of course, if there's an excess of 30 psig steam in the plant, this steam consumption is of no consequence. Also, the forced-draft fan requires an extra few inches of water head due to the extra steam versus air heat exchanger pressure drop.

Rules of Thumb for Temperature Gradients

I have several additional methods, based on air and flue gas temperatures, to identify combustion air leaks in air preheaters, without measuring the increase in oxygen in the flue gas. I'll make these temperature checks before going through the extra effort to sample the oxygen levels in the flue gas. I always like to have several methods to prove a particular malfunction. It's kind of like getting married. You can't be too careful.

The heat capacity of the flue gas is about 10% greater than the combustion air. That's because the products of combustion contain both the combustion air and the fuel. Thus one would expect that a temperature drop of the flue gas in the air preheater of 100°F would increase the air temperature by 110°F.

However, because of ambient heat losses from the hot flue gas, plus the effect of minor cold air leaks into the lower-pressure flue gas, I assume the temperature changes to both the flue gas and the combustion air will be approximately equal. Now, let's say we have the following temperature profiles in and out of our air preheater:

- Air inlet = 70°F
- Air outlet = 350°F
- Flue gas inlet = 660°F
- Flue gas outlet = 240°F

The temperature rise of the air is 280°F. The temperature drop of the flue gas is 420°F. The flue gas temperature loss is 140°F, or 50% greater, than the air temperature rise. This indicates that about 40% of the cold combustion air is leaking through the air preheater directly into the hot flue gas and quenching the flue gas temperature.

An alternate method of using the air preheater temperature profile is based on the concept of **temperature cross**. Meaning, in this case, that the hot-side outlet temperature (flue gas 240°F) is colder than the cold-side outlet temperature (combustion air 350°F). This sort of temperature cross is not very likely to be achieved in a simple heat exchanger such as an air preheater. Yet in the data presented above, the flue gas outlet is indeed much colder than the combustion air outlet temperature. Again, this is an indication that a large percentage of the flue gas is really cold combustion air leaks.

Having made these approximate observations based on temperature, I would then field check my oxygen levels in the flue gas entering and exiting the air preheater, using a portable oxygen analyzer to rigorously determine the extent of the air leakage rate through the air preheater tubes.

CHAPTER 25

Disabling Safety Systems

Norm, never start up a cat cracker on oxygen-enriched air.

—Jack Stanley, owner of the Good Hope refinery,
after we melted the regen cyclones, 1980

The main problem that I have observed with safety systems is not their tendency to malfunction accidentally, but rather the operators' propensity to disable them on purpose. For example, an overpressure safety relief valve is leaking. If there is a gate valve underneath the relief valve, it's quite tempting to block-in the leaking relief valve until it can be reseated. Having isolation valves below relief valves is legal if there are spare relief valves and the isolation valve is locked open with a chain. But locks can be opened, and the spare relief valve may be blocked-in, even though its sister valve is also blocked. Likely, the only foolproof method (or Cajun-proof, as we say in Louisiana) is not to provide isolation valves under the relief valves in the first place.

Sometimes the use of safety systems creates a safety hazard. I have in mind the terrible fire at the Piper Alpha offshore rig in the North Sea, off the coast of Scotland. I was hired to help investigate this tragic fire that cost about 150 lives. One of the factors that contributed to the fire was that the automated fire water suppression system had been turned off. The seawater intake to the fire water pumps was located rather close to where the underwater divers were doing some welding. For fear of causing a diver to be drawn into the seawater intake should the system start up on auto, the automatic feature of the fire water suppression system had been reportedly switched to manual operation. But then, nobody knows for sure, as the platform was utterly destroyed and most of the personnel were killed.

Incidentally, Occidental Petroleum did not like my report. I concluded that the accident was related to the flare system freezing up with hydrates (water plus light hydrocarbons, which can form a solid between 40°F and 60°F). The official conclusion was that two subcontractor pipe fitters had left a flange open on a spare pump. The two guilty fitters were conveniently dead, which limited Occidental's legal exposure.

But we don't have to go so far back in time for examples of safety systems being disabled on petroleum drilling platforms. A much more infamous incident has occurred right here, recently, near New Orleans, Louisiana.

In April 2010, BP had a newly drilled oil well blow out in the Gulf of Mexico, about 40 miles from my home. The well was equipped with a "blow-out preventer." This is a guillotine-type valve to cut off the well's flow by crushing the well's 20-inch casing. The valve is activated by pushing a button located on the drilling platform. The problem is that once the valve is activated, the casing crushed or cut, the cost to reestablish the integrity of the well is a very large percentage of the $100,000,000 cost of drilling the well. So what usually happens in such circumstances is that to prevent someone from accidentally pushing the button, it's disabled. Or, since it would take a great deal of self-confidence—even courage—to push the button, no one does so, even when a valid emergency situation arises.

This also reminds me of an incident that happened on my sulfuric acid regeneration plant in Texas City in 1975. Basically, air plus SO_2 feed gas were drawn into the suction of a giant centrifugal blower. The blower suction pressure operated at vacuum conditions of 5 inches of mercury.

The blower was driven by a stream turbine that could be tripped off from the control room. The turbine trip was activated by a large red button labeled, "Emergency Shut-Down." I saw that button several times a day for over a year, but never gave it too much thought. What I should have thought about were two questions:

- First, what constitutes an emergency?
- Second, what were the consequences of pushing the red button?

My failure to address these two issues ruined my career at Amoco Oil and to some real extent, my first marriage.

Acid Plant Shut-Down

Johnny Garza was the smartest guy on the unit. He worked on the panel during E shift. He was the smartest one of my 34 operators, five shift foremen, and two engineers. And certainly, he was a lot smarter than me. For example, one day a pipe fitter burst into the control room.

"Johnny, there's a big hole in the 30-inch plastic duct (actually wound epoxy glass pipe) about 20 feet upstream of the blower. I dun seen it when I climbed up the pipe rack to fix a steam leak. It's a big crack, Johnny," said the pipefitter.

John Garza now had two choices. One option would have been to call me. I would have raced down to the unit, placed a sheet of metal weatherproofing over the hole, and then wrapped it up with duct tape. After all, the system was under vacuum. Just upstream of the air leak was our air intake valve. The only effect of the leak was to cause us to run with the air intake valve mostly closed. And, when it comes to duct tape repairs, I am the Master. Anyway, that's what I think NOW, and not necessarily what I would have done THEN. But, I've had 35 years to consider the problem. And I've grown much smarter in the interim.

John's second option was to push the red button. Not having 35 years to review his choices, but 35 seconds, he selected the wrong option. As soon as the blower stopped, the pressure at the leak (which was upstream of the blower) went from negative to positive. The SO_2 generated from the oxidation of the spent H_2SO_4 blew out of the damaged 30-inch plastic pipe. A white cloud drifted across the unit. Seeing the cloud, a group of contractors took to their heels. One older fellow slipped and was engulfed by the choking white SO_2 cloud. He sued Amoco Oil for $500,000. During the subsequent proceedings, even the Amoco attorney came to dislike me. He wrote a nasty letter to Mr. Durland, the refinery manager. Apparently, the attorney did not believe me that this was all Johnny Garza's fault and that I was totally blameless.

What Constitutes an Emergency

If the option exists to trip off the plant in an emergency, the operators should be told beforehand what circumstances the management (meaning me) considered to be an emergency. One should never rely upon anyone, even your best operator, using good and proper judgment when under the pressure of time. I should have posted a note and then discussed with all the panel board operators the circumstances that would require pushing the red button. But then, one still has to allow the panel operator to exercise his or her judgment, at the moment of decision, because of unanticipated problems.

In this case, stopping the blower turned a noncritical problem into a serious incident involving personnel injury. After my poor and confused testimony was rejected by the judge, by both attorneys, and by an expert witness, I began to try to assemble a list of the circumstances in which shutting down the blower would do more good than harm. I've worked on this list for 35 years. Still, all I've got for my efforts is blank paper.

My career never recovered from the letter the company lawyer wrote to the plant manager, Larry Durland, who didn't like me anyway. Mr. Durland forwarded the nasty letter to Dr. Horner, the

V.P. for Refining, who also never liked me. After another year of struggle with my H_2SO_4 acid plant, I was demoted to a staff engineering position in Chicago. My soon-to-be ex-wife accused me of being a failure. Which was rather true, as I had never considered the consequences of pushing that damned red button before Johnny Garza enlightened me in 1975.

Sulfur Plant Hydrogen Sulfide Feed Gas Trip

In 1980, during the 130-day strike at Texas City, I worked as the chief operator at the sulfur recovery plant. The feed to the plant was hydrogen sulfide (H_2S). One evening I decided to check the zero point of the H_2S flow meter. I climbed up to the meter and opened the meter's bypass valve. Suddenly the entire sulfur plant tripped off:

- The H_2S feed gas was spilled to the flare. The flare's flame, which had been very small and yellow, became gigantic and sapphire blue.
- The air blower providing combustion air to the sulfur plant's "main reaction furnace" shut down completely.
- Without H_2S or air, flow through the sulfur plant was lost entirely.

Having thought about this incident for over 30 years, I still cannot grasp the logic of stopping both the H_2S feed and the combustion airflow because the indicated H_2S flow went to zero. If, for some reason, the H_2S flow really stopped, how would it have made the situation worse to allow the airflow to continue for a few minutes more? And, more to the point, why spill the entire H_2S flow to the flare when the indicated H_2S flow fell below a certain value? Of course, it's dangerous to continue the H_2S feed gas once the airflow stops. But why was the airflow tripped off in the first place?

Oh! I forgot to mention that Mr. Durland, the refinery manager, had placed me in charge of the night shift on the sulfur plant during the strike, because I had designed the plant. Apparently as the designer, I still had not learned my dual lesson about emergency shut-downs:

- When is it reasonable to activate such a shut-down?
- Will shutting off a flow make the problem better or worse?

Bypassing the Startup or Trip Panel

Now that the sulfur plant in Texas City had tripped off, I had a real problem to contend with. With no flow, the reactors and condensers would cool. Within a few hours, they would solidify with sulfur.

Also, the main reaction furnace would cool below the autoignition temperature of H_2S. But my immediate problem was the need to satisfy the automatic startup panel sequence. This sequence required that certain steps be followed before the H_2S feed control valve could be opened. These steps were:

- Air purge to remove combustibles from the main reaction furnace.
- Light the fuel gas pilot light.
- Reheat the main reaction furnace using fuel gas.
- Reintroduce the H_2S feed gas.

The problem with following this procedure was that the electric igniter on the pilot light was subject to malfunctions. So it was standard unit practice to rush into the shift foreman's office and grab the "key" that hung on the wall. This was the bypass key. It allowed one to bypass all the safety interlocks and trips on the panel. So the actual steps to be followed when the feed gas was lost to the sulfur recovery plant were:

- Turn the key in the panel to the trip bypass (i.e., off) position.
- Open the H_2S feed gas valve and try to relight the gas off the hot refractory.
- If this was unsuccessful, we had a portable propane torch that was used to relight the main reaction furnace.

Now, suppose I was not immediately successful in relighting the H_2S by autoignition off the hot refractory walls, and decided to wait a few moments. I could then have blown up the furnace. The dual lessons are then:

- Make sure a safety trip is actually appropriate, and that it does not cause a greater hazard than it is meant to prevent.
- Safety interlock systems must never be bypassed without written approval of a senior management representative for each individual occasion. If the fuel gas pilot light igniter malfunctioned, it should have been fixed rather than just abandoning the interlock startup panel.

Flame-Out Detector Malfunction

A malfunction in a safety trip may cause a far greater hazard than the problem it was intended to protect against. At the Huntway Asphalt Plant in California, they operated a small (20 mm Btu/hr) vacuum tower fired feed preheat furnace. There were six natural gas burners. Each burner was equipped with its own flameout detector and trip.

When the flame would blow out due to excessive burner tip pressure, the fuel gas to that individual burner would trip off—which superficially seems like a reasonable precaution. But here's what would really happen:

- The fire eye of one of the six flameout detectors would become slightly obscured with soot. At higher firing rates, the flame would very slightly lift off the burner tip. The fire eye would incorrectly interpret this as a burner flameout and trip off the fuel gas to that burner.
- The heater outlet temperature would drop by 20°F.
- The fuel gas regulator, which was on TRC, would open.
- The gas pressure to each of the remaining five burners would increase by 30%. This caused higher burner tip pressure and a greater tendency for burner flame liftoff.
- A second burner would then trip off, which within 2 minutes would cause all the burners to trip.
- An operator would then rush out of the control room with a flaming torch and relight the burners manually, with all control valves and trip valves bypassed. With the furnace temperature dropping fast, and the heater outlet dropping faster, there was no time to light the pilot light. Just open the main gas supply to each burner and hope the heater did not explode.

This is an example of the need to consider safety issues in their real-world context. It's true that if a burner flames out, it would be safer to stop the fuel gas flow to that burner. But on the other hand, if the effect of shutting one burner down due to a flame detection malfunction is to trip off all the burners, then an intended safety feature has created a positive safety hazard.

After all, even if one burner is blown out and the other five burners are still ignited, an explosion could hardly occur. But if all six burners have to be relit without first igniting the pilots, that could easily cause an explosion. I'm not one to draw general conclusions from limited experience. But in this case, I concluded that a minor malfunction with one of six flame scanners could well result in a major equipment failure. In conclusion, the scanners were disconnected from the fuel gas trip valves.

Intentionally Disabled Safety Devices and Trips

One of the areas of the American Oil delayed coker in Whiting, Indiana, that always frightened me was the top coke cutting deck. I was afraid of the 3,400 psig jet cutting water pressure. "Suppose," I thought, "The water accidentally comes out of the cutting head as

I walk by. It'll cut me in half." Then I would glance over at the massively robust 4-inch gate valve at the rail, see that it was shut, and my fears would be relieved.

Years passed. American Oil became Amoco, which became BP. The cutting water pump controls were automatically interlocked with the position of the coke cutting water head. That is, the cutting head had to be inside the coke drum before high water pressure could appear at the cutting head. I don't know if this safety interlock system was bypassed or malfunctioned. What I do know is that an operator on the cutting deck was cut in half by high-pressure water from the coke cutting head. I have always considered the numerous safety interlocks associated with coke drum operations to be appropriate. Yet I've observed how often such interlocks are bypassed to speed the coke drum cycles.

As an operating supervisor for Amoco in Texas City, I often wondered why overspeed trips on the steam supply to many turbines had been disabled. Meaning, an operator had tied a piece of wire around the trip lever to prevent it from unlatching. Typically, a turbine's maximum rated speed was 3,600 rpm. At 5% above this speed, or 3,750 rpm, the steam flow was supposed to be automatically tripped off to protect both the turbine and the pump it was driving from damage.

I found out only much later in my career why operators **wire up a trip**. It's because the more complex **governor speed control valve** (Words in bold are defined in the glossary.) is malfunctioning. Without an effective method to control the motive steam flow, the speed of the turbine tends to increase as the load on the pump it is driving diminishes. If, due to normal process fluctuations, the load on the pump becomes momentarily small, the turbine can speed above 3,750 rpm and trip off. To avoid this sort of crash shut-down of the entire process unit, the operators wire up the trip. And often by so doing they may have selected the lesser of two evils (see Chapter 31, "Steam Turbine Drivers").

Thus, when operators wire up trips, management may make one of two possible responses to this inherently dangerous practice:

- One: The Amoco method. Issue a decree forbidding the operators to deactivate or disconnect all existing interlock safety devices regardless of circumstances.
- Two: Fix the defective governor speed control valve mechanisms, the failure of which caused the operators to wire up the steam overspeed trips in the first place.

At both the Exxon Chemical plant in Baytown, Texas, and the Coastal Asphalt plant in Corpus Christi, Texas, serious injuries were sustained when a turbine oversped because the trip had been intentionally disabled. In both cases, it wasn't the trip that first malfunctioned, but the governor speed controller.

Incinerator Pilot Light Malfunction

For those of my readers who do not own an older home with a gas furnace or a gas fueled hot water heater, I'll explain how a pilot light works. First, you should understand that the purpose of the pilot light is twofold:

- It provides a source of ignition to the gas from the main burner.
- It provides electric power to keep the electrically operated gas supply valve open to the main burner and to the smaller pilot light itself.

Why is the pilot light important? Well, let's assume a gust of wind blows out the flame in a heater. If it were not for the automatic shutoff valve, which is only kept open by the electric current generated by the thermocouple wire located next to the pilot light, the heater box would fill with gas. Then any spark, perhaps from the electric motor used on the furnace fan, would explode the gas. To relight a pilot light that has gone out:

- Wait a few minutes for any residual combustibles in the heater box to dissipate.
- Manually depress the pilot light valve to start gas to the pilot. Note that the pilot gas flow rate is only about 1% of the gas rate to the main burner. Thus, the presumption is made that if you proceed reasonably quickly, not enough gas can be emitted into the heater box to be dangerous.
- Immediately light the small gas flow from the pilot light while keeping the pilot valve open manually.
- The pilot flame heats a thermocouple wire junction. This junction generates a few volts of electricity (Note: I've defeated the pilot's safety features with a 9-volt battery, but that's another sorry story).
- The low-voltage current generated by the thermocouple keeps the main gas valve open that supplies gas flow both to the pilot light and to the main burner.

Now for a tragic story. It took place at the Amoco Refinery in Whiting, Indiana, at their asphalt oxidizer, shown in Figure 25-1. The horrible incident occurred in 1989. Asphalt oxidation is a way of making road paving asphalt out of a crude unit vacuum reduced tar. It's also called air blowing. It's an archaic process in which air is bubbled through a tank operating at atmospheric pressure and about 530°F. The flue gases from the 50,000-barrel oxidizer tank consisted of:

- Nitrogen
- CO and CO_2

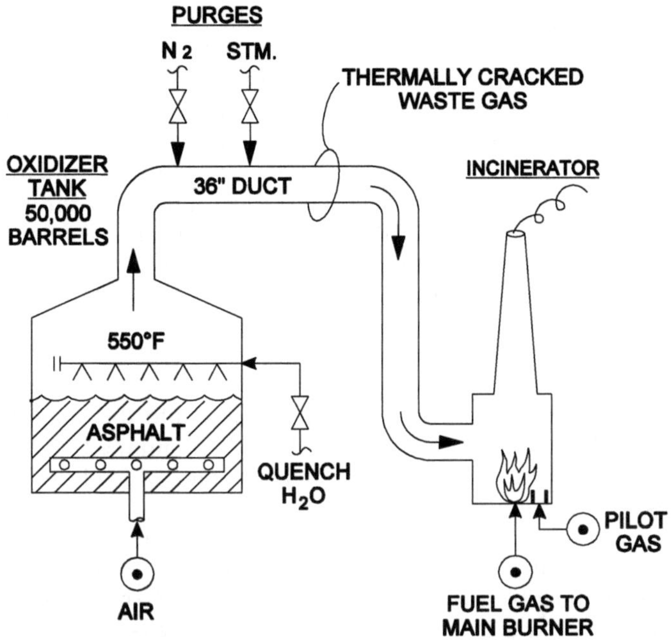

FIGURE 25-1 Failure of pilot light causes an oxidizer to explode.

- H_2S and SO_2
- Thermally cracked hydrocarbon waste gas

The flue gas, due to the sulfur compounds, had an evil odor, and was further oxidized in the incinerator, also shown in Figure 25-1.

One day, due to the operators accidentally allowing the oxidizer tank to overheat, the flow of thermally cracked waste gas increased a lot. So much waste gas was produced that the fire was smothered due to a lack of combustion air in the incinerator box. Fire was observed at the top incinerator stack, even though all combustion had been extinguished in the incinerator itself. Even the pilot light had gone out in the incinerator.

It was Sunday. The chief operator called the operating superintendent at home for advice. "Get that damn incinerator lit, pronto. We're violating our state permit. The plant manager will have a fit if he. . ."

So the operators cut the feed and the air supply out of the unit. This started to very slowly cool the oxidizer tank. Then they tried to relight the pilot with a propane torch. No luck!

So they called the operating superintendent back and he said, "You're smothering the pilot light with the steam from all the quench water (see Figure 25-1). Cut that damn quench water out. We got to get that incinerator fire back right away."

So the operators stopped the flow of the quench water. But they still couldn't reignite the pilot. So they called their supervisor back, and he said, "You need to stop the flow of the duct purge steam and purge nitrogen. They're smothering the pilot in the incinerator. We need to get that incinerator back in service right away."

Now the purpose of the purge nitrogen and purge steam injected into the duct was to prevent the propagation of fire back through the 36-inch duct and into the 50,000-barrel oxidizer tank. The tank was half full of asphalt and half full of thermally cracked gas.

With the purge streams to the duct shut off, the operators once again tried to relight the pilot. This time they were successful. Successful beyond their wildest imagination!

The pilot light reignited; combustion in the incinerator was also instantly reestablished. The fire flashed back into the 36-inch duct and exploded the oxidizer tank. The giant tank jumped 20 feet into the air, after rupturing hundreds of anchor bolts, which I saw during my inspection the following year.

The reader will note that the operators had shut off the oxidizer airflow shown in Figure 25-1. Where then did the air come from to cause a combustible mixture in the oxidizer tank? The source of this combustion air was in all probability tramp air leaks.

The 50,000-barrel tank was not a pressure vessel. It was a large tank that had been in service for a decade. As such, its roof probably had minor leaks. As the incinerator stack was creating a substantial draft, the oxidizer tank was under a slight vacuum, which apparently drew sufficient air into the tank to cause an explosive mixture to form. Note that reducing the formation of cracked gas by stopping the asphalt feed to the tank increased the probability of forming an explosive mixture in the oxidizer tank, rather than decreasing it. The reason for this is that when the pilot light first went out, the tank was probably too fuel rich to explode.

Thousands of barrels of hot asphalt poured over the four outside operators. Three were killed and one man, who was terribly burned, survived. I was retained as a consultant in the subsequent litigation. That's how I came to know all these details.

In 1979, when the oxidizer was built, I was the process coordinator for asphalt production technology for Amoco Oil in Chicago. I participated in the design of the oxidizer to the extent that I attended the P&ID and HAZOP reviews. Never once during these reviews was a rather obvious potential malfunction discussed.

"What was the correct operating procedure to follow, should the incinerator pilot light go out?"

"How could the operators safely relight the incinerator pilot light, when the oxidizer tank was still generating cracked gas?

My report summarizing my investigation of this accident concluded that the immediate cause of the explosion was the operators shutting off the flow of purge steam and nitrogen to the duct connecting the oxidizer tank to the incinerator. This is somewhat true. But the real problem is that the pilot light was not designed correctly. It should have been designed with an independent source of combustion air. Independent in the sense that if the incinerator's firebox was smothered with cracked gas flow from the oxidizer, the pilot light would continue burning. Then, with the pilot light still lit, the operators would not have had a reason to shut off the duct purge nitrogen and purge steam in the first place.

My mother would have made a great process safety engineer. She would always look at potential problems in the worst possible circumstances.

"Lou," she would scream at my dad, "Suppose we buy a house and the plumbing doesn't work. Mrs. Gold's house in Long Island burned down. Bad wiring. And how about roof leaks and foundation problems? One strong wind and your fancy new house in the suburbs will collapse into the street." Thus we continued to live in an old and decaying apartment house in South Brooklyn for 12 more years.

On the other hand, it's too bad my mom had not been advising BP before they began drilling a 13,000-foot-deep well, 5,000 feet below the surface, in the middle of Louisiana's seafood production waters.

"Listen Mr. Smarty Engineers," she would have shouted. "What are you going to do if your blow-out preventer doesn't work when your 5,000-foot-long, 20-inch casing breaks in the middle of the ocean?"

"Mrs. Lieberman," Tony Hayward, C.E.O. of BP would have explained, "Those sets of circumstances are unlikely to occur simultaneously."

And my mother would have responded, "Schlemiel! Hope for the best, but plan for the worst!"

Author's Note *The term "schlemiel" is a Yiddish term that cannot be translated into English. For many years, I thought this was my father's name. This proved not to be the case. However, my mother's propensity to worry about every conceivable possibility should form the basis for our HAZOP process review meetings and our P&ID design philosophy. Too bad the designers of the Tokyo Electric Nuclear Power Plant, who located their back-up diesel generators underground, didn't also plan for the worst.*

Furnace High-Temperature Fuel Gas Trip

Tenneco Oil in Chalmette, Louisiana, operated an aromatic fractionator with a fired, circulating reboiler. As a consequence of "mislocation" of the heater outlet TI, one of the worst accidents in the history of the refining industry occurred. Mislocation is in quotation marks because I'm not sure I've chosen the correct adjective to describe the resulting trip malfunction.

The problem began with the reboiler circulation pump shutting down for a reason I've never known. Flow through the heater stopped. The indicated heater outlet temperature began to decrease, rather than increase! But why?

Because the outlet TI thermowell was located 100 feet downstream of the heater itself. And with zero flow, the thermocouple cooled due to ambient heat loss. So the furnace fired even harder, and the tubes became red-hot. But the high-temperature fuel gas trip was not activated for the same reason. That is, the thermocouple was located too far from the heater outlet. Of course, when the circulation pump was running, this did not matter. But the designer had forgotten what my mother said about planning for the worst-case scenario.

Once flow through the heater was restored, the super-hot tubes caused a surge of vapor to blow out the trays from the aromatic fractionator. While the fractionator was being retrayed, natural gas was inadvertently admitted to the column. The natural gas exploded and killed approximately a dozen workers inside the column. The source of ignition was grinding in the tower.

Was the temperature indicator for the fuel gas trip sensor really mislocated? In retrospect, obviously yes! Would I, during a P&ID review, have added a note, "TI distance from the heater outlet to be as small as possible"? I'm quite sure I would not, as I have reviewed many such drawings and have never added such a note before I became aware of this awful incident. Now, I always try to draw upon my long experience with process plant malfunctions when I review a Process and Instrumentation Drawing for safety hazards due to equipment failures and operator errors.

Author's Comment

One of my clients, Flint Hills Refinery, recently blew up a process vessel. The source of ignition was H_2S in flare gas. On occasion, low-pressure flare gas with several percent H_2S would accidentally back into the vessel. The H_2S would react with the vessel walls to produce iron sulfide.

This vessel was normally in a caustic-propane environment. On occasion, air could also be drawn into the vessel. Normally, the mixture of air and propane vapors was too rich to explode. Meaning, the mixture was above the upper explosive limit of 12% propane in air. However, on this occasion, the ratio of propane to air dropped into the explosive region. Field investigation revealed the presence of residual deposits of iron sulfide on the inside surface of the destroyed vessel. The iron sulfide that auto-ignites at ambient temperatures provided the source of ignition that initiated the explosion.

CHAPTER 26

Vacuum Systems and Steam Jets

A theory without a test is bullshit.

—Norman P. Lieberman

The majority of my experience with vacuum systems is based on the converging–diverging steam ejector. I have also worked with several other methods of developing vacuum:

- Hogging or diverging steam jet
- Liquid ring vacuum compressor
- Liquid ejectors, using either water or diesel

I have seen all three systems working well. But 80% to 90% of the vacuum systems I have worked with, especially those servicing surface condensers, use the converging–diverging jet design. Therefore, I'll devote most of this chapter to this subject.

Converging–Diverging Ejector

This device may best be thought of as a two-stage compressor with no moving parts. The first stage of the compressor is the converging section. The second stage is the diverging section. Each section develops a separate compression ratio. By compression ratio, I mean the outlet pressure divided by the inlet pressure. For example:

$$\text{Compression ratio of converging section} = 40 \text{ mm Hg} \div 10 \text{ mm Hg} = 4.0$$

$$\text{Compression ratio of diverging section} = 100 \text{ mm Hg} \div 40 \text{ mm Hg} = 2.5$$

$$\text{Overall compression ratio} = (4.0) \times (2.5) = 10.0$$

These are typical design values for a properly performing jet operating within its design parameters of vapor loads, discharge pressure, and optimum motive steam conditions. Rarely do I observe in the field any single steam jet developing a 10:1 compression ratio. I've conducted pressure surveys on a thousand steam jets in commercial service. Perhaps 1 or 2% develop a compression ratio of more than 7 or 8 to 1. At a Conoco-Phillips chemical plant in Cedar Bayou, Texas, I've observed a single small ejector run at a 12:1 compression ratio. So anything is possible!

Let's now see how steam jets work. The first component of steam jets is the steam inlet nozzle, as shown in Figure 26-1. High-pressure motive steam flows through a specially shaped nozzle. It will help to think about your garden hose. Assume you have 40 psig city water pressure in your hose. As the water escapes through the nozzle, the 40 psig water pressure is converted to velocity. The greater the pressure of the water in the hose, the greater the velocity of the water escaping from the nozzle.

As the high-velocity steam enters the inlet of the diffuser shown in Figure 26-1, it starts to compress the noncondensible vapor drawn into the mixing chamber. I have read in some books that the motive steam "entrains" the noncondensibles. This is wrong. The noncondensible gas flows into the mixing chamber for the same reason that any gas flows into the suction of any compressor. It flows towards the inlet of the diffuser because gas naturally flows from an area of higher pressure to an area of lower pressure.

The kinetic energy required to accelerate the motive steam to sonic velocity as it enters the diffuser inlet comes from:

- The pressure of the steam
- The temperature of the steam
- The latent heat of the steam

This means that as the motive steam escapes from the steam nozzle, it cools and also partly condenses. Thus, it is normal to have water droplets blowing into the diffuser. I'll discuss the subject of

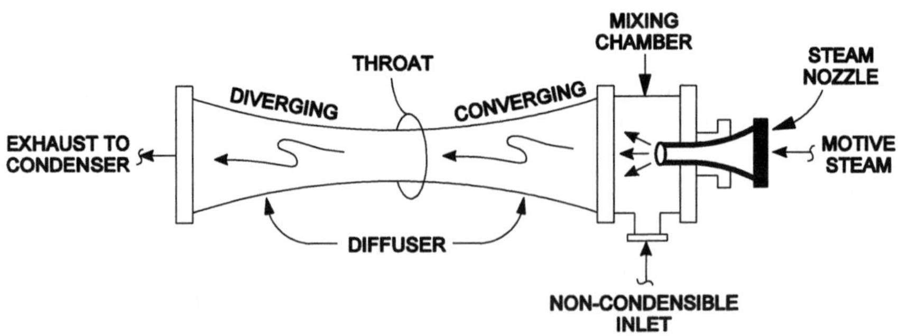

FIGURE 26-1 Components of a converging–diverging steam jet.

reduction in the steam temperature more completely in the next section, where I'll describe the effect of moisture on vacuum jet performance.

As you can see from Figure 26-1, the cross-sectional area of the diffuser diminishes as it approaches the diffuser throat. This forces the vapor velocity to increase. At some point, at or upstream of the diffuser throat, sonic velocity is, by design, supposed to be achieved. If this happens, the jet is said to be in **critical flow**. The flowing vapor has exceeded the speed of sound. This creates a pressure wave front that I call the "sonic boost." It may compress the combination of flowing steam and noncondensibles by a factor of 4 to 1. To get the sonic boost, the velocity has to be above the speed of sound. As this velocity increases, the sonic boost compression ratio does not increase. However, if this velocity falls below the speed of sound, the sonic boost compression is instantly and totally lost.

The converging section of the jet has now stopped compressing the gas. The operators will say, "the jet has broken," and they will observe a precipitous loss in vacuum. If you were standing next to the jet at this point, it would start to make a quieter sound. Then it would begin surging. More on surging later.

Velocity Boost

As the vapors pass into the diverging portion of the ejector shown in Figure 26-1, the cross-sectional area of the diffuser increases. The vapor slows down. The resulting reduction in kinetic energy is converted to pressure. I call this conversion of velocity to pressure "The velocity boost." It may compress the combination of the flowing

Summary—Ejector Performance

Vacuum ejectors are two-stage compressors with no moving parts. The energy for both the first stage (sonic boost) and the second stage (velocity boost) comes from the kinetic energy of the motive steam. The faster the steam exhausts from the steam nozzle, the larger the compression ratio in the diffuser. The kinetic energy of the motive steam is derived by converting the enthalpy (both sensible heat and latent heat), plus the steam pressure, to speed. While moisture in the motive steam upstream of the steam nozzle extracts heat, and thus kinetic energy, from the ejector, moisture downstream of the steam nozzle reflects an efficient conversion of heat to speed (an isoentropic expansion)—that is, the conversion of enthalpy to velocity. The same principles apply to steam turbines, where motive steam velocity, rather than the steam pressure, spins the turbine wheel.

steam plus the noncondensibles by a factor of 2 or 3 to 1. The velocity boost is never lost. It varies with steam pressure and the vapor load and condenser backpressure. But it's always compressing the gas to some extent. The velocity boost is essentially the second and smaller stage of a two-stage compressor, with no moving parts. When the jet "breaks," the velocity boost persists and keeps working, even though the sonic boost has stopped completely.

The most common malfunction in process facilities with vacuum ejectors is wet motive steam. See my book, *Process Engineering for a Small Planet*.

Wet Steam

When steam expands through the motive steam nozzle as shown in Figure 26-1, the velocity or kinetic energy of the steam increases. The greater the speed of the steam as it flows into the inlet to the diffuser just a few inches away, the better the job the diffuser can do in compressing the noncondensible vapor. It's the speed of the motive steam that does the compression work on these moles of noncondensible vapor.

The speed of the steam, or the kinetic energy of the steam, comes from both:

- The pressure of the steam.
- The enthalpy of the steam.

By enthalpy, I mean heat. Much of the temperature of the steam is converted to the velocity of the steam. If you don't believe me, check the temperature of the mixing chamber of an ejector. It's ambient; you can hold your hand on the body of the uninsulated mixing chamber. The motive steam is typically 150 psig, 400°F superheated steam. Why then is the steam now only about 90°F? The 310°F (400°F – 90°F), or 170 Btu per pound of steam, has not been lost. It's been converted into a different form of energy. That is, kinetic energy.

If the motive steam contains moisture, then the moisture will evaporate as it enters the lower pressure in the mixing chamber. For example, if the steam quality is very poor (as discussed in Chapter 15, "Steam Quality Problems"), it might contain 10% water. That amount of water flashing to steam at vacuum conditions would absorb 100 Btu per pound of steam. As the specific heat of steam is 0.55 Btu per pound per °F, the steam would lose:

$$100 \text{ BTU} \div 0.55 = 180°\text{F}$$

The 180°F loss in temperature would mainly be at the expense of the kinetic energy of the motive steam flowing into the diffuser. The slower steam is less able to compress the noncondensible vapors.

Thus, the vacuum produced by the jet will be degraded. If the steam gets really moist, the jet will surge. That is, it will lose its sonic boost. As the moisture content of the motive steam to the jet rises, at some point the outlet of the steam nozzle will get so cold that the droplets of water flowing out of the nozzle will change to ice. The ice will briefly (10 to 20 seconds) stop the flow of steam.

Don't think I have read this in a book. I have seen steam jets supplied with 350°F steam freeze at the Coastal Asphalt plant on the Chickasaw River in Alabama. Also, every time it rained in Norco, Louisiana, I could hear the steam jets surging at the GHR's vacuum tower. Our steam supply lines were poorly insulated and lacked heat tracing.

While superheat is not helpful in improving vacuum, reasonably dry (1% to 2% moisture) steam is vital. Steam traps do not help very much. The best solution, which I've seen several refineries use, is to have a knock-out drum on the steam supply, equipped with a demister, located in close proximity to the steam jets.

Surging

Operators typically associate a sudden loss in vacuum (vacuum breaking) with the jets making a surging sound. When a jet is working properly, it makes a steady, rather loud roaring sound. If it loses its sonic boost, it will get quieter. But only for a moment. The loss of the sonic boost suddenly reduces the vapor load to the entire jet system. The jet discharge pressure is suddenly reduced because the downstream condenser is unloaded. This raises the velocity in the jet diffuser and the diffuser throat (see Figure 26-1). The lower throat pressure and higher velocity restore the sonic boost and the compression ratio. But this draws down the moles of gas that have accumulated in the upstream vacuum system. The sudden increase in the gas flow increases the discharge pressure of the jet by increasing the flow to the downstream condenser. And then:

- The pressure in the diverging section goes up.
- The velocity (volume is inversely proportional to pressure) in the diffuser goes down.
- The velocity boost gets slightly smaller, which further raises the pressure in the throat.
- The velocity in the diffuser throat drops below the sonic velocity, and the sonic boost is therefore completely lost.
- Then the cycle repeats itself, and the operators say, "Listen. The jet is surging."

I used to think that only the first-stage jets ever surged. But I now know that I was wrong. The bigger the jet, the louder the surging sound. If a downstream (second stage or third stage) jet surges, then within 1 or 2 minutes, it causes the upstream jet to also surge. Why? Because the reduction in the downstream compression ratio increases the discharge pressure of the upstream jets. Once a jet's discharge pressure climbs above a certain value, it will lose its sonic boost, and hence it will begin to surge. The particular pressure at which the sonic boost is always lost is called the jet's **critical discharge pressure**. I'll explain this later.

For now, note that once the large first-stage jet surges, the smaller noise of the other jets surging is drowned out. You'll only hear the second- and third-stage jets surging if you are present when the train of events is initiated.

Water in Motive Steam

I'm sitting on the beautiful beach in St. Croix in the Virgin Islands. I've completed a difficult week working for the plant, the world's eighth-largest refinery, located across the island and visible from my hotel.

On one tower, I improved the vacuum top pressure from 13 mm Hg to 8 mm Hg. I had observed moisture condensing on the converging section of the primary ejector, as shown in Figure 26-2. In St. Croix, the humidity is typically 100%. Ambient temperatures are 90°F. Moisture (and in cooler climates ice) forming on the outside of the converging section of the jet is a positive indication of extremely wet steam (or liquids in the feed). The moisture in the motive steam evaporates in the low-pressure zone at the inlet to the converging section of the jet. Based on the observed skin temperatures shown in Figure 26-2, I concluded that perhaps 5 to 10 wt% of the motive steam was water. The evaporation of the water extracted a lot of heat from the motive steam.

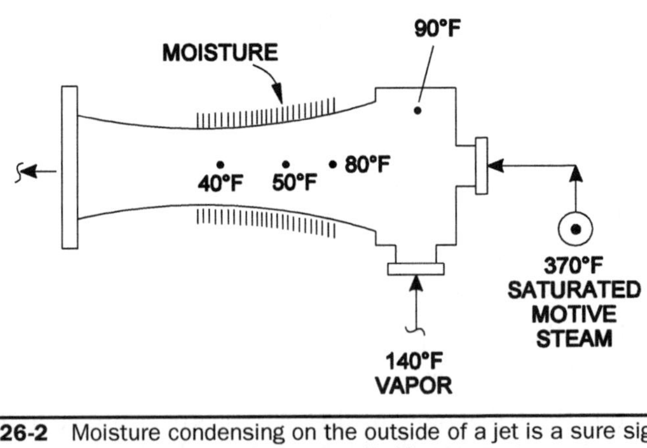

Figure 26-2 Moisture condensing on the outside of a jet is a sure sign of wet steam. The temperatures shown are all skin temperatures.

The odd thing about this situation was that the motive steam was flowing from a steam source that was superheated to 500°F. This is 130°F above the steam's saturation temperature. However, I then discovered that a **de-superheating station** had been installed on the steam supply line to the ejectors. In theory, excessive superheat in steam degrades the performance of a steam jet. In practice, the effect of superheat appears small, as I discussed in the preceding section.

However, my client, thinking that avoiding superheat in steam was critical, had an excessive amount of water injected into the motive steam. I then had the de-superheating water shut off. The pressure at the top of the vacuum tower dropped by 5 mm Hg.

Effect of Spill-Back on Pressure Stability

On a second vacuum tower in St. Croix, I noted that the tower pressure was quite erratic. It bounced between 20 and 24 mm Hg, even though the pressure control valve shown in Figure 26-3 was in service. So I climbed up to the first-stage jet. Listening to the jet, it was quite obviously surging. The surges themselves were producing the vacuum tower's pressure instability. And the cause of the surging was, I thought, that the jet was being periodically overloaded and losing its sonic boost. The cause of the overload was the spill-back pressure control valve periodically opening. It rather seemed

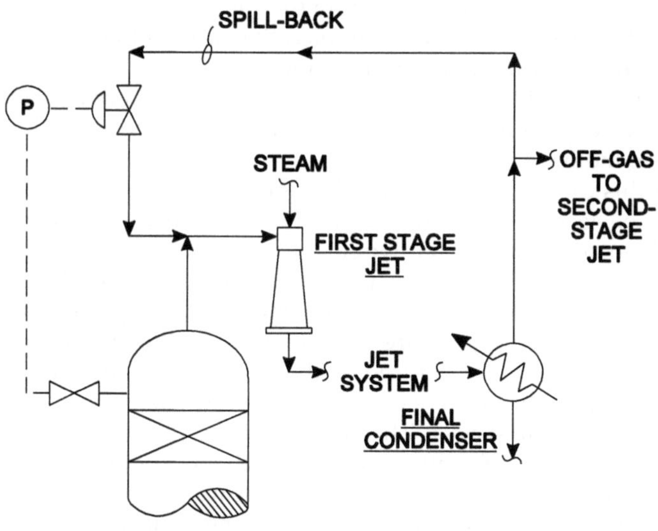

FIGURE 26-3 Spill-back valve caused jet to surge. Thus, the pressure control valve promoted rather than reduced tower pressure instability.

to me that every time the spill-back control valve swung open, the ejector would surge. Which then caused the tower pressure to jump up. Which caused the spill-back to close. Which temporarily stopped the surging.

I then had the pressure control valve switched to manual and shut. The results were quite favorable:

- The jet stopped surging.
- The tower pressure became quite steady.
- The tower top pressure declined from an erratic 20 to 24 mm Hg to a stable 16 mm Hg.

Actually, this was not that much of an improvement, compared to a bigger problem. That being the delta P across the vacuum tower's **wash oil grid**. It was a gigantic 31 mm Hg. This problem, however, did not concern my client, as they didn't sample the vacuum tower bottoms for its gas oil content. But that's another story.

Bypassing Primary Ejector

Last week I had an interesting experience at a 30,000 BSD feed vacuum tower. The first-stage jet inlet pressure was 30 mm Hg and the downstream first-stage condenser outlet pressure was 45 mm Hg. To optimize the motive steam pressure to the first-stage jet, I slowly throttled the steam supply valve. Both the inlet pressure to the jet and outlet pressure from the first condenser dropped. When I had shut off the steam totally to the primary jet, its inlet pressure had dropped to 28 mm Hg. The outlet pressure from the condenser had also fallen, to 24 mm Hg. Next, I opened the bypass line around the now idled jet and its inlet pressure fell to a satisfying 26 mm Hg.

Not only had I reduced the vacuum tower pressure by 4 mm Hg (from 30 to 26 mm Hg), but I had also saved the 6,000 lb/hr of 180 psig motive steam.

In this case, the jet was not developing much of a compression ratio (i.e., 45 mm ÷ 30 mm = 1.5). The downstream condenser was heavily loaded from the motive steam flow to the upstream jet. Then, reducing the steam flow to the jet unloaded the condenser and reduced its condensing temperature and condensing pressure. This reduced condensing pressure more than offset the loss of the small 1.5 compression ratio produced by the first-stage steam ejector.

In this case, the jet was overloaded, because only one of the two parallel jets was in service. The limited first-stage condenser capacity precluded the use of motive steam to both jets simultaneously. The malfunction with the first-stage condenser was a design error in the use of the air or vapor baffle adjacent to the noncondensible outlet nozzle, as discussed in Chapter 27.

Mechanical Defects of Jets

Sometimes, all one can say about a jet malfunction is that the problem is of an unknown mechanical nature and the jet has to be taken offline and inspected. For example:

- Referring to Figure 26-1, in one refinery the diffuser was installed in a reverse position. That is, the diverging section was upstream of the converging section. I had discovered that the motive steam was flowing backwards, out of the inlet nozzle of this jet and then into the inlet of a parallel jet. Isolating this misconnected jet (closing both the process inlet and the motive steam valves) had vastly improved the vacuum. What the actual mechanical malfunction was, I had no idea at the time. I was informed only 2 days ago by a student in one of my seminars. The incident became famous in her refinery, even though it all happened over a decade ago.
- In the same refinery, at the same time, I blocked in another one of the three parallel jets, basically for the same reason. The observed jet "inlet" nozzle temperature was hotter than the upstream vacuum tower top temperature. This is an obvious indication of reverse steam flow through the jet. However, for this jet, the mechanical malfunction was of a far more ordinary nature. Hardness deposits from poor-quality steam had accumulated in the steam nozzle (see Figure 26-1 and the two photos [Figures 26-6 and 26-7] at the end of this chapter). This is quite an easy malfunction to identify and correct. The steam nozzle (which is extremely small as compared to the whole ejector) can be unbolted or unscrewed from the back of the ejector. Some jets are equipped with a cleanout plug in the back of the motive steam nozzle for removing such hardness deposits without removal of the steam nozzle itself. These hardness deposits are not inevitable. They are a consequence of poor level control in waste heat kettle boilers, and other plant steam generation equipment, that permits dissolved salts in boiler feed water to be entrained into the produced steam. See Chapter 16, "Level Control Problems," for details as to how this problem should be corrected.
- At a refinery in Denver, I spent the longest time trying to define the problem with a primary jet on a small vacuum tower. Finally, I conceded defeat and informed my client that the jet was suffering from a mechanical malfunction. It just refused to operate anywhere near its performance curve regardless of motive steam conditions, noncondensible vapor load, discharge pressure, or the phases of the moon. Even prayer had proved ineffective. After I left, the vacuum tower was shut

down and the jet disassembled. My client informed me that the spacing between the end of the steam nozzle (inside the mixing chamber) and the inlet to the diffuser was adjustable. I had never heard of such an adjustment before. However, I do know that this dimension is critical for proper jet performance. My client, with the manufacturer's representative guidance, corrected the maladjustment and the jet then worked fine.

- At the Good Hope refinery, when I was tech manager, we found that a vacuum jet on our visbreaker residue vacuum flasher was not working. The maintenance crew disassembled the jet for inspection. They found the problem. The threads, where the jet was screwed into the back on the mixing chamber, were badly eroded. This allowed a portion of the motive steam to bypass the steam nozzle and flow directly into the mixing chamber shown in Figure 26-1. The problem was that we didn't have a spare nozzle, and a new nozzle was thousands of dollars and many weeks away. So I, your heroic author, wrapped half a roll of Teflon tape around the threads. The steam nozzle was forcefully screwed back into the mixing chamber and the jet was successfully returned to service. And if you visit me in New Orleans, you'll find that I've employed this sort of advanced technology at many locations in my house.

Diffuser Erosion

While erosion to the motive steam nozzle is a pervasive problem, erosion to the downstream diffuser is also common. For example, on our primary steam jets at the Good Hope refinery, the neutralizing amine had been injected in the wrong place. The jets were placed on top of the vacuum tower without an intervening precondenser. The noncondensibles from refinery vacuum towers contain HCl (hydrochloric acid). It can't be avoided. The HCl is formed from the hydrolysis of $MgCl_2$ and $CaCl_2$ in the atmospheric tower bottoms. That's why we should inject the neutralizing chemical or NH_3 into the overhead vapor line. But in this case, the neutralizing chemical was injected downstream of the diffuser. A good place to introduce the small neutralizing chemical flow would be into the motive steam. Which is where I switched it to, after we eroded a hole into the diffuser body. The hole had sucked air into the diffuser and caused the jet to surge.

The interesting aspect of this problem is that all we did was fix the hole to restore the ejector so that it operated on the vendor performance curve. We wrapped it with duct tape. I imagine the interior of the whole diffuser must have been badly eroded by the wet HCl attack. But this did not appear to degrade its performance. On the other hand, almost unnoticeable amounts of wear on the steam nozzle can severely affect the performance of an ejector.

Vacuum System Nomenclature

- Ejector—Term used interchangeably with Jet.
- Primary condenser—Upstream of first-stage jet.
- Wet vacuum tower—Has primary condenser.
- Dry vacuum tower—Has first-stage jet, but lacks primary condenser.
- Steam nozzle—Converts enthalpy and pressure of steam into kinetic energy.
- Converging section—Converts kinetic energy into pressure by sonic boost.
- Diverging section—Converts kinetic energy into pressure by the velocity boost.
- Sonic boost—Pressure increase due to the flowing gas exceeding the speed of sound.
- Velocity boost—Pressure increase due to the gas slowing down in the diverging section.
- Diffuser—Includes both the diverging and converging sections of the ejector.
- Motive steam—Provides the work needed to compress the gas.
- Barometric legs—Also called the seal legs. Drains liquid out of the condenser.
- Seal drum—Keeps bottom of seal legs submerged in liquid.
- Off-gas—The gas vented from seal drum.
- Final condenser—Operates above atmospheric pressure. Vents to seal drum.
- Surge—Periodic loss of the sonic boost.

Incidentally, I said I switched the injection point for the neutralizing amine. Currently, I would substitute NH_3 for the amine. While more difficult to handle, ammonia is a fraction of the cost of the very expensive neutralizing amines that are sold to us by our friendly local chemical vendor.

Steam Nozzle Erosion

I've explained in this chapter how wet steam can cause the immediate loss of the sonic boost and jet surging. Or, as the operators say, the vacuum has broken. But wet steam also has another, longer-term detrimental effect on the ejector. That is, erosion of the interior of the steam nozzle due to extremely high-velocity droplets of moisture impinging on the surface of the nozzle. The confusing aspect of this problem is that the nozzle does not appear to be damaged, when visually inspected. Apparently, the droplets of water erode the nozzle interior in a very even and smooth fashion. The relatively small increase in nozzle diameter results in the loss of jet performance.

How then can one determine, before taking a jet out of service and disassembling the nozzle, that the problem is uniquely due to an eroded motive steam nozzle?

Step 1—Place a pressure gauge on the inlet to the downstream condenser. That is, at the jet discharge.

Step 2—Place a pressure gauge on the motive steam inlet.

Step 3—Place a pressure gauge on the jet inlet.

Step 4—Slowly close the steam inlet valve. Reduce the motive steam pressure to perhaps 70% to 80% of the design motive steam pressure. This pressure is stamped on the body of the jet.

Step 5—If the pressure at the ejector inlet gets lower (i.e., the vacuum is getting better), then the motive steam nozzle is eroded and needs to be replaced.

Step 6—Don't jump to conclusions. It could be the vacuum has improved because the jet discharge pressure has dropped. Check the condenser inlet pressure. If this pressure has also decreased, then you have unloaded the condenser due to decreased steam load. Likely this may indicate that this condenser is fouled or undersized for the service.

This procedure is also used to optimize the steam flow to jets, regardless of their current physical condition. The optimum operating flowing steam pressure is not necessarily the design motive steam pressure. One has to experiment.

Critical Jet Discharge Pressure

If the jet's critical discharge pressure is exceeded, then the jet will lose its sonic boost and surge. The converse is not necessarily true. If the jet's discharge pressure is below sonic velocity, it may or may not develop the sonic boost, because there may be other malfunctions.

How do we know the ejector's design critical discharge pressure? It's listed on the jet's data sheet or performance curve. It's the same as the maximum jet discharge pressure. Above this pressure, the jet cannot be expected to operate on its vendor-supplied performance curve. No matter how low the load, no matter how dry the steam, no matter how good the jet's mechanical condition, exceeding the jet's design critical discharge pressure is going to cause the jet to surge and eventually lose its sonic boost.

The most common cause of high jet discharge pressure, other than downstream jets surging, are condenser-related problems. Chapter 27, "Vacuum Surface Condensers and Precondensers," deals extensively with such malfunctions.

Fouling of Waste Gas Burner

It's a really bad idea to mix vacuum tower off-gas into a fuel gas system. On occasion, this will contaminate the fuel gas system with air. The oxygen will form gummy sulfur deposits that plug burner tips. Any plant that has a history of having to steam out furnace burner tips to reduce plugging is certainly melting out sulfur deposits. If you don't believe me, scrape out and dry these blackish deposits. They will burn with a very pale blue flame and give off a white, choking SO_2 smoke.

To avoid the plugging problem, most of my clients will burn their vacuum tower off-gas from the seal drum in a separate waste gas burner. This cracked gas from the seal drum is rich in H_2S and requires amine scrubbing. Requires in the sense of environmental regulations to remove the 10% to 30% H_2S from fuel gas. Still, even after the H_2S has been extracted with amine, the residual gas from the seal drum will plug the waste gas burner. This creates a backpressure on the final-stage ejector which may exceed that ejector's critical discharge pressure. This jet can then surge. This raises the discharge pressure of the upstream jet, which may also start to surge. Eventually the primary ejector loses its sonic boost and the pressure in the vacuum tower jumps up. The operators say, "The vacuum has broken." It's true the jet is no longer surging, but that's only due to the complete loss of the sonic boost.

My problem is that the surging sound of the small final jets is not very loud, and I don't hear that well. So you have to be close to these small jets to hear their surging sound. Regardless, a pressure of 5 or 6 psig in the seal drum should indicate partial plugging of the waste gas burner. Divert the off-gas to flare for a brief period (which is okay for emergency maintenance), and clean the waste gas burner.

Overloading Vacuum Jets

Often both my clients and I believe that a poor vacuum is due to a mechanical problem with the jets. I just made that mistake at the Murphy Oil refinery near my home. I stupidly had them replace one of their second-stage jets because I thought the jet was not running on its performance curve. I say stupidly for two really good reasons:

- Reason One: I did not have the manufacturer's curve.
- Reason Two: The new jet didn't help.

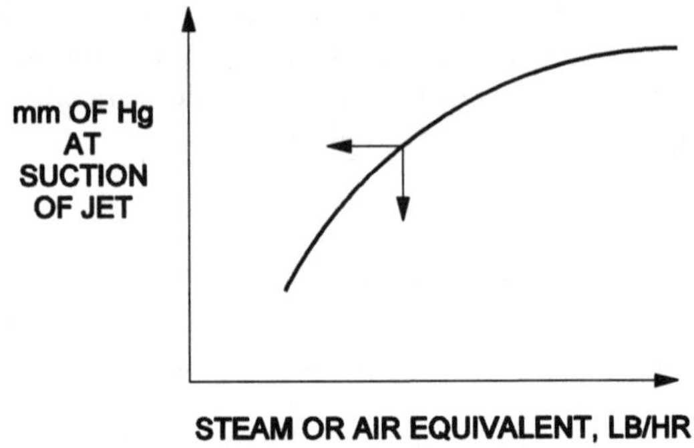

FIGURE 26-4 Ejector performance curve supplied by manufacturer.

As a result of my error, I can now cross Murphy Oil off my list of clients. And that's the way it ought to be.

The first step in checking a jet for overload is to obtain the manufacturer's performance curve. Don't waste your time looking in the project files for this curve. Unlike pump and compressor performance curves, ejector performance curves are not typically supplied with the equipment. You will have to call up the equipment vendor for the curve.

NOTE *The curve only applies at or below the critical or design discharge pressure noted on the curve. If you are above this pressure, you cannot use this curve. The vertical axis of the curve (see Figure 26-4) is the pressure measured at the mixing chamber shown in Figure 26-1. You will need to put a vacuum pressure gauge at this location. If you do not have a pressure connection at this point, but only a ½-inch plug, proceed as follows:*

- *Get a valve with a ½-inch screwed end ready.*
- *With a friend present, loosen the plug to finger-tight.*
- *Quickly pull the plug and have your friend screw in the valve. When I say "quickly," I mean 1 or 2 seconds.*

I would not write this if I hadn't done it myself dozens of times, and consider it to be safe if done quickly.

Air Equivalent

Some performance curves will be tabulated in "air equivalent" and some will be "steam equivalent." There is a rigorous method to convert different molecular weight hydrocarbons to either air or steam equivalent. Roughly speaking, divide the hydrocarbon flow by the square root of the ratio of the molecular weight. For example, I have 100 pounds per hour of butane vapors of molecular weight of 58. Air has a molecular weight of half of 58. That is 29. The square root of 58 ÷ 29 = 1.414. Therefore, I would divide my 100 pounds of butane by 1.414 to obtain an air equivalent flow of 100 ÷ 1.414 = 71 pounds of air.

For practice, see if you can calculate that 10 pounds per hour of steam has an air equivalent of about 13 pounds per hour (square root of 29 ÷ 18 times 10 pounds).

The composition of the gas is based on an off-gas sample obtained from the top of the seal drum, or the vapor outlet of the final condenser, or from the vapor flowing to the waste gas burner. Best not to try to obtain the sample from a subatmospheric pressure point and risk air contamination. Take the sample carefully in a nonmetallic container to avoid having tramp oxygen reacting with H_2S. Be careful! The H_2S concentration will be 100,000 to 300,000 ppm in crude unit vacuum tower seal drums. A fatal H_2S concentration in air is only 1,000 ppm. So wear your fresh air pack.

To obtain the sample, use a quart glass bottle. Blow the sample gas into the bottle so that you can feel a little pressure coming out of the bottle. Then, draw your sample out of the bottle. Prompt analysis is best to avoid the H_2S reacting with oxygen from air leaks.

But how about measuring the flow? Here's the method that I use to get the gas flow from the final condenser:

Step 1—Check the off-gas for nitrogen. Suppose for this case that the lab result is 5%.

Step 2—Connect a fitting with a ¼- to ½-inch opening anywhere on the first-stage condenser.

Step 3—Open the valve to this fitting all the way. If the vacuum starts to break or the jets start to surge, use a fitting with a smaller opening.

Step 4—Check the off-gas for nitrogen. Suppose the lab result is 15%. You only need to wait a few minutes between Steps 3 and 4.

Step 5—Assume the velocity of air sucked through your fitting is 1,000 ft/sec. Using the fitting open area, calculate the airflow, in ft^3 per second. Multiply by 3,600 to get 5,000 SCF/hr for the ½-inch open fitting.

Step 6—Your sample has increased its nitrogen content from 5% to 15%. The nitrogen content of air is 79%. Therefore to calculate the flow of off-gas:

$$(5{,}000\ \text{ft}^3/\text{sec})\ [79\% \div (15\% - 5\%)] = 39{,}500\ \text{SCF/hr}$$

Step 7—To convert to moles per hour, divide by 360 to get 110 mph.

Step 8—Now you have to add in the moles of condensable hydrocarbon to the first-stage jet. If there is a precondenser, then the moles of condensable hydrocarbons to this jet are small and can be neglected. If there is no precondenser, then continue on to Step 9.

Step 9—Measure the volume of liquid hydrocarbons collected in the hot well or seal drum. I do this by shutting off the hot well pump and seeing how long it takes the hydrocarbon liquid level to increase by 1 foot. Then I sample the hydrocarbon to obtain its molecular weight. Typically, in a refinery vacuum tower, this will be about 140 pounds per mole.

Step 10—For a vacuum tower with no precondenser, add in the pounds of steam used in the tower. That is, the sum of:

- In the bottom stripping section
- In side strippers
- Velocity steam or condensate used in the fired heater tube passes

Step 11—For a vacuum tower with a precondenser, you must add in the vapor pressure of water at the precondenser vapor outlet temperature. For example, the precondenser outlet temperature is 102°F. The vapor pressure of water from my steam table at 102°F is 1.0 psia or 51.7 mm Hg. The observed pressure at the inlet to the first stage jet is 0.2 bar absolute, or 152 mm Hg. Therefore the moisture content of my vapor is:

$$(51.7) \div (152) = 34\%$$

Since your lab sample of seal drum gas was likely reported on a dry basis, this will be a big additional vapor load to the first-stage jet.

Now, you can see if you're on the jet performance curve shown in Figure 26-4 (which is good), or above the curve (which is bad). Being below the curve is quite improbable. If you are substantially above the curve, the malfunction is a mechanical defect or poor motive steam conditions, as I've already described. Perhaps you're on the curve, but way out on the right. Or worse yet, you're really far out on the right, so that the curve has stopped. Meaning, you have overloaded the jet with gas. Now what?

Well, you've got trouble. The noncondensible load to the jet is excessive. The main areas of excess gas are:

- Air leaks in the hot part of the tower.
- Air leaks in the cold part of the tower.
- Excess cracked gas in the feed.
- Excess cracking in tower bottoms.
- Feed contamination with lighter components.
- Heat exchanger leaks.
- Naphthenic acid decomposition.
- Seal leaks on idle pumps.

I could probably write a whole book on this subject. But I've restrained myself and just written Chapter 28, "Excess Gas Overloads: Vacuum System Ejectors."

Relative Jet Efficiency

Recently, I was working at a refinery in Coffeyville, Kansas. The operators had been running two jets in parallel. They wished to take one of the two jets out of service. I checked the mixing section skin temperature on both jets. Basically, I was measuring the exit temperature of the steam nozzles (see Figure 26-1):

- Jet A = 90°F
- Jet B = 110°F

Which jet is less efficient and thus should be removed from service? As you consider this question, please note:

- The heat content or enthalpy of the steam is converted to velocity in the steam nozzle. The more kinetic energy the motive steam has, the better the jet will work in compressing the off-gas from the vacuum tower.
- Both jets were supplied with the same quality (moisture content) motive steam.

The answer is jet B is less efficient, because the steam is warmer and hence is presumed to be moving more slowly. If the steam supply to both jets was different, I would most likely have thought jet A suffered from poor-quality steam.

High Discharge Pressure

Often my clients think I'm conducting field observations on their process facilities. But in reality, I'm playing with their equipment. For example, I was working in a refinery in Arkansas City, Texas, on a vacuum tower problem. I found that I could cause the vacuum to break by increasing the level in the hot well (see Figure 26-5):

- Raising the level in the hot well drum backed liquid up into the condenser.
- This caused the liquid drain temperature to go down but the condenser vapor outlet temperature to increase.
- The high vapor outlet temperature increased the vapor pressure of the water in the condenser.
- This increased the discharge pressure of the upstream ejector above its "critical discharge pressure." This critical pressure is shown on the jet system specification sheet issued by the manufacturer.

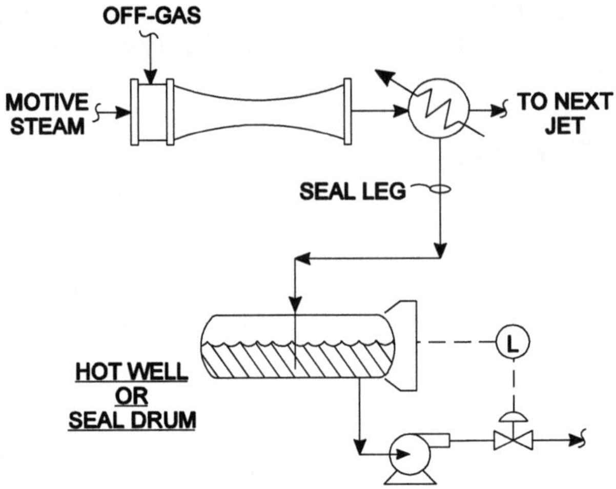

FIGURE 26-5 Raising level in the hot well causes jet to surge.

- The jet was forced out of its "critical mode of operation." That means that the jet lost its sonic boost.
- The jet began to surge and the vacuum broke in the upstream vacuum tower.

Lowering the level in the seal drum allowed me to restore the sonic boost within a few minutes, and thus recover the vacuum. And here's the best part. My wife Liz delivered a sandwich to me on the sixth-level platform, so I wouldn't have to interrupt my important field test for lunch. I have discussed other factors that cause high jet discharge pressure, due to condenser problems, in Chapter 27, "Vacuum Surface Condensers and Precondensers."

The final condensers of refinery crude vacuum tower jet systems are subject to extreme fouling rates and high delta Ps. I have seen pressure drops of 2 or 3 psi (100 to 150 mm Hg) through these final condensers. However, because the final-stage jets are often designed to discharge to a high (5 psig) pressure, cleaning these condensers has only a small effect on vacuum. Unless the fouling has caused the final-stage ejector to exceed its critical discharge pressure; then, cleaning the final condenser has a huge beneficial effect.

Hogging Jets

Hogging jets are used in several services:

- To extract air that leaks through the mechanical seals on shafts in condensing steam turbines.

- To air-free large vacuum systems rapidly to speed startups. Hogging jets are not intended for continuous services, even though they are habitually used for that purpose.
- To draw noxious vapors out of storage pits or tanks for disposal at a safer location. My experience is with the use of hogging jets on sulfur plant product sulfur pits. The H_2S that's dissolved in the sulfur flashes slowly out of the sulfur product in the underground concrete pits.

The hogging jet is just an ordinary steam jet, but it lacks a converging section and thus never develops a sonic boost. It discharges directly to the atmosphere rather than through a condenser. Thus, there is no recovery of noncondensible gas and no seal drum. Other than poor motive steam conditions and steam nozzle wear, there isn't much that goes wrong with a hogging jet. I saw one working on a cracker wet gas compressor steam turbine surface condenser last month. It was discharging to the atmosphere with a suction pressure of about 180 mm Hg. Atmospheric pressure was around 720 mm Hg in Kansas that day. So the hogging jet was developing a compression ratio of about 4:1. This is really good for such a jet. On the other hand, the hogging jet is relatively inefficient, when compared to an ordinary converging-diverging jet, as far as motive steam consumption is concerned.

Liquid Seal Ring Compressor

A liquid seal ring pump is a sort of rotary compressor. Gas is squeezed between the turning vanes of the rotor and a pool of liquid. The gas and liquid are squeezed out of the compressor or pump case together and into a vapor–liquid separation drum. The liquid is circulated via a small pump through a cooler, where the heat of compression is removed. I have a better description and photo of this device in my book, *Troubleshooting Process Operations*.

Liquid seal ring pumps circulate either diesel oil or water. The water, being more volatile, limits the vacuum that they can develop. Meaning, their lowest suction pressure corresponds to the vapor pressure of water at the temperature in the compressor or pump case. Diesel oil, which has a lower vapor pressure, is circulated when a lower suction pressure is required.

Texaco refineries routinely favored the use of two-stage liquid seal ring pumps, which worked just fine. For deep vacuums, they used a first-stage steam jet, followed by two in-series seal ring pumps.

The only malfunction that I experienced with the seal ring compressors at the Good Hope refinery over a two-year period was organic fouling of the circulating water cooler. H_2S-consuming bacteria accumulated in the water circulated through the compressor case. We cured this problem by biocide injection into the inlet to the cooler.

From the designer's perspective, these vacuum pumps are more energy efficient than vacuum jets, but they are also more costly to purchase and install.

Liquid Ejectors

My experience with liquid ejectors is limited to one large system in Lithuania, built by the Russians in the 1980s. The vacuum tower had a precondenser, and the pressure in the precondenser closely corresponded to the vapor pressure of water at the precondenser vapor outlet temperature. Thus, the two-stage liquid ejector system was never an issue, and was working properly. Diesel was circulated through a cooler and back to the ejector system. The fact that no one in the plant knew much about the system is ordinarily a good indication of its reliability.

High-Pressure, Superheated Motive Steam

A word of caution about steam conditions: Often, reducing the motive steam pressure below the design pressure improves vacuum. This needs to be determined by trial and error. Of course, if you get the steam pressure too low, you will cause the jet to surge and break the vacuum.

On the other hand, I've never seen the vacuum improved by exceeding the jet nameplate motive steam pressure. Usually, excessive motive steam pressure causes a loss in vacuum.

Same for superheated steam. While dry steam is critical, as I've discussed in this chapter, superheated steam does not help, for reasons that I do not know. If you have higher-pressure, superheated steam, you can modify (or replace) your existing jets for the new steam conditions. This will result in a reduction of the pounds of motive steam consumed for each pound of noncondensible gas compressed in the ejector system.

Fouled Steam Nozzle

Allow me to complete this chapter with a nice letter I received via e-mail today from a client. It's an exact copy. The notes in [ ] are mine:

Hi Norman:

I just wanted to say thanks for the course recently—my beads [i.e., a gift I award for smart answers] hang over my computer to remind me that this machine is good at wasting time.

Figure 26-6 Fouled nozzle.

I wanted to share a recent success with you. I got stuck into the ejector system of surface condensers in one of our plants [the author of this letter has a problem with his English writing skills]. Through various testing I was convinced it was the ejector nozzles that were a problem. I found this (picture of fouled nozzle) when we opened up. [See Figure 26-6.]

Lots of gunk which I filed out and returned the nozzles [see Figure 26-7]. Saving about 90 T/H superheated steam—20 MW electricity!

Figure 26-7 Clean nozzle.

Thanks for inspiring me to overcome the inertia of the office and bureaucracy.

Warwick Hayes
Process Engineer
OPI—Gas Circuit
Sasol Technology—Secunda
South Africa
Warwick.hayes@sasol.com

The energy savings due to the improved vacuum in the surface condenser is equivalent to five billion Btus per day (that's billion; not million). I was kind of depressed today. I ruined a new expensive alloy blade on my bandsaw. Cut into a nail. But this e-mail has filled my day with sunlight. So I thought I'd share it with you.

CHAPTER 27

Vacuum Surface Condensers and Precondensers

When I hear, I forget. When I see, I remember. When I do, I understand.

—Confucius

I've forgotten the names of my ex-girlfriends, and the Spanish verbs I learned in school, and what I had for breakfast. But I remember everything about surface condensers in vacuum service. By some mental quirk, I can recall all my encounters with process equipment. In case you don't believe me, I'll prove it.

What is the origin of the term **surface condenser**? The original condensers in vacuum service were called **barometric condensers**. Cold water was sprayed into an empty chamber. Exhaust steam from a steam engine flowed into this empty chamber where it was condensed by direct physical contact with the cold water. At a condensation temperature of less than 100°C, the steam condensed at a subatmospheric pressure (or a vacuum). Exhausting steam from an engine or turbine under vacuum conditions is the best way to extract more work from each pound of steam.

The cold water and the steam condensate drain out of the condenser through a pipe called the **barometric leg**. For this to work, the condenser must be elevated above grade. If the pressure in the condenser is 0.5 bar absolute (7.3 psia), then the condenser has to be elevated 5 meters (17 feet) above grade for drainage.

The problem with the barometric condenser is that it mixes and contaminates the steam condensate with the cooling water. The surface condenser is a shell-and-tube heat exchanger, where this contamination is avoided. That is the only advantage for a modern surface condenser over the archaic barometric condenser. If you don't believe me, phone the CVR Refinery in Kansas. They replaced their ancient barometric condenser with a modern, multimillion-dollar surface condenser and lost capacity on their vacuum system.

Condensate Backup Due to Air Leak in Barometric Drain Line

In 1987 I had an assignment at the Mobil Oil refinery in Coryton, England. The problem was poor vacuum in the surface condenser shown in Figure 27-1. The sketch and the story are in foreign metric units, rather than American units (see Table 27-1).

Historically, the surface condenser had been able to develop a vacuum of 0.1 bar (i.e., 76 mm Hg). Now the best vacuum that could be developed was 0.3 bar (i.e., 228 mm Hg). What was the malfunction that caused this limitation?

I noted a flange on the barometric drain leg, 7 meters below the bottom of the surface condenser. Seven meters equals 0.7 bar absolute. If I add the head of water equivalent to 0.7 bar to the 0.3 bar absolute surface condenser pressure, I obtain 1.0 bar absolute or atmospheric pressure (note that Mobil Coryton is at sea level).

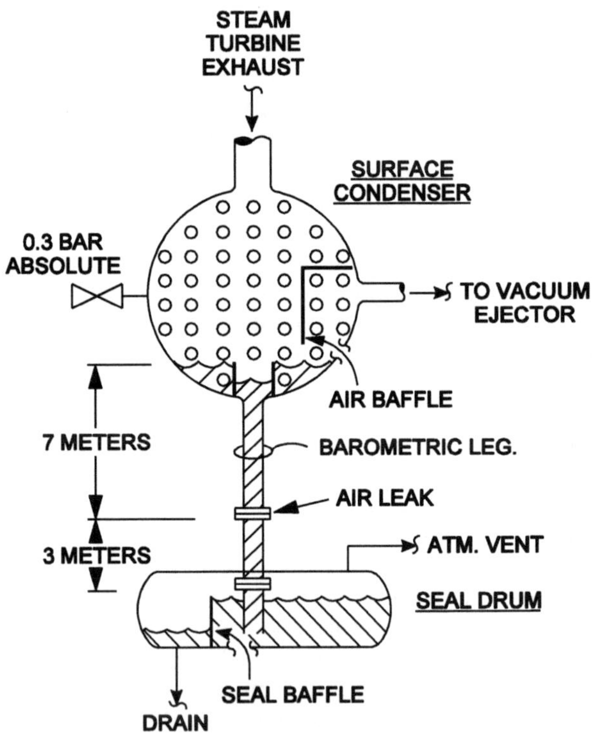

FIGURE 27-1 Air leak limits vacuum in surface condenser.

Foreign	American
10 meters of water	34 feet of water
One bar absolute	14.5 psia
760 mm Hg	0.0 inches Hg vacuum

TABLE 27-1 One Atmosphere of Pressure

I cannot say that I observed that the flange was actually drawing air between the flange faces. However, I reasoned as follows:

- If the pressure in the surface condenser dropped from 0.3 bar absolute to 0.2 bar absolute, then the flange would have drawn air into the barometric leg.
- With air bubbling up through the leg, the pressure head of water required to promote condensate drainage would have been reduced.
- Water in the surface condenser would have backed up and covered a portion of the tubes.
- Tubes submerged in water will not contribute to steam condensation, as the steam cannot contact the submerged tubes.
- The incremental uncondensed steam escapes from the side vapor outlet nozzle on the surface condenser (see Figure 27-1). This increases the vapor load to the downstream vacuum ejector.
- The extra moles of steam flowing to the ejector will cause its inlet pressure to rise. This will increase the pressure in the surface condenser.
- The surface condenser pressure increases, so that at the elevation of the leaking flange on the barometric leg, the head pressure of water equals 1 bar absolute or the pressure at sea level.
- This pressure stops the air leaking into the flange and restores drainage through the barometric leg. But any reduction in the surface condenser pressure below 0.3 bar absolute would reestablish the leak, until the system reached equilibrium.

Now what? I could write up my theory in a report to Mobil Oil. Or I could take some aluminum duct tape and wrap up the suspect flange. After I wrapped the flange, the vacuum in the surface condenser slid down to 0.1 bar absolute. The interesting part of this incident was that I could not actually observe that the flange was noticeably leaking. I drew the correct conclusion based on my calculations, rather than on field observations.

Incidentally, I have presented this example many times in my seminars. When I explain the calculations using American feet and psi, no one understands. When I explain my calculations using meters and bars, everyone understands.

Leaks Inside Seal Drum

I can go back a lot farther than 1987. I can reach back in my mind to the Amoco refinery in Sugar Creek, Missouri, in 1969. This 100,000 BSD refinery was located a few streets over from the former home of America's best president ever: Harry Truman (1945–1952). My visit to the plant was to help them improve the performance of their vacuum tower. Referring to Figure 27-1, I noticed that the level in the seal drum was showing 100% on the control panel. I then reasoned as follows:

- If I reduced the level in the seal drum, then the level in the surface condenser would also be reduced.
- If the level in the surface condenser dropped, the effective area on the condenser would be increased. This would reduce the amount of uncondensed steam and hydrocarbon vapors flowing to the downstream jet.
- The jets would be unloaded and pull a better vacuum on the vacuum tower.

But the operators said, "Mr. Lieberman, we know you're an engineer. We know that what you're saying makes sense. But we know that don't work that-a-way here. But, if that's what you want, we'll give it a try."

And here's what then happened:

Step 1—The operator lowered the level in the seal drum from 100% to 50% on the left side of the seal baffle (Figure 27-1).

Step 2—The internal flange in the seal drum was uncovered.

Step 3—The internal flange, which was evidently leaking, sucked gas out of the seal drum and up into the barometric leg.

Step 4—Drainage through the barometric leg was retarded and (I guess) the level in the surface condenser rose.

Step 5—The vacuum broke. And this is not a guess, but a fact!

Step 6—A piping flange on the suction of the vacuum tower bottoms pump blew its gasket. Meaning, as the vacuum broke, the pressure on this flange went from a negative to a positive pressure.

Step 7—The 690°F tar blew out from the defective gasket and autoignited. Fortunately, the resulting fire did not burn down Harry Truman's home.

Now, 43 years later, I would never ignore such a comment by an operator. Now I always expect process equipment malfunctions. Now I attempt to reconcile operators' observations with possible equipment malfunctions before making operating changes to process units. But you can see why certain of my encounters with process equipment have created such lasting impressions.

Sludge Accumulation in Seal Drum

In all the years I've worked in the identification of vacuum system malfunctions, I never encountered the problem shown in Figure 27-2. Never, until this week at the Sinclair refinery in Wyoming. Sludge in the bottom of the vacuum system's seal drum restricted drainage from the vacuum interstage condensers. I obtained a sample of the black, gritty material. It was plain iron sulfide.

Corrosion in the condensers due to wet H_2S, plus a few ppm of HCl, is common. Such corrosion deposits will be flushed into the seal drum. Why, I wondered, had I never encountered this malfunction before? Restricting condensate drainage will surely cause liquid backup that floods the condenser. This raises the condenser pressure and causes the jets to bog down and the vacuum, as the operators say, "will break!"

The reason for this unusual malfunction at the Sinclair plant was that the bottoms of the seal legs were only 6 inches above the bottom of the 6-foot ID seal drum, as shown in Figure 27-2. As the overflow baffle was 4 feet high, the 6-inch clearance of the seal legs makes no sense. The depth of the liquid seal only needs to be a few inches, not 42 inches (i.e., 4 feet minus 6 inches).

I suppose that the problem was aggravated by the failure of my client to wash out accumulated sludge from the seal drum during last year's turnaround, even though the plant operators insisted that this

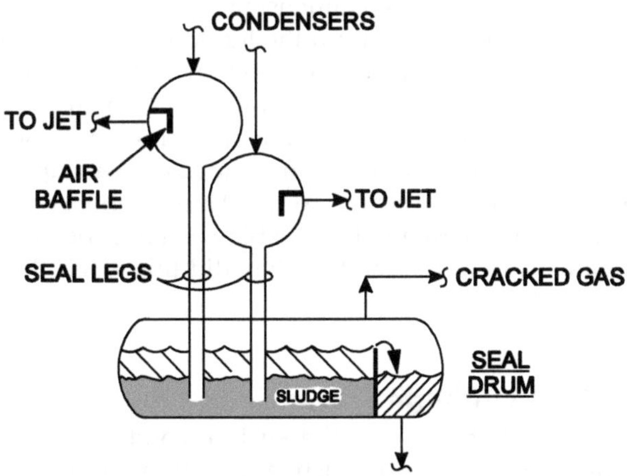

FIGURE 27-2 Sludge accumulation in a seal drum floods the condensers and causes the vacuum to break suddenly.

was necessary. So now, when the vacuum breaks, the same operators must rush out to madly blow out the plugged seal legs with 200 psig steam.

I suggested that the bottom 2 feet of the seal legs be cut off at the next opportunity.

Detecting Condensate Backup

For a surface condenser commonly used on steam turbine exhausts, the chances of restricting condensate drainage due to fouling are remote. But in refinery vacuum tower operations, fouling deposits on the shell side of surface condensers are certain. The cracked gases combine with O_2 leaks to form gums. The HCl and H_2S, in an aqueous environment, form water-insoluble $Fe(HS)_2$ corrosion products. The gums and iron sulfide particulates are the basis for shell-side fouling deposits and restrictive flow in the barometric legs.

If liquid starts to back up in the surface condenser, you will observe that:

- The barometric leg gets cooler.
- The vapor outlet to the downstream jet gets warmer.

If condensate was draining freely from the condenser, the vapor and liquid outlet temperatures will be roughly the same. However, as we will see, there is another problem with the surface condenser that causes the vapor outlet to be hotter than the liquid outlet. That problem is a defective air baffle seal to the shell. But I'll come back to this later on. For now, what I wish to describe is how to prove the condenser malfunction is condensate backup.

The method is the opposite of what I tried in Sugar Creek in 1969. That is, raise the liquid level in the seal drum above the overflow baffle shown in Figure 27-1. If condensate is not already backed up in the surface condenser, pushing up the liquid level in the barometric leg a few inches will not have any effect on vacuum or the surface condenser vapor outlet temperature. However, if vacuum is lost, and if the surface condenser vapor outlet temperature does rise, then the condenser was already suffering from condensate backup before you raised the level in the seal drum. At the Amoco refinery in Whiting, Indiana, the operators used to have to blow waxy deposits out of the seal legs with steam in the winter. They did this when the seal legs became cool to the touch.

Air Baffle Internal Leakage

The air or vapor baffle inside the surface condenser is critical to proper performance of the downstream vacuum ejector or vacuum liquid ring seal pump. Figure 27-3 illustrates the function of the air or

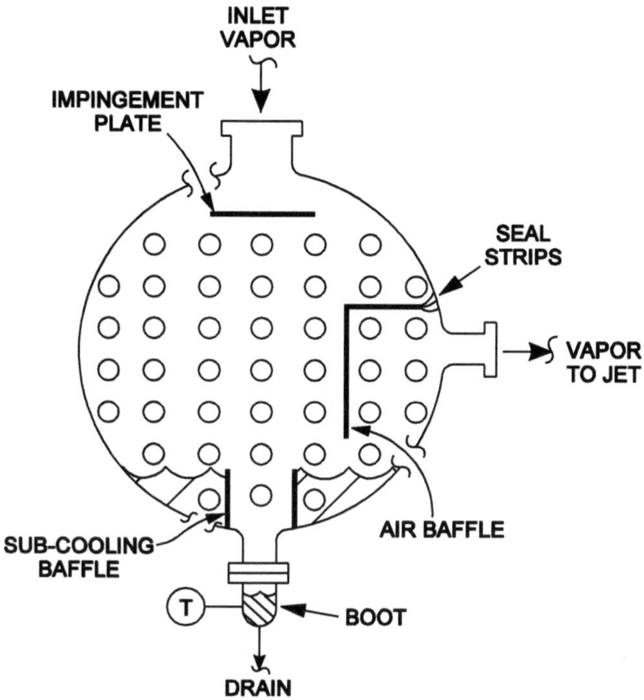

FIGURE 27-3 Internal components of a surface condenser. Seal strips retard leakage around air baffle.

vapor baffle. The inlet vapor flows down across the tubes and underneath and then back up inside the air baffle. The objective is to minimize the temperature and the moisture content of the residual vapor flow to the downstream vacuum jet. The usual malfunction is leakage of hot vapor between the edge of the air baffle and the shell ID. The cause of this malfunction is defective seal strips that ought to seal the baffle, which is part of the tube bundle, to the shell ID. The causes of the malfunction to the seal strips are identical to that which I described for the horizontal baffle in the two-pass shell configuration (see Chapter 10, "Shell-and-Tube Heat Exchangers in Sensible Heat Transfer Service"). The leaking seal strips permit the hot, moisture-laden vapor to bypass the tube bundle and blow directly into the downstream jet. This overloads the jet and diminishes vacuum. Often these air baffle seal strips fail in refinery vacuum tower service because they are made of copper alloys subject to attack by the HCl and NH_3 in the vacuum tower overhead vapors. On occasion, I've extracted bits of such degraded seal strips from the suction of surface condenser boot condensate pump suction screens.

It's simple to identify that the air baffle seal strips have failed. Check the skin temperatures of the surface condenser shell at the

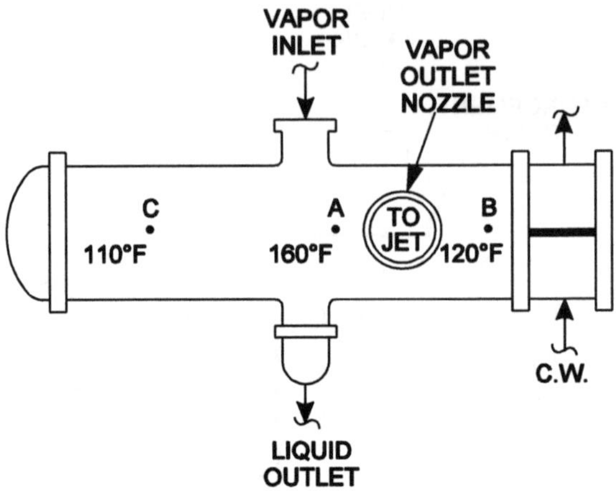

FIGURE 27-4 Skin temperatures indicate an air baffle leak inside the surface condenser.

elevation an inch or so below the seal strips with your infrared gun. I've shown the correct location to check these temperatures on Figure 27-4. If point A is close to the temperature of the jet inlet, and if A is 40 or 50°F hotter than B and C, then the air baffle seal strips are leaking.

At the Coastal refinery in Aruba, we reduced the vacuum tower flash zone pressure from 75 mm Hg to 62 mm Hg by renewing the seal strips on the second-stage condensers. The temperature profile I observed prior to renewing the seal strips is shown in Figure 27-4. After the seal strip repairs, all the temperatures were between 125 and 130°F, including the jet inlet (or the condenser vapor outlet) temperature.

Oversized Impingement Plate

At the Texaco refinery in Delaware City (now Valero), I made a significant improvement in vacuum by reducing the delta P in the first-stage condenser. This was done by reducing the size of the impingement plate shown in Figure 27-3. The impingement plate was far larger than the condenser inlet nozzle diameter, which introduced unnecessary shell-side pressure drop. A delta P measured on a condenser in vacuum service above 6 or 8 mm Hg should be considered a sign of a malfunction due to either fouling or a flawed mechanical design.

The largest surface condenser I ever worked with was at the Good Hope refinery (now also Valero) near my home in New Orleans. It served the 10,000 horsepower K-805 supplementary FCU catalyst air blower steam turbine exhaust. This machine had been idle for decades when I tried to commission it. One of the problems I encountered was condensate backup. How did I know that the condensate was backing up in the condenser? As we increased the motive steam flow to the turbine to increase the blower speed:

- The boot temperature dropped (see Figure 27-3).
- The vapor outlet temperature to the downstream jet increased.
- The vacuum at the steam turbine exhaust became worse.
- Further increases in steam flow to the turbine failed to increase the blower's speed or discharge air pressure.

But the level indication in the boot seemed to be working fine. Reducing the level from 60% to 30% in the boot's gauge glass failed to help. Now what?

Joe Petrocelli, an old Navy man, came to my rescue. "Norm," he said, "You see that the boot is just bolted onto the shell."

"Yeah! But why is that, Joe?" I asked. "It would have been cheaper to just weld it on."

"Cause we learned in the Navy that crud from the steam or condenser shell can gum up the bottom coupla rows of tubes. This kinda restricts the drainage of steam condensate out of the shell and into the boot. So we got taught in the Navy—I worked in an aircraft carrier in the engine room during the battle of Leyte Gulf—that you've got to drop the boot. Then hose down as best you can the bottom coupla rows of tubes."

"Joe, that doesn't make a lot of sense. Can't the steam condensate just run along the bottom of the shell and into the boot? The bottom rows of tubes are a few inches above the shell floor," I asked.

"No, Mr. Norm. You're wrong. There are a set of **subcooling baffles** [see Figure 27-3] that run along the shell ID on either side of the central water boot. They're to chill the condensate off below its boiling point. Makes it easier to pump a subcooled liquid. Keeps them condensate pumps from cavitation and slippin'," explained Lieutenant Petrocelli. "But those baffles also make it harder for the condensate to drain into the boot."

So I had the boot dropped and had the whole shell side chemically cleaned, in addition to hosing down the tubes we could reach from the open boot nozzle. When we started back up, the turbine ran much better. Also, the surface condenser vapor outlet and boot drain temperatures were quite similar. However, Joe Petrocelli was not happy.

"Norm, that chemically cleaning stuff is wrong. It may have worked, but it damn sure ain't the Navy way."

Unfortunately, the high vibration trip on K-805 failed to function. The next day, the operators tried to run the turbine at its **critical speed** and thus destroyed the machine. But ladies and gentlemen, that's a story for another day (see my book, *Troubleshooting Process Plant Control*).

Impingement Plates as Vapor Distributors

Using an impingement plate below the vapor inlet of a surface condenser is intended to protect the upper row of tubes from erosion. Caution is advised! At the Huntway asphalt plant in San Francisco, my client installed a new tube bundle with too large an impingement plate. The precondenser delta P increased from 5 to 50 mm Hg, and asphalt product specs could no longer be met. Rather similar to my Texaco story.

However, an extra-large impingement plate, if designed properly, can improve the overall heat transfer efficiency of a surface condenser or precondenser. Let's assume that the bundle is 20 feet long. Also that it has four tube support baffles, a floating head, and channel head tubesheets. The vapor inlet nozzle is presumed to be in the center of the shell. The impingement plate should then be designed as follows, to promote vapor distribution:

- Maximize the width consistent with the shell dimensions at the upper row of tubes.
- The length of the impingement plate should extend between the farthest tube support baffles. This will cover 60% to 70% of the horizontal distance between the two tube sheets.
- Drill 1-inch or 1½-inch holes, equally spaced, through the impingement plate, including the area beneath the vapor inlet nozzle.
- The sum of the area of these holes should be equal to or slightly greater than the area of the vapor inlet nozzle.

I've copied this design from exchangers I've seen in service. I can't testify as to how much benefit is derived from such an impingement plate vapor distributor. Incidentally, if your condenser has the shell inlet located at the opposite end of the exchanger from the vapor outlet, this concept does not apply, and a conventional impingement plate is required to protect the tubes from erosion.

Cooling Water Pressure to Surface Condensers

Because the surface condenser must be elevated for condensate drainage, the cooling water flowing up to the condenser can easily, and often does, fall below atmospheric pressure. For example, if the cooling water supply pressure at grade is 2 bar (30 psig), and the condenser is elevated 23 meters (75 feet) above grade, the water supply pressure at the condenser inlet will be 0.7 bar absolute (9 inches of Hg vacuum). As the hotter, lower-pressure water exits from the condenser, small quantities of dissolved air are evolved. The air chokes off the flow of water in the cooling water return line. The reduced water flow through the condenser raises the vapor outlet temperature. The flow to the downstream jet will increase. Vacuum will consequently get worse. Note, however, that the pressure head lost by the water as it flows uphill is entirely regained when the water flows back to grade.

To suppress the evolution of the dissolved air, throttle back on the local isolation gate valve on the cooling water return line. Partly closing this valve will, strange to say, increase the water flow to the surface condenser. I know this sounds weird—but it's true. But a theory without a test is bullshit. So here's the test:

Step 1—Monitor the cooling water outlet temperature from your condenser bundle.

Step 2—Don't worry about the water outlet pressure. Throttle back slowly on the water outlet gate valve and watch your cooling water outlet temperature.

Step 3—If air evolution is a problem, the cooling water outlet temperature will decrease by 1°F or 2°F.

Step 4—Play with the outlet valve until you have minimized the cooling water outlet temperature.

I learned this trick in the Aruba refinery from Leo Chago, shortly before he passed on. A friend lost, but not forgotten.

The problem of low cooling water pressure to the elevated surface condensers may cause two other types of malfunctions: air leaks and inability to back-flush the tubes. If a valve is leaking on the cooling water return line at the elevation of the condenser, air may be continuously drawn into the flowing water. As in the previous example, the airflow reduces the flow of water by displacing some water flow from the return water piping. If you think a fitting or flange is leaking, spray a water hose over the suspected leak. Water, not air, will be sucked into the leak. If the cooling water outlet temperature from the condenser drops, you have located the leak.

The ability to **back-flush** cooling water tubes in all services is critical. Wood chips, sand, and shells (but not calcium hardness deposits) can be removed by flowing water from the cooling water return header backwards through the tube bundle, and draining the water to the sewer. Of course, this is quite impossible if the return header pressure at the elevation of the surface condenser is below atmospheric pressure.

We had this problem in Aruba on eight giant vacuum tower overhead precondenser shells. To provide high-pressure water for routine back-flushing, we tied a 10-inch line from the refinery firewater circuit into a new back-flush header above the eight condensers. The fire water pressure was sufficient to back-flush the condensers. That was in 1998. Now in 2011, I'm older and wiser. Now I would never use plant fire water for such a routine process service. In an emergency, I might not have sufficient pressure in the fire water system to control a fire.

Tube Leaks

I never had a tube leak in a surface condenser. But when I checked for one, the procedure was pretty simple. I would get a sample of steam condensate from the condensate recovery pump. As we used chromates in our cooling water treatment program, I would just have our lab check for chromates in the condensate. The concentration of the chromates would then correlate with the magnitude of the exchanger leak. I do not believe that a relatively small water leak could seriously impair the efficiency of the surface condenser. Your cooling water chemical vendor will have a portable test kit that can be used for whatever treatment chemical is used in your cooling water system.

Air Leaks

Leakage of air into the surface condenser is a more serious matter. A small leak could easily load up the ejectors. To see if jet capacity and/or air leaks are limiting the surface condenser vacuum, you should proceed as follows:

Step 1—Check the temperature at the vapor outlet of the condenser (not the liquid outlet or boot temperature).

Step 2—Compare the measured pressure at the surface condenser vapor outlet to the vapor pressure of water (from your steam tables) at the temperature determined in Step 1.

Step 3—If the measured pressure is more than 5 mm Hg greater than the calculated vapor pressure of water in Step 2, the limiting factor for the surface condenser performance is a combination of an air leak and a lack of required ejector capacity.

Cooling Water Flow Configuration

The largest vacuum tower precondensers I ever saw were the eight giant exchangers at the former Esso Largo refinery in Aruba. I sketched the nozzle and tube support baffle configuration in Figure 27-5. Can you see Esso's mistake? It required 9 years and 50-odd trips to Aruba for me to notice the design malfunction.

First, observe that there is no air baffle in the condenser. Also, note that the vapor outlet is on top of the exchanger. The cooling water outlet, and therefore the final rows of tubes, are also on top. This means that the warmest cooling water is contacting the condenser vapor effluent.

It's true that for water coolers in general, it's best to have the cold water entering the bottom of the channel head and the warm water leaving at the top of the channel head as shown in Figure 27-5. But in this condenser, the effluent vapor, the temperature of which we are

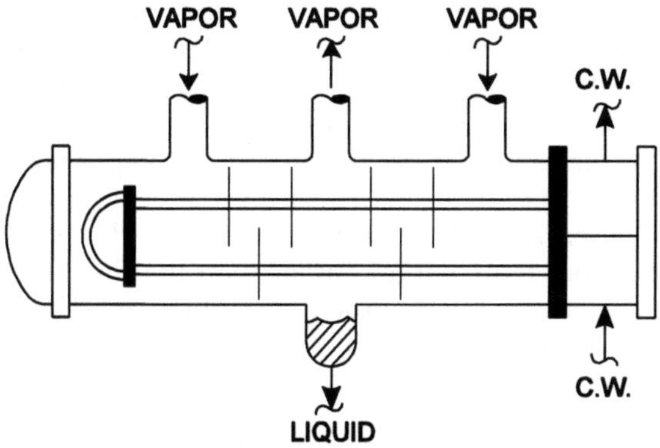

Figure 27-5 Cooling water flowing in the wrong direction increases precondenser outlet temperature.

trying to minimize, is also exiting from the top of the exchanger shell. So I had the cooling water flow reversed (i.e., the opposite of what is shown in Figure 27-5) in the channel heads of all eight shells by a relatively simple change in the cooling water supply and return piping headers. The result of this reversal in water flow direction was to reduce the precondenser outlet temperature by an average of about 2°F. This resulted in a reduction of the vacuum tower flash zone pressure of about 3 to 4 mm Hg. Meaning, the flash zone pressure dropped by an average of 4% to 5%. This further resulted in a significant increase in gas oil recovery from the vacuum tower bottoms residual tar (i.e., coker feed).

One could argue that a better solution to this design malfunction would be to retrofit the precondensers with air baffles (Figure 27-1) and side draw-off nozzles for the vapor. After all, that's the standard and correct engineering design for precondensers. But that would have required new bundles and new nozzles. My client, Coastal, would never have approved such a major project. But that's the nature of working with process equipment that suffers from design errors. You have to do the best you can with what you have and learn from other engineers' mistakes.

Management Malfunctions

One day, an energy consultant visited the former Gulf refinery in Port Arthur, Texas. He advised Gulf's management that in order to save steam, the surface condenser outlet temperature connected to **condensing steam turbines** should not exceed 100°F. This limitation was then conveyed to the plant operators as a "standard operating instruction."

But there was a problem. The cooling water supply temperature in Port Arthur in July is around 93°F. This makes it rather difficult to reduce the surface condenser vapor outlet temperature to below 100°F. However, the surface condenser outlet temperature for every steam turbine that I have seen is not measured at the vapor outlet, but in the hot well or boot, as per Figure 27-3. This measures the steam condensate drain temperature, rather than the vapor temperature to the downstream jets.

So, inadvertently, the Port Arthur operators had been instructed to minimize the boot temperature, rather than the flowing vapor temperature to the jets. And that's exactly what they did. They backed the steam condensate level in the boot up into the surface condenser itself.

Precondenser Delta P

Liz and I were working in the Calumet refinery in Shreveport, Louisiana, last week. The plant manager told me their main problem was high vacuum tower pressure caused by a high delta P in the tower's precondenser. The precondenser was a fixed tube-sheet design, meaning that the bundle could not be pulled for shell-side cleaning. We agreed that the best option was to purchase a new $500,000 (installed) condenser.

In a precondenser, the vapor comes in the top, the noncondensibles exit from the side, and the liquid drains from the bottom boot. There was a ½-inch connection on the boot. I checked each of the three points for pressure:

- Vapor inlet: 75 mm Hg
- Noncondensible outlet: 25 mm Hg
- Boot: 110 mm Hg

How could the boot pressure be 35 mm Hg higher than the inlet pressure? Obviously, the precondenser was suffering from condensate backup.

Next, Liz and I checked the level in the vacuum tower's overhead seal drum. On the panel it was shown at 80%. But all the level taps were plugged with scale. Eventually, I determined that the seal drum was actually 100% full of hydrocarbon liquids.

The liquid had backed up into the shell side of the precondenser and restricted the flow of vapor. Hence the high delta P. When we reduced the seal drum level, the precondenser delta P declined to a more normal 5 mm Hg, and the vacuum tower excessive pressure was corrected.

Just another of the thousands of lessons I've learned illustrating the importance of direct field observations in identifying equipment malfunctions.

This promoted a greater degree of subcooling of the condensate. It also promoted:

- A reduction in surface area exposed to the condensing vapors.
- Which raised the vapor outlet temperature and the vapor flow to the downstream jets.
- Which caused the pressure in the surface condenser to rise.
- Which reduced the amount of work that could be extracted from each pound of steam (see Chapter 31, "Steam Turbine Drivers").
- Which then required more steam flow to drive the turbine.

Thus, steam was wasted rather than saved by giving the operators an arbitrary guideline for the surface condenser outlet temperature. And whose fault was this? Well, it just illustrates the importance of having knowledgeable technicians or engineers working with the operators in the field to accomplish management objectives.

Half the problems that we have with poor vacuum in the surface condensers are not related to the condensers themselves, but to downstream malfunctions in the steam jets or ejectors. This is the subject of Chapter 26, "Vacuum Systems and Steam Jets."

Fixed Tube Sheet Condensers

For condensers in refinery vacuum tower overhead service, the ability to clean the shell side is critical. Both for the precondenser and jet interstage exchangers, fouling is certain. The cracked olefinic gas contains diolefins. In the presence of oxygen from the inevitable air leaks, free radicals will form. These free radicals will polymerize to form gums. The gums will mix with corrosion products produced from HCl and H_2S and with the sulfur compounds produced from the partial oxidation of H_2S. This results in a sticky, somewhat waxy mess on the shell side. I have not had any particular success in chemically cleaning such deposits.

Fixed tube sheet design exchangers are often offered by the ejector vendors as a less expensive alternative to an ordinary pull-through bundle with a floating head (TEMA designation AES). Such exchangers can only be cleaned on the tube side, which is fine for a surface condenser handling a steam turbine exhaust. A fixed tube sheet exchanger cannot be cleaned on the shell side for the simple reason that the bundle cannot be extracted from the shell.

If you ignore this advice, you will eventually replace the fixed tube sheet exchanger anyway with an exchanger that permits the bundle to be pulled for cleaning. All of my clients, sooner or later, come to this unpleasant realization. So you might as well save some money and avoid the lost gas oil recovery due to poor vacuum, and order the correct sort of condensers in the first place. Just specify "AES" on your bid specs.

Incidentally, the condenser that is most prone to fouling on the shell side is the final condenser on the discharge of the last-stage jet. Why this is so, I cannot say, but the problem is almost universal.

Author's Comments

I was teaching a seminar at the P.B. refinery in Australia last month.

"Look, ladies and gentlemen," I explained, "You all are working for a criminal enterprise. Your management has likely escaped from a home for the confinement of the insane."

"Norm, kindly cut the crap," Matthew O'Brien interrupted. "You're supposed to be teaching us about troubleshooting surface condenser malfunctions in vacuum systems, not criticizing our management, who, incidentally, is paying your extravagant seminar fee. Anyway, our management are not criminals."

"Then why, Matt," I asked, "Are they trying to destroy our planet?"

"They're not! Our management is doing everything within reason to reduce CO_2 emissions," Matthew responded.

"Haven't you seen the P.B. television ads?" Natasha added. "It explains our Green Corporate Policy. Our management is actively engaged in supporting environmental excellence."

"Natasha, dear girl. What does that actually mean?"

"Norm!" Matthew interrupted again. "It's all management B.S."

"Okay then, Matt! What are you doing about environmental degradation?"

"There's nothing an engineer at my level can do. Anyway, let's go out and look at my condenser. That's what you're being paid for. Remember, Mr. Lieberman?"

Matthew, Natasha, and I studied the spare condenser tube bundle lying in the mud.

"Look, Norm! There are no windows cut in the tube support baffles," observed Natasha. "How can noncondensibles reach the vapor outlet nozzle?"

"That's a good question. The vapor will have to bypass most of the tubes. A window for vapor flow in each baffle is sure required," I answered.

"And there's no impingement plate. And the vapor baffle you described in class isn't an L-shaped baffle at all. It's just a skinny strip of steel." Natasha's pale face flushed with excitement.

"Yeah, Natasha," I agreed, "That vapor baffle strip will never work. What do you think, Matt?"

"Oh! Natasha's right. It's all screwed up. I'll order a replacement bundle ASAP."

"Well, I was thinking you could modify this bundle. Being that it's a spare. Just laying here idle," I said.

"No. It will be a lot simpler just to order a new surface condenser bundle. It's a real pain dealing with our maintenance department. They're gonna want all kinds of instructions and dimensions and mechanical details. Natasha's right. This bundle is designed completely wrong. We'll replace it with a new, correctly designed bundle," Matthew concluded.

I watched the setting sun flatten itself against the eastern horizon, as Matt and Natasha discussed expediting the delivery of the new surface condenser for the P.B. refinery in southern Australia. It reminded me of a book I read a while back, *On the Beach,* which also takes place in southern Australia.

CHAPTER 28

Excess Gas Overloads: Vacuum System Ejectors

I saw the lightning and waited on the thunder. How far off, I thought and wondered.

—Bob Seger, "Night Moves"

As described in Chapter 26, "Vacuum Systems and Steam Jets," excessive noncondensibles or gas flow will overload a jet system with the result of:

- Loss of the sonic boost.
- Surging of the jet.
- The "breaking" of the vacuum in the upstream vacuum tower or the surface condenser.

The most common, if not the most serious source of excessive noncondensibles, is air in-leakage.

Air Leaks

For an overloaded vacuum system serving a steam turbine, there is only one possible source of noncondensibles. That is an air leak. Most of the normal air in-leakage occurs along the turbine's shaft seal. When the turbine is not running, air leakage along the shaft, which is drawn into the surface condenser, is normal. But when the turbine starts to spin and the shaft begins to get hotter, most of this seal leak stops. My observation is based on the vacuum in the surface condenser improving once the turbine is up and running.

To a large extent, the air that leaks past the turbine's mechanical seal should be extracted by the small hogging-type steam jet on the side of the turbine, located near the coupling (see Chapter 26). Thus, the small steam jets located downstream of the surface condenser should ideally only be extracting air leaks in the surface condenser itself.

The ejected air is exhausted out of a vent discharging from the final (second) stage condenser. If there are no air leaks, there will be no flow from this vent. It is not at all unusual for me to set a piece of cardboard on top of this vent without changing the pressure in the surface condenser or having the cardboard blow off. On the other hand, you can hold a strip of cloth over this vent and get a qualitative idea as to how much air is being exhausted. Then, as I reduce the air in-leakage, I can easily observe if the vent is blowing with reduced velocity.

How to Find Air Leaks

When air expands, it cools. All gases (other than H_2 and CO_2) cool upon expansion. Atmospheric air when drawn through a small hole into a vacuum system will cool. Not only because the air has a positive **Joule–Thompson expansion coefficient**, but also because a lot of its enthalpy (heat) is being converted into kinetic energy (speed). This will produce a cold spot on a leaking pipe or fitting. In New Orleans, the land of humidity, this is going to cause localized condensation to form on the pipe. If you touch such a spot and it's cold, there is an air leak.

To confirm your air leak, I would wrap it with tape. If it's the sort of leak that can't be wrapped (like a screwed fitting), pour water over the suspect leak. You will instantly see if the airflow from the atmospheric vent diminishes. Within a few minutes, vacuum will improve if your ejectors were limiting the performance of the surface condenser.

Not uncommonly, the air leak could be in a flange. Wrap the suspect flange with tape. Punch a hole in the tape and spray aerosol shaving cream over the hole in the tape. If the cream is sucked in, there is a leak.

A more expensive method (which I have never used, but heard about from my wealthy clients) is:

- Spray helium gas over a suspected air leak area.
- Install a portable He monitor on the seal drum noncondensible vent.

You can do the same thing yourself without hiring an expensive contractor. Use Freon and a portable Freon leak detector available at your local air conditioner supply store. This is a good general method to detect leaks in many heat exchangers as well.

Air Leaks on Vacuum Towers

Finding air leaks in refinery vacuum systems is more difficult than in turbine exhaust surface condensers. Not only is the system far bigger, but the air leaks could be associated with the hotter parts of the system. So you will need to get a sample of off-gas in a nonmetallic container. This is critical. You must prevent the reaction:

$$H_2S + O_2 \rightarrow H_2O + \text{sulfur}$$

A metallic container would catalyze this reaction. The O_2 content must be preserved. For example, you obtain a sample of the vacuum tower tail gas. Suppose it's 40% N_2. You would suspect that you would then have 11% O_2:

$$(40\%) \times (21\% \div 78\%) = 11\%\ O_2$$

But you don't. You have 1% O_2 and 10% CO_2! What happened to the other 10% O_2? And where did the 10% CO_2 come from? Well, the air leak was in the vacuum heater transfer line, which ordinarily runs at a slight subatmospheric pressure. In the Coastal refinery in Aruba, we once identified a hole in a vacuum heater tube based on the CO_2 and N_2 content of the vacuum jet off-gas. That is, the O_2 was burning the hydrocarbon content of the feed, and thus producing CO_2. Of course, some of the O_2 is converted to CO and H_2O. So my main point is, if the O_2 content is disproportionately low compared to the N_2 content, then the air leak is in the hot part of the vacuum tower, and not in the overhead system.

Incidentally, if you have the following sort of analysis of your noncondensibles:

- $N_2 = 20\%$
- $O_2 = 5\%$
- $CO_2 = 0$
- $CO = 5\%$

the carbon monoxide content is not due to an air leak, but is caused by the thermal degradation of naphthenic acids. There is nothing you can do about this reaction, so don't worry about it.

Exchanger Leaks

I have two examples of this nasty problem. One from the now-defunct Pacific refinery north of Berkeley, California. And the second from the currently idled Valero refinery in Aruba.

In California, the leaking exchanger was a vacuum tower bottoms quench exchanger versus crude, as shown in Figure 28-1. But how

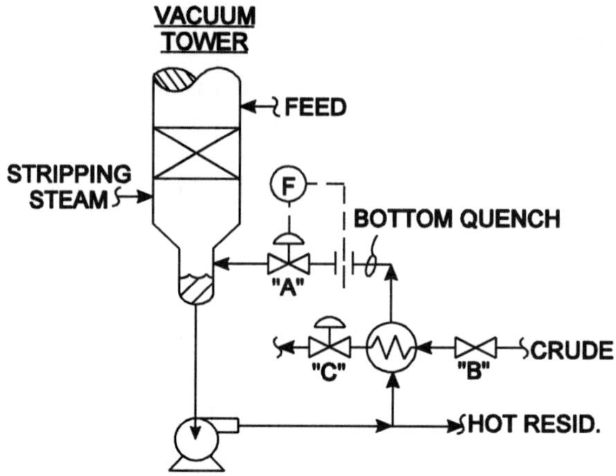

FIGURE 28-1 Leaking quench exchanger overloads vacuum system with virgin light gas (ethane, propane, butane).

did I know that the exchanger was leaking crude oil into the circulating quench resid stream?

I had noted that every time the tower bottoms quench flow increased, vacuum was lost. That is, every time valve A (Figure 28-1) opened, the vacuum degraded. One possible explanation was because of increased noncondensible flow to the jets. I therefore reasoned as follows:

- Opening valve A reduced the shell-side pressure in the quench heat exchanger.
- The pressure in the crude side (i.e., the tubes) would then exceed the pressure in the shell. Local pressure gauges supported this hypothesis.
- As the shell pressure fell below the tube pressure, crude charge would leak into the vacuum tower quench oil return stream.
- Light ends in the crude (ethane and propane) would load up the overhead jets and degrade the vacuum.

But a theory without a test is bullshit. So what was my test for this theory? What I could have done, but did not do, was to sample the off-gas for excessive propane. This would have taken too long to obtain the sample results. What I did do was (referring to Figure 28-1):

- Partially close isolation gate valve B.
- This caused control valve C to open from 40% to 60% valve position.
- The tube (crude)–side pressure in the heat exchanger dropped by 20 psi.

- The tube side–pressure was now lower than the shell (bottom quench oil)–side pressure.
- The metered vacuum tower off-gas rate dropped, which unloaded the jet system, and the vacuum improved.

My story from the Valero (at that time, Coastal) refinery in Aruba is one of the few occasions when I identified a malfunction by phone. In Aruba, hot vacuum tower feed was used to preheat crude. The reason for this strange process configuration was that the vacuum tower feed flowed through a storage tank operated at 300°F. That's not important. What was important was that the vacuum tower feed versus crude preheat exchanger was leaking.

But how could I identify this malfunction in Aruba from my office in New Orleans? I asked for a sample analysis (gas chromatograph) of the vacuum system off-gas collected from the hot well or seal drum. This indicated:

- Mole percent propylene = 2%
- Mole percent propane = 12%

Normally, the composition of the vacuum tower off-gas reflects the origin of the gas. That is, **thermal cracking**. In thermally cracked gas, we find that the ratio of olefins to saturates (i.e., propylene to propane), is about 40:60. But in my sample of off-gas from Aruba, the ratio was 2:12. That is, I had about four times the expected concentration of propane, based on the propylene concentration.

The engineers in Aruba decided to put my leaking heat exchanger theory to a test. They reduced the pressure on the crude oil side of the suspect exchanger a few psi below the pressure on the vacuum tower feed side of the exchanger. As the vacuum tower feed went through a large intermediate tank, the vacuum tower pressure and vacuum off-gas rate slowly declined over the next day. Finally, the off-gas analysis showed that the propane content of the seal drum off-gas had fallen from 12% to 3%. Incidentally, the amounts of ethane and butane (iso and normal) had also declined.

Poor Vacuum Tower Feed Stripping

I hate Canadians. It's not their fanatical passion for hockey. It's that they live in a cold country and refuse to admit it. For example, I was working at the Suncor Oil Sands Upgrading plant in Ft. McMurray, Alberta, one January. The problem was a poor vacuum on their new, gigantic, vacuum tower. The off-gas analysis showed the following composition:

- Saturated butanes (iso plus normal) = 20 mole%
- Unsaturated butanes and butylenes (C_4 olefins) = 2 mole%

If the C_4s were just a product of thermal cracking of the vacuum tower feed, the saturated and unsaturated compositions would have been approximately equal. In this unit, there were no heat exchangers located on the vacuum tower feed. There was no circulating bottom quench, as in the prior example. If the vacuum tower pumparound heat exchangers had been leaking, then the vacuum tower gas oil products would have been contaminated with black tar sands oil (i.e., bitumen). Therefore, I decided that the problem was not a leak of black bitumen into the vacuum tower, as in my prior example.

The problem of excess saturated butane in the feed could only be due to poor steam stripping efficiency in the upstream tower. That is, the butanes were not being efficiently stripped out of the vacuum tower feed. So I increased the stripping steam rate and the malfunction got worse. Meaning the pressure in the downstream vacuum tower went up.

Most unfortunately, my increasing the stripping steam rate to the upstream tower also caused this tower's stripper to pressure up. Why? Because we lacked sufficient overhead condenser capacity to handle the extra stripping steam. And this was on a new unit, using air coolers, in the middle of the winter, with ambient temperatures of minus 20°F!

And when I complained that my nose, feet, and fingers were numb from the cold, and that at minus 20°F, they really should not be limited by air cooler condenser capacity, the unit supervisor, Probkar Reddy, said, "But Norm, it's been a warm winter. I guess that's why we don't have sufficient stripper condensing capacity, to strip the vacuum tower feed properly and prevent the butanes from overloading the first-stage vacuum jets."

And that's why I hate Canadians. They live in a minus 20°F environment and claim it's warm. And the worst aspect of the story is that Probkar comes from a subtropical region of India and isn't a native Canadian, anyway!

Evolution of Cracked Gas

The rate of formation of cracked gas is a function of time and temperature. By time, I mean the residence time that liquid is held at a particular temperature. The official way of stating this is that thermal cracking is a **zero-order reaction**. In my book, *Troubleshooting Process Operations*, 4th ed., I published a chart correlating:

- Observed gas flow from jets
- Barrels of vacuum tower feed
- Vacuum tower flash zone temperature

This chart is based on empirical observations and thus is a useful design tool. But this chart and the theory of thermal cracking being a zero-order reaction is suspect. The problem is that cracked gas can contain 40% H_2S on a dry basis.

The term **dry basis** means that I have condensed out the moisture in my sample of off-gas before checking the H_2S content with my Drager tube or Gas-Tec tube. So before you accidentally kill yourself with H_2S (fatal concentration being 1,000 ppm or 0.1% in air), here's how to check for H_2S and safely obtain a sample of gas from a seal drum or hot well:

- The hot well will have a slight positive pressure of 1 to 4 psig.
- Put on your Scot Air Pak or B.A. equipment.
- Get a pint-size can or bottle and two lengths of plastic tubing.
- Blow gas into your bottle from the bleeder on the off-gas line from the hot well drum. Open the valve just enough so that you can feel a little gas blowing out of your bottle.
- Draw the sample up into your Drager tube (this is a standard, portable lab tool, used to measure, by chemical reaction, the concentration of various gases such as H_2S, SO_2, CO_2, NH_3, etc., on a dry basis).
- To retain a sample for the lab, I inflate a bicycle tire using a hand pump (looks like a rubber ball).

In Aruba where the vacuum tower feed might be 4 wt% sulfur, I measured 40% H_2S in such a sample of cracked gas. In other places, with the sulfur content of vacuum tower feed about 1%, I've measured about 15% H_2S in the off-gas from the seal drum. So obviously, a large amount of the cracked gas rate is related to the sulfur composition of the feed. Hence, the evolution of cracked gas is actually dependent on things other than time and temperature.

In order of concentration (on a dry basis), a typical off-gas from the vacuum tower seal drum would contain:

- Methane (up to 50%)
- H_2S (10% to 40%)
- Hydrogen (maybe 10%)
- C_2s (maybe 10%)
- C_3s and C_4s (maybe 10%)
- Nitrogen (about 5%)
- CO_2 and CO (about 2%)

This analysis excludes air leaks. The 5% N_2 is due to air contamination on sampling.

Sources of Cracked Gas

There are three areas in vacuum tower operation where cracked gas is generated. All three are under our direct control:

- Vacuum heater passes or tubes
- Transfer line and flash zone
- Bottom boot of the vacuum tower

The vacuum tower heater operates differently than other process heaters. That is, the maximum temperature that is experienced in the heater is not at the heater outlet, but back upstream in the heater passes—perhaps two, three, four or more tubes upstream of the outlet. That's because as the flow nears the outlet, it undergoes a big delta P. A lot of flashing occurs as the heater outlet pressure falls to subatmospheric pressure near or at the outlet. All this evaporation or flashing requires lots of latent heat of evaporation. Some of this heat does come from the radiant heat transfer in the tube. But most of it comes from the flowing oil itself. That is, the rapid reduction in pressure causes the sensible heat content of the flowing oil to be converted to latent heat of evaporation. Thus, the oil temperature drops, even though radiant heat transfer continues. Hence, the heater outlet temperature might be 40°F to 50°F lower than the peak tube temperature. This value of 40°F or so is not only a calculated number. At the Mobil refinery in Coryton (U.K.), I observed just such an effect on their vacuum heater, which had many process side TI points through the heater passes.

I have represented this concept of peak heater temperature in Figure 28-2. The lower curve shows the effect of using steam in the heater passes. The steam reduces the hydrocarbon partial pressure. This promotes the earlier vaporization of the gas oil in the heater

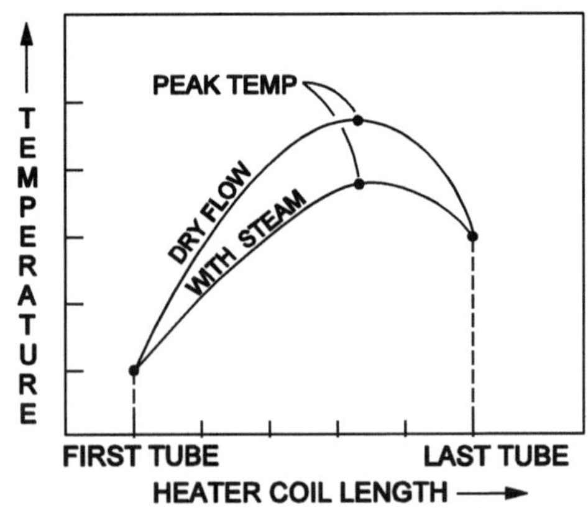

FIGURE 28-2 Effect of steam on peak vacuum heater coil temperature.

passes and thus suppresses the peak temperature. The rate of cracked gas evolution is strongly related to this peak temperature, rather than to the heater outlet temperature, which is cooler.

Unfortunately, this bit of knowledge once cost me $65,000. You see, I had a contract from the Unocal refinery in San Francisco to revamp their vacuum tower for increased gas oil recovery. Rather than just carry out my assignment like a sane and sensible engineer, I decided to run a plant test:

- I increased the steam in the heater passes from zero to 1½ pounds of steam per barrel of vacuum heater feed. The heater inlet pressure increased from 60 to 100 psig. I would have used even more steam, but the heater feed inlet pass valves were almost wide open.
- The cracked gas flow went down a lot.
- The vacuum became better as I unloaded the jets.
- I gradually increased the heater outlet temperature until the vacuum started to slip back down.
- Next, I optimized the steam pressure to the jets as described in Chapter 26. Then I cranked up the cooling water flow to the vacuum tower overhead precondenser and added a little more stripping steam to the vacuum tower bottoms stripping section.
- Then I added a few more pounds of steam to the heater passes and a few more degrees to the heater outlet and back-flushed the vacuum tower precondenser as discussed in Chapter 27, "Vacuum Surface Condensers and Precondensers."
- And then gas oil production increased to the required design value of 9,000 BSD. And then my former friend Paul Schraeder canceled my $65,000 contract and issued a $2,000 purchase order for my one day's field services. Only then did I remember what my Dad used to tell me, "Son, you are selling knowledge, not time. Go dig ditches if you want to sell time."

Effect of Heater Outlet Temperature

I recall standing in the parking lot of the Fina refinery in Big Springs, Texas. This plant is located in a desert 40 miles east of Odessa. Warning! There are no springs in Big Springs, Texas. It's bone dry and quite cold in the morning. And so is the cooling water to their vacuum tower precondenser. As the heat of the day comes on, the precondenser vapor outlet temperature rises. As does the vapor pressure of water in the precondenser vapor flow to the first-stage jet. This jet starts to surge due to loss of its sonic boost as the vapor load to it escalates due to the extra moles of steam in the precondenser vapor effluent. I could hear this surging sound in the parking lot every day about 9:00 A.M. Also, the vacuum on the tower would increase from about 70 mm Hg to about 110 mm Hg. As a result of the degraded tower vacuum, heavy vacuum gas oil production would drop by 2,000 or 3,000 BSD.

To mitigate this problem, I had the operators reduce the heater outlet by 5°F. This relatively small move would usually unload the primary jet enough to restore the sonic boost. If not, and I still heard the surging from the parking lot, I would knock off another 5°F from the heater outlet. The loss of 5°F or 10°F flash zone temperature had a minor detrimental effect on the recovery of heavy gas oil, but was nothing as compared to the 40 mm Hg drop in vacuum due to the loss of the sonic boost in the first-stage ejector.

I'm usually cautious when raising the heater outlet temperature, especially in vacuum service. Certainly, if raising heater outlet temperature by 5°F produces a loss in vacuum of 5 mm Hg (0.2 inches of Hg), I feel that gas oil recovery will be adversely affected. But even if it's not, and I gain some gas oil production, I would still be concerned about increased coke formation in the heater tubes. The reason I was confident at the Unocal San Francisco refinery in raising heater outlet temperature was because I had suppressed coke formation rates with the coil steam. So, regardless of any design criteria or rules of thumb pertaining to the flash zone or heater outlet temperature, let me give you some sage advice:

- If the metered off-gas flow increases by 10% to 20% as you increase the heater outlet, back off, because you will shorten the heater run length due to coke formation in the tubes.
- More steam in the heater tubes will significantly increase heater run length as limited by coke formation and high tube skin temperatures in the heater tubes.
- Don't marry girls who cannot get along with their own families.

Incidentally, if you have a precondenser, you may be able to reduce the load to the first stage by raising the tower top temperature. I noticed this at the Good Hope refinery vacuum tower. Raising the top temperature flashes more naphtha overhead, which absorbs some lighter hydrocarbons out of the gas to the first-stage jets. Indeed, one of my clients injected some heavy naphtha into the precondenser feed for the same purpose. Of course, if the precondenser outlet temperature goes up more than a few °F, you may not gain anything by increased tower top temperature.

And then there's the incident that occurred in the Ecopetrol refinery in Colombia. The engineers woke me from my siesta one

Saturday afternoon. "Señor Lieberman! Bueno news! We have found that adding mucho neutralizing NH_3 improves vacuum. We abrieto the NH_3 feed valve 100%, which flows to the discharge of the first-stage jet. Mira, Señor Norm! Vacuum has greatly improved. Porque?"

Because, I explained:

$$NH_3 + H_2S = NH_4HS$$

which is soluble in water in the downstream condenser. In effect, they had extracted the H_2S out of the vapor to the downstream second-stage jet. If this jet was overloaded and was running without its sonic boost, a great improvement in vacuum could be expected. But by the time we all arrived at the plant, the pressure had risen back to normal.

"Muy malo, Señor Norm! Que pasa?"

The cylinders of ammonia had all gone empty. They had consumed a week's worth of ammonia in 5 hours. However, if one could do the same with amine, and then regenerate the amine, why then, one could . . . But I'm writing this book based on hands-on experience only. However, it's an idea I've often had. Maybe you would like to try it yourself.

Cracked Gas Evolution from Boot

Please refer to Figure 28-1. Large amounts of cracked gas can be evolved from the vacuum resid in the tower bottoms. On one unit at Murphy Oil in New Orleans, we found that 50% plus of the evolved gas was due to cracked gas from the bottoms boot. You can see if this is a problem by lowering the boot quench temperature. Or, if you do not have such quench facilities, reduce the tower bottoms level.

If the cracked gas rate goes down, then your vacuum is going to get better. Typically, keeping the tower bottoms temperature below 680°F avoids the whole problem. But, if the boot's residence time is low, boot temperatures above 700°F are also okay.

Air Leaks in Heater Transfer Lines

This is a safety issue of some importance. I have been involved in the apparent sudden failure of vacuum heater transfer lines at the American Oil refinery at Sugar Creek, Missouri, and at the Good Hope refinery in New Orleans. I say "apparent" sudden failure because the leaks in the heater transfer lines had not developed suddenly at all. If we had been tracking the N_2 and CO_2 content of the off-gas, we could have seen the leaks developing for months. It's just that these leaks were under the insulation covering in an area of the heater transfer line operating under a subatmospheric pressure. An operator in one of my seminars said he once removed the insulation from a vacuum heater transfer line and could see the feed rushing past him. We found the leaks at Sugar Creek and in Louisiana in a more dramatic fashion. A minor problem with the steam jet systems caused a sudden loss of tower vacuum. The defective section of the heater transfer line went to a positive pressure. The hot oil blew out, autoignited, and large fires resulted.

Why are transfer line leaks so common (transfer line is the term used to describe the pipe connecting the heater to the tower) in vacuum services?

- Naphthenic acids inherent in many crudes.
- High (750°F to 800°F) temperature.
- But mainly, high velocity. Especially if the velocity exceeds 50% of sonic and the transfer line is constructed (God forbid) out of 410 chrome steel.

So the gradual increase in the CO_2 in off-gas, even if it's not greatly affecting the vacuum, should still be considered an indication of a malfunction that you should worry about.

Purges and Seal Gas

Often the top tap of level-sensing instrumentation is purged with natural gas or nitrogen through a restriction orifice. In my designs, I don't use level control, but pump **suction pressure control** (as described in my book, *Troubleshooting Process Plant Control*). So for me it's not a problem. But natural gas is methane, and any purge gas on level taps will contribute to the off-gas handled by the jets. Sometimes pressure-sensing connections and flow orifice taps are purged with refinery fuel gas or natural gas. Packing (i.e., filling) these connections with a light gas oil or diesel oil is sufficient, without gas purging, provided the clean gas oil is changed periodically. I recall that my friend who works for Chevron in El Segundo said they did this change-out perhaps once a month.

In newer pumps in black oil service, we now use **nitrogen barrier (dry gas) seals**, rather than seal flush oil. I don't really have any idea as to how much N_2 would be used on an idle vacuum tower pump. But whatever the flow, it will contribute to the off-gas rate to be handled by the ejector system. So when the pump is on standby, block-in the nitrogen flow to the pump seal. I suppose some clever mechanical engineer will call the seal vendor and tell us the N_2 flow is tiny. But that's only true if the seal face tolerances are as per the manufacturer's specs. And this is a book about malfunctions. So block-in the N_2 to idle pumps.

Reminder about Water Partial Pressure

After you have identified all your various sources contributing to your off-gas flow, don't make the same mistake that I sometimes do. That is, I forget to back-calculate the moles of H_2O. Under vacuum conditions, the vapor pressure of water at the condenser vapor outlet temperature can easily be a very large part of the total pressure. It is not at all uncommon for the partial pressure of H_2O to be 50% to 80% of the total pressure. The extra moles of water may greatly increase the vapor flow to the downstream ejector. Then convert the moles of H_2O to its air equivalent, as I described in Chapter 26. An accurate field measurement of the condenser vapor outlet temperature is quite critical in this service to avoid misstating the moisture partial pressure.

Reminder about Water Partial Pressure

After you have identified all your various sources contributing to [illegible]

CHAPTER 29

Centrifugal Pump NPSH Limitations: Cavitation, Seal, and Bearing Failures

Anything bad that can happen, will happen.

—Author's creed for being a refinery operating superintendent

Poor pump suction conditions will cause cavitation and will damage the pump's mechanical seal. Many of the fires and explosions in hydrocarbon process plants are caused by pump mechanical seal failure. Seal failure is the leading mechanical equipment malfunction in almost every process plant. The pump mechanical seal is the device that keeps the process fluid from escaping around the spinning shaft where the shaft goes through the pump case.

Most persistent or chronic seal failures are caused by cavitation. Pump cavitation is a consequence of the process fluid flashing to vapor as it enters the pump case. This happens because the process fluid loses pressure as it flows from the larger-diameter suction line into the smaller eye of the pump impeller. To prevent this from happening, the liquid has to be above its boiling point pressure before it enters the pump. Or, another way of saying the same thing is that the liquid has to be below its boiling point temperature before it enters the pump case. I can quantify this idea by defining the term **Available NPSH** as:

- The physical pressure at the suction of the pump, minus the vapor pressure of the liquid at the suction of the pump.

In mathematical terms this means:

$$\text{Available NPSH} = (P_1 - P_2) \times 2.31 \div (\text{SG})$$

where Available NPSH = Net positive suction head, feet

P_1 = Physical pressure measured at the pump suction, downstream of the pump's suction screen, in psia (not psig)

P_2 = Vapor pressure of the liquid, at the observed pump inlet temperature in psia

SG = Specific gravity of the liquid pumped relative to water at 60°F at the flowing temperature

The amount of pressure the liquid loses as it enters the pump case increases with flow. The pump manufacturer provides a curve that informs us as to how much pressure, or really feet of head, is required by the pump. As the flow to a pump increases, this required feet of head also increases. This is called the **Required NPSH** of the pump.

When the Required NPSH equals the Available NPSH, the pump will cavitate or slip.

Symptoms of Cavitation

1. The pump discharge pressure is erratic. It varies from normal to low. Alternately, a variation of normal to high pump discharge pressure is never an indication of cavitation.
2. The pump flow becomes erratically low.
3. The pump suction pressure is drawn down slowly but evenly, and then jumps back up.
4. The motor amps are erratically low.
5. The pump may (but not always) make a rattling sound.

Unfortunately, reality is often more complex than this conventional definition of cavitation. Many larger, high-head pumps exhibit none of the preceding symptoms when they are limited by marginal suction conditions. They just pump less with a reduced, but quite stable, discharge pressure. This is contrary to what you'll read in other books, but that's the way it is. A pump is operating with a stable discharge pressure and flow. I then increase the suction pressure from 1 to 3 psig and the discharge pressure increases from 80 to 150 psig. This only seems to happen for large pumps working with a low suction pressure of a few psig.

Pumps suffering from this sort of malfunction do not destroy their mechanical seals, as do pumps that are cavitating in the more conventional erratic mode.

Seal Failure Due to Piping Alignment Problems

Once I had a centrifugal pump that suffered from repeated seal failures. The pump connections were blinded-off and the pump was sent to the maintenance shop. On one occasion, I watched as the pipe fitters reinstalled my pump. After bolting the pump back onto its base, they reconnected the flange on the suction of the pump. At this point I noticed the discharge flange on the pump was 8 inches out of alignment with the flange on the discharge piping. The pipe fitter was forced to use a come-along to force the two flanges together (a come-along is a crank and chain used to forcefully pull a pipe into a new position).

The stress introduced by the out-of-alignment discharge piping also tended to put a stress on the pump's mechanical seal. This promoted seal failures. Of course, you can't expect the pump and piping flanges to line up exactly. The two flanges need to be forced into alignment with the pointed end of a wrench inserted through the bolt holes. Then the first two bolts can be inserted and the gasket placed between the flanges. The pointed end of the wrench is removed and the flanges bolted up. This amount of manually forced realignment is acceptable. As I learned, cutting and rewelding the piping to avoid more forceful flange alignment is an effective way to reduce repeated mechanical seal failures.

A similar problem can exist due to piping stresses transmitted to a pump by uneven heating of common piping between parallel pumps. There are, for example, three pumps in parallel, circulating 700°F slurry oil. One pump is on standby and thus is relatively cold. Special care must be taken to accommodate the differential rate of the thermal expansion of the piping when designing the piping expansion loops. This is done by a computer method that checks the piping stresses under a wide variety of operating parameters. It's a complex calculation, which is the job for a mechanical engineer, and not process engineers and plant operators.

Causes of Lack of NPSH

The required NPSH is not only that stated on the pump manufacturer's performance curve (see Figure 29-1 as an example). You also have to add in two additional head losses:

- Frictional loss through the pump's suction piping and fittings.
- Upstream vessel nozzle exit loss, including the effect of those nasty and often unnecessary vortex breakers.

If the design has used a normal suction line velocity of 1 to 2 ft/sec, these losses will be negligible, provided the outlet nozzle from the vessel is not partly plugged. Only too often the vessel

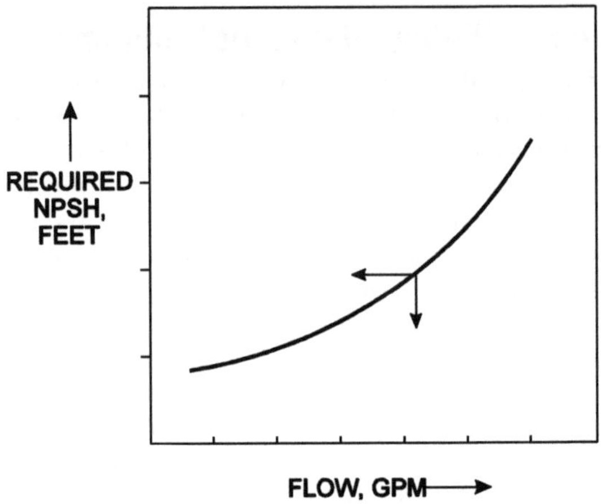

Figure 29-1 Typical NPSH required versus flow curve.

internal bottom's outlet screen becomes fouled. Or, the fractionation trays collapse on the outlet nozzle. In Texas City, the maintenance department once removed a totally dead contractor from a xylene splitter pump suction nozzle. Equation (29.1), in a subsequent section of the chapter details how to calculate the suction head of liquid, which is then compared to the actual head of liquid observed.

One useful method to discriminate between a restricted outlet nozzle and frictional loss in the suction line is to proceed as follows:

Step 1—Place a pressure gauge on the suction of the pump.

Step 2—Increase the flow until the pump just starts to cavitate.

Step 3—Reduce the flow by 1 or 2% until cavitation just stops.

Step 4—If the suction pressure comes up by 1 or 2 psi, the problem is frictional loss in the suction line.

Step 5—If the suction pressure comes up by 5 or 10 psi, the malfunction is a partly plugged vessel outlet nozzle.

Frequently, it's the vessel outlet nozzles that cause the malfunction. Pump suction "Y" strainer screens also commonly restrict flow. That's why we have spare pumps, so that the strainers may be cleaned.

Starting NPSH Requirements

I have written quite a bit about this subject in my book, *Troubleshooting Process Operations*, 4th ed. The basic idea is that when we start a pump, the liquid in the suction line is stagnant. To accelerate the mass of liquid takes energy. The energy cannot come from the pump. It comes from the available NPSH. This starting NPSH is subtracted from the NPSH needed to run the pump. So, to avoid cavitating the pump on startup, the good operator opens the discharge valve slowly—which is fine, up to a point. The difficulty is that running the pump at a very low rate for a long time causes the barrier fluid, or seal flush, between the seal faces to dry out. This also will damage the seal faces.

In *Troubleshooting Process Operations*, I provided the lengthy calculation procedure to determine the additional NPSH requirements needed to avoid cavitation on startup. But with existing equipment, what options are available to overcome this malfunction? I list them here:

- Maximize the upstream vessel level.
- Spray water on the bare suction piping and on the pump case. A few degrees of subcooling reduce the fluid vapor pressure a lot, especially for lighter components.
- Raise the pressure in the upstream vessel by 5 or 10 psig. The pressure at the suction of the pump will increase instantly. However, the lighter liquid formed at the vapor–liquid interface will take 5 or 10 minutes to migrate down into the pump suction. For that short period of time, you will have quite a bit of extra available NPSH to start your pump. At equilibrium, this extra NPSH is of course gone.
- If there is a spill-back line, this will make it slightly easier to start the pump. However, excessive use of the spill-back will overheat the pump itself and cause cavitation.

There's a widespread misconception that starting a centrifugal pump with the discharge shut is a good idea. This too will overheat the liquid inside the pump case. The discharge valve needs to be opened one-quarter to one-half turn, and not shut on startup.

Leaks into Pump Suction

Water leaking into the suction of a hot hydrocarbon pump will flash to steam. The evolved vapors will choke off the flow and cause cavitation.

For pumps in vacuum service, the pump suction may be under a subatmospheric pressure. This is also true for pumps at an elevation above the liquid level in a vessel (for example, a sump pump). Tank field pumps, which are removing the heel from a tank, typically develop a vacuum at their suction when the tank level drops to a few feet. A flange leak of very minor proportions in the suction piping of such tanks will cause the pump to cavitate and lose discharge pressure, because air is drawn into the pump suction flow.

As such pumps' mechanical seals are also under a subatmospheric pressure, a minor seal leak will also draw air into the pump case and promote cavitation. Especially common are packing leaks on isolation gate valves, which draw air into the pump's suction past the valve stem.

To identify the particular leak, I take a water hose or a can of the appropriate heavy oil and pour it slowly over the area where I suspect a leak. The fluid will temporarily displace the air. And, for a few seconds, the pump discharge pressure will be restored. My friend Jerry recently found a leaking flange on the suction of a vacuum gas oil pump using this simple method.

Anything that causes the liquid in the suction line of a pump to partially vaporize will cause a loss of NPSH and cavitation. However, as the following incident illustrates, this sort of malfunction is often not as simple as the problems just described.

Mixing Two Dissimilar Streams in a Pump Suction

I recall at the Coastal refinery in Eagle Point, New Jersey, that I had a pump cavitation problem. My client had joined two streams at a tee-junction.

- 100°F light naphtha
- 150°F heavy naphtha

Both streams were at their saturated liquid (i.e., at their boiling or bubble points) conditions at the pump suction conditions. When the two streams were blended together on my computer simulation model at the pump suction conditions, a pumpable saturated liquid with no vaporization was predicted.

But in reality, the existing pump was subject to periodic cavitation. I suppose my assumption of perfect mixing and equilibrium was flawed. It's okay to mix two subcooled liquids at the pump suction. But I have learned not to do this with two saturated liquid streams.

As I write this I'm returning from the Ras Tanura refinery in Saudi Arabia. They have a chronic problem on a debutanizer bottoms reboiler circulation pump. They have been destroying mechanical seals on these giant pumps for 10 years. Pump conditions are (see Figure 29-2):

- Suction pressure = 225 psig
- Suction temperature = 550°F
- Flow = 7,000 GPM

It's a nasty problem because the material being pumped is above its autoignition temperature. So a seal failure results in a fire. The giant size of the pump increases the hazard. The hazard is real enough. Two guys were killed a few years ago during such a fire on this unit.

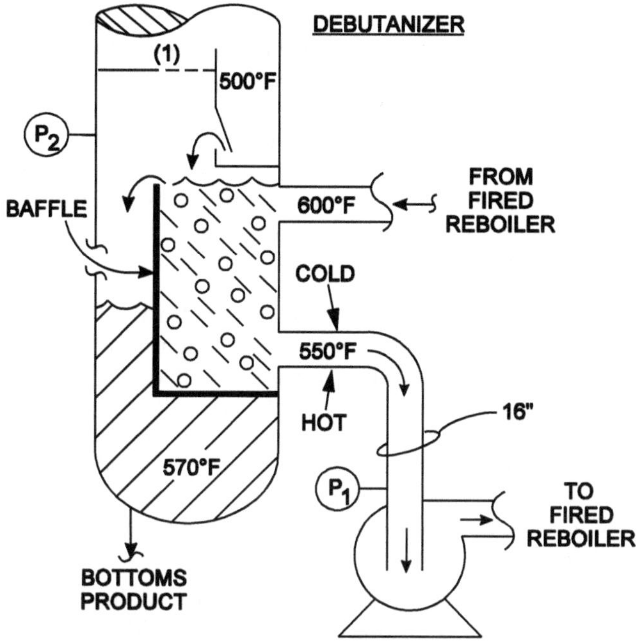

FIGURE 29-2 Mixing lighter and heavier components in a pump suction will promote cavitation.

My refinery contact, Raj Malik, has felt that marginal suction head has been the real malfunction causing the pump seal failures. I call this lack of available NPSH. Reduction in the tower operating pressure makes this tendency to cavitate worse.

If you will study Figure 29-2, you may see the similarity with my experience at the Coastal Eagle Point refinery. The difference is that the two liquids are comingled in the bottom of the debutanizer, rather than in a tee-junction. I'll explain:

- 500°F light liquid overflows the bottom tray's seal pan to the right of the vertical baffle.
- The 600°F heavier liquid from the reboiler return also enters to the right side of the vertical baffle.
- As the two different liquids mix, vapor is generated throughout the pump draw-off sump, so that a static head of liquid does not exist in the draw-off sump.
- The feet of head lost through the draw-off nozzle (the nozzle exit loss due to liquid acceleration) then promotes vaporization in the 16-inch pump suction line.
- Formation of vapor in the suction line reduces the NPSH available to the pump.

"But Norm," Raj objected, "That all sounds like a theory. Where's the test that proves your theory? We already have 10,000 engineers in our Dammam headquarters generating theories. We don't need another one."

"Raj, there are two tests. First, put the pump in a marginal cavitation mode. That is, carefully increase the flow until the motor amps slip and the discharge pressure becomes slightly erratic. Next, reduce the flow by 1 or 2% and make the following field observations:

- Along the horizontal section of the draw-off nozzle, at the elevation of the draw, check the skin temperatures at the top and bottom of the 16-inch pipe. If my theory is correct, the top of the pipe will be 50°F to 100°F cooler than the bottom of the pipe. The reason for this is that vapor conducts heat much less efficiently than liquid to the pipe wall. Incidentally, I don't think this measurement can be meaningful for a small (4- to 6-inch) draw-off line, but it should be definitive in your 16-inch line.
- Referring to Figure 29-2, measure the delta P between P_1 and P_2. Convert this delta P to feet of liquid:

$$(\text{Delta P}) \times (2.31 \div \text{SG}) = \text{feet} \tag{29.1}$$

- Compare this value to the elevation difference between P_1 and the top edge of the baffle. If my theory is correct, the feet of head from the elevation difference will be 10% to 20% more than the feet of head calculated from the differential pressure between P_1 and P_2."

Correcting Vaporization in a Draw-Off Sump

The next day Raj reported to me that both measurements indicated the formation of vapor in the draw-off sump.

"And then Norman, my friend, what shall we do?"

"My advice is to only permit the 500°F material from tray #1 seal pan to flow into the draw-off sump. The 600°F vapor-liquid reboiler effluent should be returned into the product side of the baffle. [That is, on the left-hand side of Figure 29-2.]"

"Yeah! Okay. However, the pump's rated for 10,000 GPM at its best efficiency point," Raj complained. "If I follow your advice, my friend Norman, I'll only be running the pump at 3,000 GPM. Is that okay?"

"No, Raj! It will not be okay. Running a high head, large volume pump at 30% of its **best efficiency point** will likely also cause seal failures due to vibration. It's just not safe," I concluded.

"So what then?" Raj asked.

"Look, Raj. You should have had the tower and pump designed properly in the first place. Now you have created a really bad problem."

"Ah, Norman, my good friend. With such friends like you, one doesn't need enemies," Raj concluded sadly.

Low NPSH Pumps

At Amoco Oil, we tried to resolve a pump NPSH problem by the use of a "pig tail." That's a corkscrew-type device attached to the center of the eye of the impeller. The spinning screw was supposed to develop a few feet of head to supplement the available NPSH. It didn't work, though. The pig tail broke off soon after startup.

On the other hand, at the Texaco Waste Oil Recovery Plant (see my book, *Process Engineering for a Small Planet*), we had a vacuum tower far too close to grade for reasonable NPSH requirements to protect the bottoms pump's mechanical seal from cavitation. So Texaco purchased a very low NPSH pump. Required NPSH was only 4 to 6 feet. It was the only pump I have ever seen where the eye of the impeller was much larger in diameter than the suction line. Only one problem: This pump was super expensive. Which didn't really matter, compared to the $4,000,000,000 legal judgment that Texaco had lost to Penzoil over their Getty Oil acquisition in the 1980s.

In Aruba, we also had severe cavitation problems on some giant (100,000 BSD) hot crude oil residue pumps. The plant manager, Mr. Crusher, suggested a solution to this problem that involved digging a 6-foot-deep sump and relocating the pumps further below the tower. My idea was to remove the fallen trays lying across the tower's outlet nozzle vortex breaker. The refinery project manager ruled that Mr. Crusher's solution was a more cost-effective solution to this lack of NPSH. The opinions of the maintenance people and operators who had to work in this pit are best not recorded in this text. Anyway, it rarely rains in Aruba.

Plugged Suction Screens

I was once almost fired from my job as an operating superintendent in Texas City for discarding a pump suction screen on a multistage (i.e., five wheels) LPG product pump. The screen kept plugging. The resulting cavitation repeatedly caused blown pump mechanical seals and pump outages. I suppose I was wrong and Amoco was right about that screen. But there is some optimum mesh size for such a screen. Too fine a mesh is worse than too large a mesh. I believe that the maximum permitted particle size that a pump can safely pass is defined by the clearance between the impeller and the pump's case. Anyway, it's best to size the screen's mesh using the maximum particle size that the pump manufacturer has specified for their pump. Being overly conservative in this area compromises mechanical seal integrity and is not actually a conservative design.

Some pump suction screens are located inside the vessel. This may be both good and bad. In 1980, we retrofitted our trayed crude fractionator at the Good Hope refinery with 2-inch Cascade minirings. This in itself was an error. But that's another story. The thin-gauge rings were crushed on installation and slipped through the support grid shown in Figure 29-3. Even 30 years later, I still become angry at Roddy, the slick salesman who sold us those wretched rings. But that too is another story. The rings migrated into the pump suction and caused severe cavitation. When we tried to isolate the cavitating pump to repair its seal, we found the suction valve would not seat-off properly. The same malfunction affected both spare pumps. So we shut the unit down.

We quickly found the problem. The crushed rings had backed up above the Y strainer and above the elevation of the suction gate valve. Then, when our operators closed the valve, the gate smashed the rings against the seat.

"Lieberman," the owner of the refinery, Jack Stanley, screamed at me, "Keep your rings out of my pumps!"

Figure 29-3 Tower internal screen plugs causing pump cavitation.

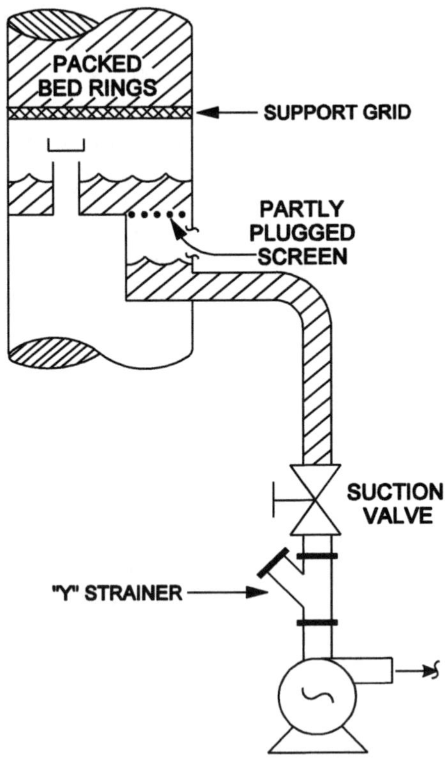

"Mr. Stanley, they're not 'my rings.' You wanted to replace our perfectly good trays with those thin rings. Not me."

"Okay, Norm. Kindly do something to keep my rings out of your pumps."

So I had the maintenance guys install a screen over the draw-off sump shown in Figure 29-3. After all, I reasoned, the top of the sump was 24 by 24 inches. The draw-off pipe was only 6 inches ID. I had specified that the screen be constructed from:

- 317 (L) s.s. wires (L stands for low carbon. Low carbon is better for welding.)
- Number 9 wire (about ⅛ inch).
- Wire spacing 1 inch.

Obviously the wire spacing needed to be less than the 2-inch rings. But I carelessly failed to inspect the installation. The three Fates (Faith, Hope, and Charity) noticed my laxity and decided to intervene. They placed a No. 8 mesh screen over the sump. The word "mesh" means the number of wires per inch. The screen worked great for a few months. Then the pump began to cavitate again. My pressure survey on the pump's suction line indicated with certainty that this time the entire suction line was empty. The new internal screen above the draw-off sump had plugged with iron sulfide ($Fe[HS]_2$) corrosion products. The publisher has deleted Jack Stanley's comments to me about my screen design from this text.

The details and ultimate fix of this problem are to be found in my book, *Process Design for Reliable Operations*, 3rd ed. My idea was good, but the execution was bad.

Preserving Mechanical Seal Integrity

As a pump loses NPSH, its suction pressure goes down. By reconfiguring the controls to operate on correct suction pressure control rather than on incorrect level control, you can avoid the whole problem. I have discussed this sort of control modification, intended to preserve the integrity of mechanical seals, in Chapter 16, "Level Control Problems." The intent of suction pressure control is to avoid mechanical seal failures due to a low pump suction pressure.

Effect of Increased Vessel Pressure

At equilibrium or steady state, increasing the vessel pressure does not alter the available NPSH. That is because the vapor pressure of the liquid pumped increases at the same rate as the pressure increases at the pump suction. The light ends in the vapor space dissolve in the liquid phase due to the greater pressure. Yet many operators believe that increased operating pressure does relieve cavitation. And this observation is partly true, for two reasons:

1. If a pump is cavitating, a rapid increase in tower pressure will simultaneously increase suction pressure. But it may take 10 minutes for the lighter liquid formed at the vapor–liquid interface to reach the pump suction. Then, cavitation will be restored.
2. Centrifugal pumps do pump small amounts of vapor. I see it happen every day on my pool pump which has a transparent intake filter. The official number is 1 volume percent. Of course, on a weight percent basis, that's only 0.005 or less percent, depending on vapor density. So if a pump is just at the edge of cavitating anyway, a reduction in the vessel's operating pressure can push it over the edge. In my previous story from Saudi Arabia, that was one of the symptoms that indicated to me that there was vapor in the draw-off nozzle and suction line.

Effect of Lighter Liquid

There is nothing to indicate that specific gravity affects the available NPSH. After all, head is expressed in feet. Yet many operators believe that lighter hydrocarbon liquids have a greater tendency to cause pumps to cavitate. The reason for this duplicates my previous explanation for the effect of reduced vessel pressure. That is, if a pump is in marginal cavitation, reducing the molecular weight of the liquid being pumped will increase the volume of vapor entering the eye of the impeller. Meaning, lighter liquids flash to bigger volumes of vapor. If you had a reasonable amount of NPSH to start with, you would not have noticed the effect of the lighter lower molecular weight liquid on your pump's tendency to cavitate.

Effect of Lower Temperature on Pump Cavitation

At equilibrium or steady state, increasing or decreasing the vessel temperature does not alter the available NPSH. That's because the vapor pressure of the liquid pumped does not change at a lower temperature. The light ends in the vapor phase dissolve into the liquid due to the colder temperature. Yet many operators believe that cooler liquid temperatures suppress cavitation. And to some extent, this too is true—that is, if the liquid can be cooled off after it has been in contact with the vapor. As, for example, spraying cold water onto a pump case or the pump suction line. That subcools the liquid. Which means the vapor pressure of the liquid is reduced before it enters the pump. Hence the available net positive suction head is increased.

I can use this idea to increase the available NPSH to a pump in two ways. Sometimes I place a small cooler between the pump and the upstream vessel. But this creates a bigger delta P on the pump suction. Also, it's expensive and takes a long time to implement.

A simpler method is to recycle some colder subcooled liquid back to the suction of the pump. I find this to be a rather effective way to suppress cavitation.

Sump Pumps

Pumps drawing water out of a ditch require NPSH just like any centrifugal pump. It's just that we subtract, rather than add, the elevation difference between the vessel and the pump when calculating the available NPSH. The suction head is provided by atmospheric pressure.

But when the pump is idle, the case of the sump pump is under vacuum. Water may then drain by gravity out of the pump case and suction line. Air is slowly drawn through the dry seal faces. Then, when the pump is restarted, it cavitates and vibrates. I had this common problem on an oil–water recovery pump in Port Arthur, Texas. To correct the malfunction, we connected an external source of flushing water to the seal. Then, when the pump was not in service, instead of air being drawn through the seal faces, water flowed into the pump case. Not only did this stop the pump from losing its prime when not running, but it also reduced the rate of mechanical seal failures.

Prior to my intervention, the engineer on the unit had attempted to correct the problem by changing the 25 horsepower motor driver to a 40 horsepower motor. No great wonder that some operators hold some engineers in disdain. This engineer had never set foot on this process area.

Other Causes of Seal Damage

Compared to cavitation, other causes of seal damage are less important. I'll summarize a few of these malfunctions:

- Ice crystals forming between the seal's faces. Occurs in wet propane or butane service. A small leak autorefrigerates and promotes ice crystal formation, which forces the seal faces apart and adds to the leak. Keep the seal a little warm with a wisp of steam to keep the ice crystals melted.
- Poor pump alignment with process piping. Watch your pipe fitters when they reinstall a pump. Do both the suction and discharge flanges line up? Or are they pulling the flanges into alignment? If so, you need to modify the suction and discharge piping or install flexible expansion joints. This last

option is exceptionally hazardous, as expansion joints are subject to catastrophic failure.

- Scratches on the seal faces indicate contamination of the seal flush. For self-flushed pumps, you will have to add a filter on the recirculation line.
- Excessive seal flush pressure. Forces the seal faces apart. Then the seal leaks at the normal seal oil pressure. Do you have a pressure indication on your externally seal flushed pumps?

Bearing Vibrations

- *Bearing damage.* Vibrations transmitted to the seal faces cause excessive wear on the soft, carbon seal stationary face. Keep your pump bearings properly lubricated.
- *Lack of exercise.* Idle pump bearings lose lubrication after a while. Then, when you start the pump, it runs rough until the bearings pick up lubrication. Seals may be damaged during this run-in period. Don't leave your pumps idle for months. Run them once a month, not just once a year.
- *Water may contaminate the bearing housing.* Water enters the bearing housing by being drawn into the housing along the shaft in the form of humid air. The moisture can then condense. The carbon seals (also called dust covers) on either side of the bearing housing should exclude air in-leakage. But these simple seals need replacement with time. It's best to check for moisture at least once a week. Install a small drain valve on the ½-inch drain on the bottom of the bearing housing. Then drain a few ounces of lube oil into a small glass and see if there's water or dirty lube oil.

The correct oil level in the bearing housing is up to the bottom of the oil slinger ring. That's essentially up to the bottom edge of the bearing. Oil levels any higher cause the bearing to overheat. In the normal wet sump system of bearing lubrication, the oil level is controlled in the oiler glass bulb. When you drain the oil down off the bearing housing once a week, your level in the oiler glass bulb should drop a bit and a bubble of air should be seen in the glass. If it is not, the small connection between the oiler glass and the bearing housing is plugged. There's an adjustment device inside the oiler bulb that the unit machinist needs to set to adjust to the correct lube oil level in the bearing housing.

Many pumps are lubricated by the oil mist system, rather than the wet sump system. For the oil mist system, the bearing housing is kept drained, and none of the problems I've just discussed ought to apply. I've never worked with an oil mist system, and its use is becoming less common in refineries.

One thing you can be sure of: If you permit bearing vibration to persist, you will eventually blow out your pump's mechanical seal. And this is potentially dangerous.

Pumping Aerated Liquids

I have recently encountered a very nasty pump malfunction that promoted severe vibrations to the suction and discharge piping and damage to the pump's mechanical seal. The high viscosity of a stripper's bottoms product (2,000 centipoise) retarded the separation rate of vapor and liquid. In this case, the pump was a positive displacement screw-type pump, rather than an ordinary centrifugal pump. Once viscosity rises above 100 centipoises, a positive displacement pump is needed.

As the bubbles of vapor flowed into the pump, they caused the pump to run rough. The compressed vapors collapsed at the pump discharge and released energy. This promoted severe vibration in the discharge piping and, to a lesser extent, vibrations in the suction piping. Cracks developed in the welds of the suction piping.

So far, there has not been a catastrophic failure of the piping. But the facility has been in operation for less than a year. My recommendation to my client was to run the tower bottoms temperature hotter. Increasing the pumped liquid temperature by 25°C will reduce its viscosity by half (to 1,000 c.p.). Experience has shown that reduction in viscosity appreciably increases the escape rate of vapor bubbles from the liquid phase. Hopefully, they will follow my advice of adding a side reboiler to the stripper. This ought to increase the pump suction temperature from 250°C to 280°C.

In the preceding case, the suction pressure appeared adequate to provide the pump's required NPSH. However, aerated liquid was still drawn into the pump suction. While the pumping capacity was marginally reduced, my overwhelming concern was vibration, seal damage, and potentially catastrophic piping failures.

Incidentally, positive displacement pumps, such as screw, gear, piston, or reciprocating pumps, have a required NPSH just like an ordinary centrifugal pump. If too much of the flowing pressure is converted into acceleration, then the liquid will partly vaporize inside the pump's case. The pump will then cavitate; mechanical damage is certain to follow. The real difference between the positive displacement pumps (used in high-viscosity services) and the centrifugal pumps is that:

- For a centrifugal pump, the discharge pressure is a function of the pump itself.
- For a positive displacement pump, the discharge pressure is a function of the downstream pressure resistance of the process piping, heat exchangers, and vessel operating pressure.

Benefits of External Seal Flush

Most of our smaller pumps in clean service are self-flushed pumps. This means that the seal's barrier fluid comes from the pump discharge. However, when the pump cavitates, the seal flush flow becomes erratic. Also, if the pumped fluid contains occasional particulates, the seal flush filter may plug or the seal flush flow can become erosive.

An engineer working for John Crane suggested at one of my seminars that the use of an external source of flushing medium can mitigate seal failure due to these problems. I cannot refute or support this theory from my experience. However, it likely explains why some of my clients use external seal flush on light, clean hydrocarbons, such as diesel oil and jet fuel, which are subject to occasional particulate contamination.

Pressure Control and Pump Cavitation

If a centrifugal pump is limited by its available NPSH, then the pressure control of the tower or vessel is critical. A pressure loss of a few kPa at the top of the vessel will instantly lead to a corresponding change in pressure at the suction of the pump. But the vapor pressure of the liquid, at the suction of the pump, will drop far more slowly, due to the residence time of liquid at the bottom of the vessel. Hence, the available NPSH is instantly diminished as the vessel pressure falls.

I'm sitting at the airport in Edmonton as I write these words. I have been working for Nexen in Fort McMurray, who is tearing up a fractionator bottoms pump's mechanical seal due to cavitation. The pump normally has an adequate NPSH to run properly. But when it begins to rain or the sun goes behind a dark cloud:

- The fractionator overhead air cooler duty increases.
- The fractionator tower top pressure falls.
- The pressure at the suction of the bottoms pump also falls, just as quickly.
- The pump starts to make a cavitating, rattling sound.
- The discharge piping jumps off its supports by half an inch or more.

So I have designed for Nexen a proper split-range pressure control system, to maintain a constant pressure in the overhead reflux drum. (See my book, *Troubleshooting Process Plant Control*.) As pressure rises, the noncondensible vapors are automatically vented to the flare. As tower pressure drops, nitrogen is introduced into the top of the reflux drum, to restore the tower operating pressure and the bottoms pump suction pressure.

But things progress quite slowly at the giant Nexen Tar Sands Synthetic Crude Oil Recovery Plant in Alberta. So I'm rather glad to be here, at the airport in Edmonton, rather than standing next to that rattling, cavitating, 500°F hot asphalt pump in Fort McMurray waiting for a piping flange to rupture and to possibly be engulfed in burning asphalt.

CHAPTER 30

Centrifugal Pump and Driver Capacity Limits: Lubrication Failures

He who sows the wind, reaps the whirlwind.

—Our current hydrocarbon production
policy for our little planet

There are three possible constraints on pump capacity:

- NPSH (see Chapter 29); this is cavitation
- Horsepower or driver limits
- Pump size or impeller constraints

An NPSH (net positive suction head) limitation will cause a pump to cavitate as flow is increased. Cavitation is potentially damaging to the pump's mechanical seal and other parts of the pump. Other limitations simply constrain capacity but cause no damage. For example, if a pump is limited by driver horsepower, then:

- For a constant-speed AC motor, the pump's motor will trip off on high amps.
- For a variable-speed turbine driver, the speed control governor will open 100% and the pump and turbine will slow below their set point speed.

Combating Excessive Motor Loads

If I have a motor that is approaching its FLA point **(full limit amp)**, any increase in load will trip the motor off and shut down the pump. Motor amps at constant pump speed will increase with:

- Specific gravity—If the SG increases by 10%, so will the motor amps. A centrifugal pump produces that same feet of head regardless of the SG of the liquid pumped. But work (or amps) is proportional to foot-pounds. If the SG increases by 10%, so does the pounds, and so does the amperage load, or the electrical work.
- Viscosity—You can ignore this effect for a viscosity of less than 40 centistokes (c.s.). Almost all applications in petrochemical plants or refineries are less than 40 centistokes viscosity. Above 70 c.s., there's a rapid increase in driver horsepower. Pumps in such service should be some type of positive displacement pump (like a gear pump), not a centrifugal pump.
- Flow—Typically, opening the discharge valve of a pump increases the flow faster than the head declines, as shown in the pump curve in Figure 30-1. But that's only true on the flat part of the curve (point A). On the steep part of the curve (point B), if flow is increasing at a slower rate, then the head is dropping. This means that opening the discharge valve at a pump will:
    - Increase amps near point A
    - Decrease amps near point B

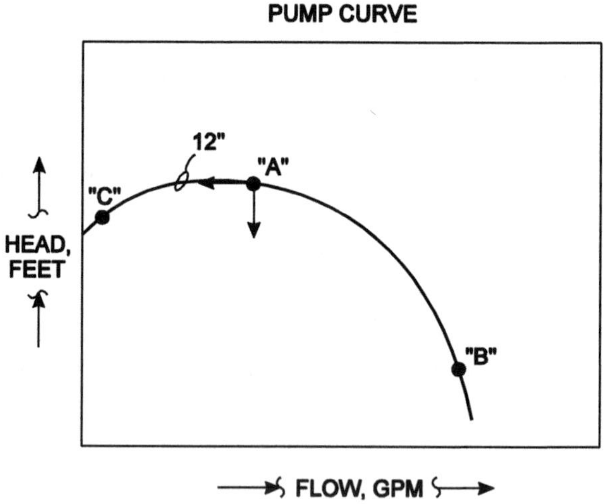

FIGURE 30-1 At point "B," flow is increasing slowly and head is declining rapidly. Thus, the motor amperage will drop between "A" and "B."

In practice, pumps are designed to operate on the flat part of their performance curve (point A) and never on the steep part of the curve (point B). I've tried to run pumps too far out on their curve. As long as the downstream control valve is 100% open, the flow is stable. But even a small movement of the control valve causes the flow to swing over a wide range. It seems from the control console that the pump is cavitating. But it is not.

Pump Operation on Flat Part of Performance Curve

Typically then, operating on the flat, or normal, portion of the pump performance curve will increase the motor amp load by about 10% for each 15% increase in flow. Thus, the motor will trip off on excessive amperage load if the rated amperage or the full limit amperage (FLA) point of the pump's motor is exceeded. What then could be done to overcome this limitation?

The first and simplest step is to clean the back plate or grille of the motor. Fouling of this grille causes the cooling flow of air to be reduced. The motor windings then become hotter, which increases the motor amperage. This will typically affect the motor amps by just a few percent.

Next, check the trip point of the motor. Such "tripping" of motors used to confuse me. At home, the overload trip on a motor is activated by excessive winding temperature. If the motor gets too hot, it opens a thermal switch, which shuts down the motor until the motor cools off. Usually, it takes several minutes of running at excessive amps before this happens.

But at work, things are different. An amperage trip point is set and the motor will trip off when the set amp load is reached. Often, the unit electrician sets this amp trip point too conservatively. It's an attempt to extend the service life of the motor, before its electric coils have to be rewound. On many motors, I have had this trip point reset 10% to 15% higher with no short-term effects. I suppose that I'm trading reduced motor life for increased pumping capacity.

Trimming the diameter of the impeller is another way to increase pump capacity. This may seem backwards, as one normally increases impeller size to get more flow. But just the contrary is true if a pump is limited by motor horsepower. Reducing an impeller diameter by 10% reduces the amp load on the motor by about 27%. This allows the pump to run out farther on its performance curve—with a reduced discharge pressure, of course. You need to study the pump's vendor performance curves for a range of impeller sizes to determine the extent that you will trim the pump impeller diameter.

My favorite method of overcoming motor limitations is to combine a reduction in impeller size with a restoration of the pump's internal impeller-to-case mechanical clearance. Typically, a pump loses efficiency in two ways:

- The impeller wear ring needs to be replaced. The wear ring surrounds the eye or inlet of the impeller. It is very easy to replace.
- The clearance between the vane tip of the impeller and the pump case has eroded out. This is very difficult to fix. The interior of the pump case has to be built back up with weld metal. Then the case has to be remachined on a lathe.

Both malfunctions cause the process fluid to internally bypass from the pump discharge to the suction. A pump suffering from internal bypassing will have a noticeable increase in its normal temperature rise. On one pump at a refinery in Aruba, after I had the pump overhauled properly, the temperature increase of the liquid was too small to measure. Before the overhaul, I measured a 10°F to 20°F increase in discharge versus suction temperature. For this pump, I eliminated the motor as a bottleneck by:

- Reducing the pump clearance to the manufacturer's specifications.
- Reducing the impeller size.

I've written about a similar incident at the Fina refinery in Texas in greater detail in my book, *Process Engineering for a Small Planet*. In both instances, my clients' solution to this process bottleneck was to buy a larger horsepower motor. My solution was to use applied engineering—which is often in short supply, as compared to the purchase of new motors.

Turbine-Driven Pumps

Steam turbine-driven pumps will run slower, but will not trip off, if the driver horsepower capacity is exceeded. The comments I've made about checking impeller-to-pump case internal clearances and reducing the impeller size applies for these variable-speed drivers. If an alternating current motor actually slowed due to overload and it failed to trip, the motor would burn up and be destroyed. For a turbine driver, the **governor speed control valve** will open to 100% as the machine slows, but no mechanical damage will occur.

To see if any malfunction is occurring with the turbine itself, first check the position of the **horsepower valves**. These are also called:

- Nozzle port valves
- Hand valves
- Power valves

- Star valves
- Speed valves

I'll discuss the use of the port valves in Chapter 31, "Steam Turbine Drivers." For now, check that these valves are both open. (Most machines have two port valves. About 20% have three.) Sometimes turbines are purchased without external port valves. Then you'll have to shut down and open the steam chest to see if the ports are plugged (i.e., closed with a steel-threaded plug) internally. The port valves typically are located a little below the governor and are operated with 3-inch hand wheels. Make sure you open them all the way for maximum horsepower availability.

Next, check the motive steam pressure in two places. The supply pressure upstream of the governor—let's say this is 400 psig. Now check the steam pressure of the steam chest. If you don't have a pressure gauge or connection to the steam chest, use the low-point condensate drain from the steam chest to attach your pressure gauge. If the turbine is not in operation, there is always at least a plug on the top of the turbine case for a pressure gauge. If the pressure on the chest is 380 psig, then there's a 20 psi delta P across the governor speed control valve. That's okay. But if this pressure is 300 to 350 psig, then the governor is not actually opening 100%. The "linkages" on the governor are not set correctly. Your machinist should be able to correct this malfunction.

I've had nasty arguments with plant maintenance people on this subject. To win the argument, at the Irving Oil refinery in St. John, New Brunswick, I pushed against the governor with my foot.

"See!" I screamed above the whining turbine, "The turbine's running faster, the steam chest pressure is rising, and the crude charge pump's discharge pressure's also increasing."

"Lieberman," replied the unit machinist, "Are you crazy, or stupid, or both? Get your G-d foot off of the governor."

Steam Quality Issues

Moisture and hardness deposits in the motive steam supply to a turbine will diminish the work available from the turbine driver. The hardness deposits consist of entrained salts that accumulate on the turbine blades. The moisture reduces the enthalpy content of the steam. I have discussed these problems in Chapter 15, "Steam Quality Problems," in greater detail. I'll add here, though, that many problems with steam-driven pumps originate on the waste heat boilers on the very process units that are complaining about poor steam quality. Improper level control in the waste heat steam generators is the usual malfunction. This too I've discussed in detail in Chapter 16, "Level Control Problems" and in my book, *Troubleshooting Process Plant Control.*

Effect of Specific Gravity on Pump Capacity

A centrifugal pump develops the same feet of head regardless of the density of the liquid. However, the pump discharge pressure is a linear function of the SG. If the density goes down by 10%, so does the discharge pressure. This is of no consequence if I'm pumping a fluid up a hill. That just requires more feet of head. But if I'm trying to pump liquid into a 2,000 psig reactor, a 10% loss in density means I'll need an extra 10% in head. And if I don't have the extra feet, then the pump gets pushed back on its curve (Figure 30-1). Then I'll lose a lot more than 10% of my flow. What can I do?

If possible, reduce the temperature of the hydrocarbon. Every 60°C or 100°F decrease in temperature will increase the SG by 5%. But if this causes the fluid viscosity to increase above 40 or 60 c.s., you may gain little or nothing from this temperature reduction. If viscosity rises to the 80 to 100 c.s. range, pump capacity will likely decline as the flowing temperature is decreased.

The preceding discussion does not apply to aqueous solutions. For water, density varies above 100°F by about 3% for each 100°F.

Marginal Cavitation Limits

Once, I was working at the CVR refinery in Coffeyville, Kansas (site of the famous 19th century Dalton brothers' bank robbery). A crude charge booster pump had a discharge pressure of 80 psig with a suction pressure of 5 inches Hg vacuum (that is, about −2 psig). I partially bypassed an upstream preheat exchanger to raise the suction pressure to a positive 3 psig. The pump discharge pressure jumped to 150 psig. I call this "marginal cavitation." The pump discharge pressure and flow were steady. During normal cavitation, the flow and discharge pressure would both be erratically low. Also, the pump would make an unpleasant sound. But in marginal cavitation, a pump operates smoothly, but on an inferior performance curve. Raising the suction pressure by 5 psi caused the discharge pressure to increase by 70 psig, a sure indication of pump marginal cavitation.

I have only seen this happen on larger pumps operating at a low suction pressure. Operating in this marginal cavitation mode does not seem to damage the pump's mechanical seal as there is no increase in vibration. Normal cavitation, as indicated by erratic pump discharge pressure, will certainly damage the pump's seal. See Chapter 29 on "Centrifugal Pump NPSH Limitations."

If a pump is not operating on its performance curve, raise the liquid level in the pump's suction drum by a few feet. If the pump jumps back onto its operating curve, your problem was a lack of NPSH. This proves the pump was suffering from marginal cavitation.

Note that I've written level, not pressure. Raising the pressure on the suction changes the vapor–liquid equilibrium in the drum. Thus, the liquid becomes more volatile and no net increase in available NPSH results. In my previous example at the CVR refinery, I raised the suction pressure of the pump by reducing the upstream pressure drop. This did not alter the composition of the liquid flowing to the pump, and thus the available NPSH was increased.

Checking Performance on Pump Curve

Centrifugal pumps operate on the manufacturer's performance curves. I've shown such a curve for a 12-inch impeller in Figure 30-1. The size of the impeller refers to the outside diameter of the impeller. A 4-inch impeller is used for a small pump. A 20-inch impeller is used for a giant pump. If you think a pump is malfunctioning, the first step is to see if the pump is on its curve.

In the U.S.A., most process unit flows are reported at standard conditions, meaning the volumetric flow has been corrected for temperature at 60°F. If you record a standard flow of 100 GPM at 360°F, you will need to correct the actual flow for expansion due to temperature. Perhaps 5% expansion for each 100°F for hydrocarbon, 3% for each 100°F for aqueous system. That defines the horizontal axis of the pump curve.

The vertical axis is in differential feet. Measure the head by:

$$\text{Feet} = (P_2 - P_1) \times (2.3 \div \text{SG})$$

where P_2 = Observed discharge pressure, psig

P_1 = Observed suction pressure, psig

SG = Specific gravity corrected for flowing temperature

But suppose there's no trustworthy flow meter? Perhaps you can measure the inventory change in a downstream or upstream storage tank and calculate the flow. Adding a tracer chemical to the flow at a known rate and checking the downstream concentration of the tracer chemical is another method. More often, I can calculate flows based on heat and material balances.

One technique that I've now found is worthless in checking a pump performance, is the "deadhead" method. That is, block-in the pump's discharge and note the discharge pressure at zero flow. It's true that if the pump deadhead pressure is below the curve, the pump's internal clearances are worn. But if the pump is developing its design head at no flow, this does not indicate that it will develop its design flow further out on the curve as the discharge valve is opened.

If a pump is running below its performance curve, the malfunction can be, starting with the most common:

- The impeller is smaller than the data sheet indicates.
- The impeller wear ring needs replacement.

- The internal surface of the pump case needs to be rebuilt (a major overhaul).
- The impeller itself is worn.
- The pump impeller is spinning backwards.
- The vanes between the inner and outboard walls on the impeller are not recessed correctly. You will need to locate a new impeller to see how the vanes should look.
- The pump is running slow. Ordinary AC motor-driven pumps cannot run slow. So this comment applies to turbine drives, motors with variable frequencies, or direct current (DC) drives. Most pumps (90%+) that you'll work with are constant-speed machines (AC motor drives, at 60 Hertz, 3 phase, 3600 rpm).
- A discharge-to-suction bypass valve is leaking through.
- The pump is in marginal cavitation, as described above.

Never accept the following excuse from the maintenance department, "Yeah, the pump's not running on its curve. But it's 40 years old. We did everything we can. You'd better get a new pump!" That's just low-grade B.S. Probably the internal clearances are excessive.

Pump Spinning Backwards

Many pumps will develop 70% to 80% of their design flow and pressure when the impeller is spinning backwards. This can happen in two ways. The polarity of the electric leads can be reversed. Then the impeller will spin backwards because the motor is rotating backwards. There is an arrow stamped on the pump case showing the correct direction of rotation. Touch your pencil to the spinning shaft and you can feel the direction of rotation. Or, see if the air is blowing out of the back of the motor case. If so, the motor—and the pump—are spinning in the wrong direction.

For larger pumps that have both inboard and outboard suctions, the impeller itself can be installed backwards. (Inboard means the side of the pump facing the driver.) The impeller may be spinning in the correct direction of rotation, but the impeller vanes are spinning backwards.

I well remember in 1975 the cooling water pump on my H_2SO_4 regeneration plant in Texas City. I had the pump sent in for maintenance to increase cooling water flow to my acid coolers. The Amoco Texas City maintenance shop specialized in the nonrepair of centrifugal pumps. When my pump was put back in service, its discharge pressure was running lower than before it was nonrepaired.

But old Zip, my ancient shift foreman, rescued the situation.

"Look, Mr. Norm, the air's a-blowing out of that motor. They done wired up my motor backwards again. Them scab Yankee contractor electricians from New York City likely done it on purpose!"

Undersized Control Valves

Increased pump discharge pressure will reduce a centrifugal pump's flow. One of the ways many operators employ to combat this limitation is to open the bypass around the downstream process control valve. I've done it a thousand times myself. But experience has taught me that this is in general a bad operating practice. For example, how is the console operator going to remember which control valves are partly bypassed? Just imagine, during an operational upset or plant emergency, the consequences of having a control valve bypassed! The idea of bypassing control valves contradicts the concept of central control.

It's quite true that many control valves are vastly undersized. Either the valve itself is too small, or the valve's internal trim is undersized. To see if the control valve is bottlenecking a pump's capacity, proceed as follows: With the control valve 100% open, open the bypass valve by two or three turns. If the flow increases by 10% to 20% or more, the control valves need to be resized. If the flow only increases by a few percent, consider increasing the size of the pump's impeller.

Incidentally, increased discharge pressure should not greatly affect the volumetric capacity of a positive displacement pump, such as a reciprocating pump, if the pump is in good mechanical condition.

Running Pumps in Parallel

Increasing a flow by turning on the spare pump may be fine, or it may be dangerous. To run centrifugal pumps safely in parallel, both pumps have to be operating on a similar performance curve. If the spare pump is a turbine drive, or is driven by a variable-frequency AC motor, then the spare pump needs to be adjusted to operate at the same speed as the primary pump. Also, both pumps must be in a similar mechanical condition. Otherwise, the strong pump will rob the weak pump of most of the flow.

For example, if I put a second pump online, and the total flow goes up by 20%, I will not worry. But if I place a second pump online, and the flow goes up by only 5% or less, I may have committed an unsafe act. It certainly is not necessarily true that the second pump is running at a low flow rate. The second pump may have stolen the flow from the first pump. There are several methods available to determine which pump may be suffering from too low a flow:

- The motor amps will be lower.
- The pump will make a low, moaning sound. This is hard to hear above the general roar of most process plants.
- The pump discharge temperature may be a few degrees hotter.

For large, high-head process pumps, running the pump at less than 60% of its "best efficiency point" (which is usually similar to its design point) is considered by pump experts to be potentially damaging to the pump's mechanical seal and other pump internal components.

Sometimes, placing a second pump online will not generate more flow because the pumps are running on the flat portion of their curves. If so, they are sharing the flow, and their motor amps will be similar. This is not dangerous.

In reality, I do not bother consulting the pump curve. I don't wander around the plant with a bunch of curves in my overalls. If I hear a pump making a soft, surging, moaning sound, I start to worry. If I see an operator running two identical pumps in parallel, and one pump is pulling half the amps of its sister pump, I start to worry. Then, I'll call up the Inspection Department to have the **pump's displacement** checked. That's a measure of the amplitude and frequency of the pump's vibration. They can then advise me if I'm really running the pump at too low a flow.

Negative Pump Curve

Oscar Wyatt, the founder of the Coastal Corporation, is currently in prison. Supposedly he was purchasing crude oil illegally from Iraq. More likely, he omitted a pay-off to some politically connected official in the federal bureaucracy. Anyway, Mr. Wyatt, who I consider a great American, was not that great at sizing pumps. When he purchased the old ESSO Lago refinery in Aruba in 1990, he decided to replace the existing, rather small vacuum tower bottoms pumps with some truly gigantic pumps. The problem with these pumps was cavitation. The suction piping and draw-off nozzle were restricting the flow to the pumps. The resulting cavitation was damaging the pump's mechanical seals.

I remember standing on the pump's concrete pad and manually throttling on the discharge gate valve, trying to stop the pump's cavitation, which I was able to do. But as I did this, the pump's discharge pressure started to slip down.

To my dismay, the discharge pressure fell from 320 to 295 psig as I reduced flow by slowly restricting the discharge valve. And the pump base started to vibrate in a most alarming fashion. It did not require a maintenance expert to realize that the pump could not safely be run with such extreme vibrations. On the pump curve (Figure 30-1), I've shown operating point C. This is not the manufacturer's performance curve, which did not show the "negative" portion of the pump curve—negative in the sense that the pump discharge pressure declined as I throttled on the pump discharge gate valve to reduce flow! Life is full of surprises.

Bearing Lubrication

The purpose of the ball bearing race is to support the weight of the pump's shaft. For big pumps, with very heavy impellers, there will be both an inboard bearing (i.e., closer to the driver) and an outboard bearing. I've shown a cross-section of a ball bearing in Figure 30-2.

Note that the lube oil level is kept as low as possible, while still in contact with the very bottom of the bearing race. Too high an oil level causes the bearing to churn around in the oil. This causes the bearings to operate at an undesirable elevated temperature, which shortens the bearings' life. The oil level in the container that holds the lube oil (called the bearing housing) is maintained by a device called the oiler glass. The unit machinist (not the operators) adjusts the level in the bearing housing by making an internal adjustment to a screw-type mechanism inside the oiler glass.

If the bearings are not lubricated correctly, the bearings will wear out and cause the pump to vibrate. This vibration will then damage the pump's mechanical seal. Process fluid will begin to squirt out from the pump, where the shaft passes into the pump case. The hot fluid will then autoignite and a huge fire will result. You can imagine how I've become so smart on this particular subject.

Then how do we check for proper pump lubrication? First, please note that there is a totally different sort of bearing lubrication system than what I have just described. That's called the oil mist system. An oil mist is sprayed into the bearing housing, but the lube oil level is drained out of the bearing housing. I've never used the oil mist system, and therefore I will not discuss it.

The more common system, which I will discuss, is called the wet sump system. For this wet sump system we verify proper lubrication as follows:

Step 1—There is a small bull's-eye sight glass in the lower portion of the bearing housing. That defines the correct level for the lube oil.

Step 2—There is a ½-inch case drain plug on the very bottom of the bearing housing. Replace that plug with a small drain valve.

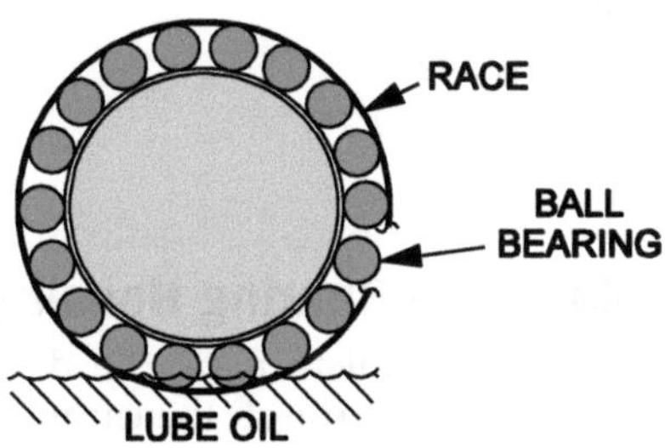

Figure 30-2 End view of a bearing showing the correct lube oil level. A higher level will cause the bearing to over-heat.

Step 3—Once a day, or once a week, drain a half cup of lube oil out of the bearing housing.

Step 4—You ought to see the oil level in the oiler glass drop a bit, and a bubble of air should rise through the oiler glass. If you do not see this, the connection between the oiler glass and the bearing housing is plugged, which happens quite frequently.

Step 5—Look at the sample of lube oil. Is it clean or dirty? If dirty, shut the pump down and change the lube oil. Just like you would on your car.

Step 6—In New Orleans, our big problem in bearing lubrication is water. Moist air may be inducted into the bearing housing, due to the rotating action of the pump's shaft. Not a problem in Fort McMurray, in northern Alberta. There the air is bone dry. But in south Texas, along the Gulf, moist air causes water to accumulate inside the bearing housing if the dust cover or carbon seals are worn out. These **carbon seals** are a simple device intended to prevent external air from being drawn into the bearing housing. If you are repeatedly draining water from the bearing housing every day, then your carbon seals have to be renewed. Unfortunately, water is not an acceptable lubricant for ball bearings.

During the 1980 strike at Texas City, I discovered a way to manufacture lube oils from air. One day I noticed on a turbine-driven pump that the lube oil level in my oiler glass was going up, not down! Actually, I later discovered that I was not really making any lube oil. It's just that the steam from the turbine was blowing up against the inboard bearing carbon seal and forcing steam into the bearing housing. So if you see an increase in the lube oil level in the oiler glass, that's a good indication that water is accumulating in the bearing housing. Such water intrusion in the bearing housing is, in my experience, the most common reason for shortened bearing life in process plant pumps.

Don't think, as you all read this stuff, that I check bearing lubrication for centrifugal pumps every day. I haven't done it since the 1974 and the 1980 strikes at the Amoco Oil refinery in Texas City. But, since I've done it myself, with my own hands, I really am an expert:

"When I hear, I forget
When I see, I remember
When I do, I understand."

Pressure Accumulation in Bearing Housing

I previously noted that moisture can be drawn into the bearing housing with air inducted through the carbon seals on each side of the bearing housing. This air is not usually a problem. It is simply released through a vent on the top of the bearing housing. However this vent, which is also designed to retard the loss of lube oil in the vented air, is subject to fouling and plugging. This will cause an accumulation of pressure in the bearing housing. The resulting excess air pressure may also push up the oil level in the oiler glass, and it may also push down the oil level in the bearing housing itself. Note that a heightened level in the glass may also be due to water, as described earlier. The lubrication oil level in the bearing housing may then become too low due to this excess pressure. Or the increased lube oil level in the oiler glass may indicate to the outside operator that the pump has a satisfactory lube oil level in the bearing housing, and thus the operator fails to add required lubrication oil. Periodic cleaning of the vent cap should prevent this sort of malfunction.

Pump Rotation Schedule

Very slowly, lubrication oil will drain from the surface of an idle pump's bearing. Then, if the centrifugal pump has been idled for an extended period of time, the bearings dry out. Therefore, when you run the pump, at first the bearing has no lubrication. The pump will vibrate and run rough for a few moments. Turning the pump by hand before startup will not help. A pump needs to be run at speed for the bearings to pick up lubrication. You cannot expect a piece of rotating equipment to sit idle for a year and avoid this problem of rough running on startup.

Run your equipment! Switch pumps every few weeks, not every few months. Not only for the sake of the mechanical integrity of the bearings, but also for the integrity of the pump's mechanical seal. Remember that when you turn a pump by hand, you're turning a dry bearing, which only makes the situation worse. The same logic applies to centrifugal compressors. Rotating equipment, unlike fine wine, does not improve with age as it sits idly on standby.

Motor Lubrication

Motors have bearings, too. At home, most of our motors are sealed and the bearing cannot be lubricated. At work, many of our motors can and should be lubricated. Typically, there is a small grease fitting on a motor. The motor should be lubricated by just a few squirts with a grease gun. Too much grease can be bad. Also, motors should definitely not be lubricated once a month. Something like once or twice a year is typical. This is a job for the unit electrician and not for the untrained unit operators or people like your author. It might even be best to follow the manufacturer's instructions.

Origin of Pump Vibrations

This is a purely empirical observation. It seems to me, based on the relative severity of vibrations, that pump vibration is sometimes caused by vibrations originating in the discharge line. The collapse of bubbles in the discharge line escaping from the pump impeller (i.e., cavitation) may be one such cause. Alternately, improper discharge line supports, or the geometry of the pump discharge line, may cause it to vibrate. The resulting vibrations may then be transmitted back to the pump itself and result in mechanical seal damage.

I have observed this sort of vibration pattern on process pumps both in asphalt service and in cooling water pumps. I've suggested to my clients that dampening out the vibrations on the discharge line of a centrifugal pump would aid in reducing their pump mechanical seal failure rates.

CHAPTER 31
Steam Turbine Drivers

God sees the truth but waits.

—Count Leo Tolstoy

The idea that steam can do work formed the basis for our modern industrial society. Even today, more energy is consumed in generating steam than is consumed by all the cars in the world. Mainly I'm referring to the steam used to spin turbines that generate electric power.

How, though, does a steam turbine work, and what are its potential malfunctions? An entire branch of science is devoted to answering these dual questions. I'm referring to thermodynamics. The reason you, dear reader, failed to understand Thermo was because it was taught by a group of professors who didn't understand it, either. I'll correct this unfortunate situation right now.

- The steam that drives a turbine is called the motive steam.
- The steam that leaves the turbine is called the exhaust steam.

The exhaust steam has less energy than the motive steam. It's the difference in energy between the motive and exhaust steam that spins the turbine wheel. But in what form is this difference in energy transferred to the turbine wheel?

Energy appears in several different forms:

- As heat (enthalpy)
- As speed (kinetic energy)
- As pressure (the repulsive force of gaseous molecules)
- As elevation (potential energy or gravity)

The form of energy that spins the turbine is speed. Consider the form of energy that causes a windmill to spin. It's the wind. That's why it's called a windmill. If it was the pressure of the air that spun the windmill, it would be called a pressure-mill.

As motive steam enters a turbine, it's not moving particularly fast—maybe 40 or 50 ft/sec. But when the motive steam strikes the turbine wheel, its velocity is perhaps 1,000 ft/sec. The kinetic energy of the steam really increased. But where did that extra kinetic energy come from?

1. Some came from the reduction in the pressure of the steam.
2. A lot came from the reduction in the temperature of the steam.
3. And a lot came from the latent heat of the steam!

Let's say I have an ordinary steam turbine. The motive steam is saturated 100 psig steam. The turbine is exhausting to the atmosphere. I'll assume that the turbine is 100% efficient. Actually, many turbines are really rather efficient. On this basis:

1. About 10% to 20% of the energy converted to the kinetic energy of the steam came from the reduction in steam pressure of 100 psi.
2. About 30% to 40% of the energy converted to the kinetic energy of the steam came from the reduction in the temperature of the steam.
3. All the rest, maybe 40% to 50% of the energy converted to the kinetic energy of the motive steam, came from the reduction in the latent heat of the steam. Meaning, some of the steam condensed.

The idea that gases move faster as they cool is easy to prove. Allow the air in your car tire to blow past your fingers when you depress the tire's valve. When I say that half of the energy needed to accelerate the motive steam comes from the reduction in latent heat, what I'm really saying is that the steam is partially condensing. About 10% of the motive steam is converted to water before the steam imparts any energy to the turbine wheel. Further, the wetter the steam, the better!

Is Wet Steam Good or Bad?

We all know moisture or low-quality motive steam is bad. I've discussed this in Chapter 15, "Steam Quality Problems." But that's the moisture in the motive steam supply. Moisture that is produced as a consequence of accelerating the motive steam (i.e., as a consequence of increasing the kinetic energy of the steam) is good. The faster the steam moves, the more work we can extract from each pound of steam.

I have assumed that my turbine is 100% efficient. But no machines are really 100% efficient. As the conversion of the heat content of the steam to kinetic energy becomes less efficient, the moisture content of the exhaust steam is reduced. And that is bad, because it's the kinetic energy of the steam that spins our turbines. And the greater the amount of power we supply to the turbine, the more horsepower or kilowatts we can extract from the turbine.

I can quantify these effects by reference to a Mollier diagram that you will find in the back of your steam tables. I've discussed all this in great detail in my book, *Process Engineering for a Small Planet*. But to repeat part of what I reviewed in this book:

- If I expand steam through an orifice in such a way that I maximize its kinetic energy, this is called an **isoentropic expansion**.
- If I expand steam through an orifice in such a way that I minimize its increase in kinetic energy, this is called an **isoenthalpic expansion**.

The word *isoenthalpic* means that the heat content and kinetic energy of the steam are kept constant. The word *isoentropic* means that I've done all I can to convert the heat of the steam to kinetic energy. And since it's the speed of the steam that spins the turbine wheel, we say:

- An isoentropic expansion preserves the ability of the expanding steam to do work by minimizing the enthalpy of the steam.
- An isoenthalpic expansion wastes the ability of the expanding steam to do work by preserving the heat content of the steam.

Interaction of Speeding Steam with a Machine

Note that an isoentropic expansion has nothing do with machines. But now that I've accelerated the steam, I'm ready to use it to do work. Mechanically, how shall I accelerate the steam? That's simple. I'll expand the steam flow through a nozzle. Now, rush into your garden. Turn on the water to your garden hose. Adjust the nozzle until the water squirts out of the hose with the greatest force. Next, turn your bike over on its seat. Squirt water on the front wheel of the bike.

Did you observe the discharge pressure from your hose? Was it not atmospheric pressure? Did you see that it's the velocity of water, not its pressure, that caused your bicycle wheel to spin?

I've summarized these ideas in Figure 31-1. As V_2 becomes progressively greater than V_1, we can extract more work from the steam.

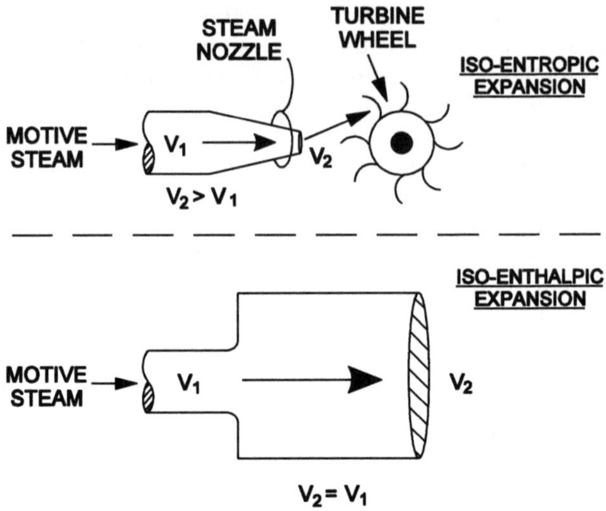

Figure 31-1 Everything you need to know about thermodynamics, but were afraid to ask. An Iso-entropic expansion is good. An Iso-enthalpic expansion is evil. An Iso-entropic expansion is reversible.

Turbine-Driven Pumps

The turbine shown in Figure 31-2 has a constant supply of motive steam and thus is running at constant horsepower. I'm starting out at point B on the head-versus-flow curve. As I close the level control valve on the discharge of the pump, what happens to the speed of the turbine? Does the turbine now spin:

- Faster?
- Slower?
- Same?

The correct answer is that the turbine will run faster. Faster because the work required to spin the pump has decreased. Work is best understood not by the concept of force times distance. Look at it my way; work is carrying gallons of water (pounds) up a hill (feet). Meaning work is foot-pounds (ft-lb).

From the pump curve shown in Figure 31-2, we can see as we move from point B to point A that the head (ft) goes up by 10%, but the flow (lb) drops by 50%. So I conclude:

- The work required to spin the pump is going down by about 40%.
- The work generated by the turbine is the same, because the turbine horsepower is constant—because I've kept the motive steam flow the same.

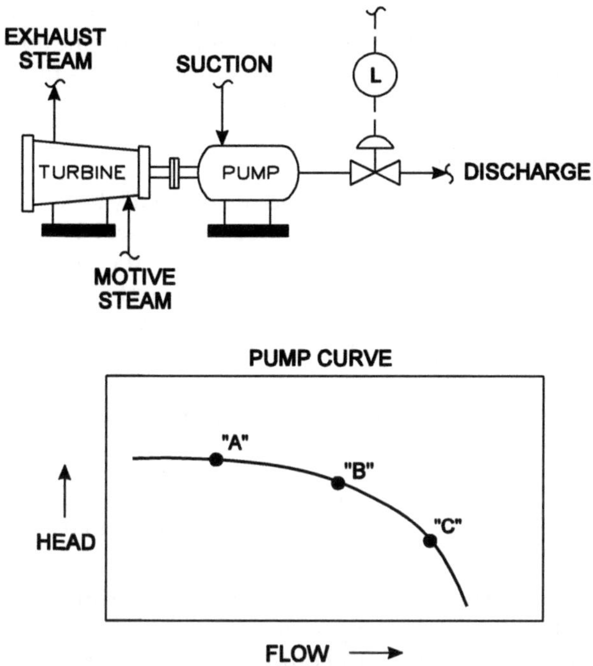

Figure 31-2 Throttling on the pump discharge causes the turbine to run faster.

- The turbine and the pump now run faster to consume the extra available horsepower. (Note that at the increased speed we're now going to jump up to a different curve—but I'll ignore that complication for simplicity's sake.)

Let's start over. I'm starting out at point C on my pump curve. As I close the control valve on the pump's discharge, what now happens to the speed of the turbine? The correct answer is that it slows. It slows because the work required to spin the pump has increased. From the pump curve you can see, as we move from point C to point B, the head goes up by 100%. But the flow drops by only 30%. So the work required to spin the pump goes up. But as the turbine is generating a constant horsepower output, it slows down. And if I didn't want it to slow, I would have to increase the motive steam flow rate.

If I had a motor-driven pump, the horsepower output of the motor driver would vary with the pump's requirements, but the motor speed would remain constant.

A turbine driver is different. The turbine itself controls the horsepower that can be consumed by the pump. At a constant steam flow to the turbine, the pump speed will self-adjust to match the turbine's horsepower output.

If a pump is limited by the motor horsepower, the motor trips off and the pump shuts down. If a pump is limited by turbine horsepower, the pump slows down and develops less head and less flow.

Governor Speed Control

The flow of steam to a turbine should not normally be kept constant. It's supposed to be controlled by the **speed control governor**. I say it's "supposed" to be because only too often the governor is not operational. The governor is like the cruise control and gas pedal on your car. If the turbine is running below its set speed, the governor will open to supply more horsepower and increase the pump speed.

But how does the operator know what speed to select? There are two possible answers:

- The Truth: The turbine is set to run at its maximum rated speed the day the unit starts up and runs at that speed for the life of the unit.
- Wishful Thinking: The turbine is set to run at that speed that keeps the pump's downstream control valve in a mostly open, but still controllable position. My "wish" is to minimize the parasitic energy loss across the process control valve.

If we ran with the control valve wide open, or bypassed the control valve, this parasitic loss would be close to zero. This is good. But how then would we manipulate the variable that we need to control, such as an upstream level, or a flow, or a pressure, or a temperature? My answer is in the original way such variables were controlled. That is, by varying the pump speed directly! For example, a level transmitter output would directly control a steam turbine's governor speed control valve. Why then do we have control valves on the discharge of turbine-driven pumps? The answer is that there is no answer. I suppose one answer is that if a turbine-driven pump is working in parallel with a motor-driven pump, both pumps will be lined up through a common discharge control valve. Still, that's not a very good reason. We could still run the turbine on speed control and the downstream control valve would drift 100% open if we tuned the control loop with that as our intent.

Strange to say, but turbine-driven centrifugal compressors historically have run on speed control. Even stranger: Do you drive your car to achieve a constant engine speed, and then control the car's forward progress with the brake? The entire concept of throttling a turbine-driven pump's flow with a discharge control valve should be classified as an engineering aberration.

Steam Chest Port Valves

Figure 31-3 represents the components of a steam turbine. It does not in any way represent the real mechanical construction of a turbine. It does explain the process components of a steam turbine. The motive steam enters the steam chest. The steam then flows through three ports into the turbine case. The pressure inside the turbine case is the exhaust steam pressure.

The ports are really steam nozzles, where the enthalpy and the pressure of the steam are converted to velocity. The ports may be opened or closed by means of the port valves. When an operator closes a port valve:

- Initially the turbine wheels slow.
- The speed controller opens the governor valve to emit more steam to the steam chest.
- The delta P through the governor valve drops and the pressure in the steam chest increases.
- More steam flows through each of the remaining two open ports.

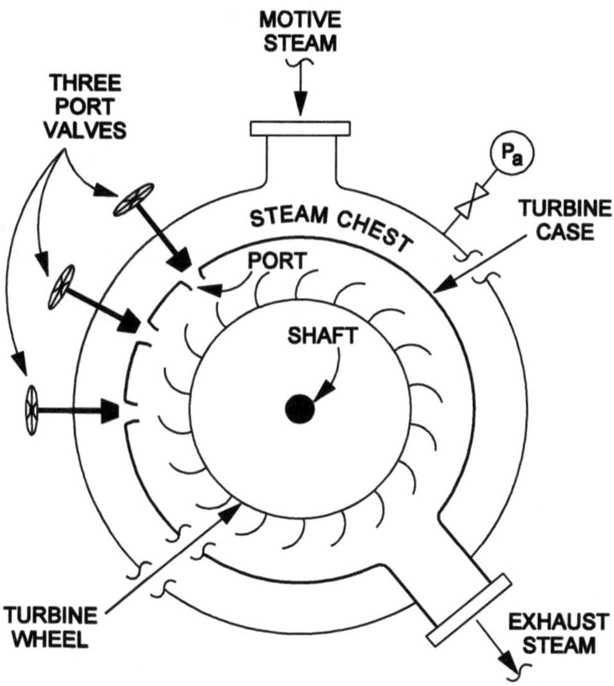

FIGURE 31-3 Closing a port valve will cause the pressure across the steam inlet control valve to drop and save steam.

- The velocity of the steam escaping from the ports or nozzles increases.
- The steam now strikes the turbine wheel with more kinetic energy.
- The turbine's speed is restored to its set speed, but using less motive steam than before a steam port valve was closed.

To summarize, it's the momentum (mass times velocity) that spins the turbine. By closing a port valve, and forcing the governor speed controller to open, I have traded more velocity for less mass. Thus, I saved steam. I can continue to close port valves until the governor valve is in a mostly open but still controllable position.

Port Valve Malfunctions

Thus you see that there are two ways to improve turbine efficiency. One is to close the port valves and force the governor to open. The second is to slow the turbine and force the pump discharge valve to open.

Often, the seat on the port valve has been damaged. This happens when an operator leaves the port valve partly open for an extended period of time. Then even after the port valve is externally shut, it still passes steam. A port valve must be either completely open or completely shut to avoid damage to the valve due to the erosive velocity of the steam.

Very few operators try to optimize a turbine's speed. Even fewer know the function of a port valve. Yet in Texas City, during the 1980 strike, I was able to reduce both the 400 psig motive steam consumption and consequently the 100 psig exhaust steam rate by about 30,000 pounds by reducing turbine speed and shutting port valves on a dozen or so steam turbines.

Incidentally, many operators refer to port valves as hand or power valves. Many of the port valves do not have handles. They are intended to be turned with a small wrench supplied with the turbine and attached to the turbine by a thin chain.

Low Steam Chest Pressure

A common malfunction with steam turbines is a low steam chest pressure. The chest pressure is shown as gauge P_a on Figure 31-3. The steam chest pressure should be within 5% of the steam supply pressure. Otherwise, both turbine power and efficiency will suffer. The vast majority of turbines do not have a gauge connected to the steam chest. The reason for this omission is that there is no reason. If you do not have a valve on the steam chest, then you can temporarily place a pressure gauge on the condensate drain line from the steam chest. Raising the steam chest pressure is accomplished by adjusting the linkages on the governor. That's a job for the unit machinist.

If everything is working correctly, slowing the turbine down by 10% saves about 28% of the motive steam. Closing a port valve will save 5% to 10% of the steam for each port valve shut. Doesn't this all sound like something you'll like to try? Doesn't it seem like it would be great fun?

Overspeed Trip Malfunctions

There are two sorts of trip malfunctions that I've seen:

- Natural—Caused by hardness deposits from the supply steam.
- Man-made—Caused by a piece of wire wrapped around the trip mechanism.

I learned about the first type in Texas City in 1974. I was in charge of a 6,000-horsepower isobutane refrigeration centrifugal compressor. It consumed 400 psig steam and exhausted to 30 psig. The overspeed trip on this machine was a spring-activated valve that cut off the 400 psig steam flow in the event of the following emergency circumstances:

- Excess speed
- Excess vibrations
- Loss of lubrication oil pressure to the bearings

The overspeed trip was invented by James Watt in the 18th century to protect steam-driven pumps from self-destruction. In Texas City I discovered one morning that the trip had become unlatched—meaning, the trip valve should have shut down the steam flow to my turbine, which it did not. The trip valve was stuck due to carbonate deposits from the poor-quality 400 psig steam (see Chapter 15 regarding steam quality causes and cures). This incident, which happened 36 years ago, is still vivid in my memory. I dutifully informed my boss, Frank Citek, about the stuck trip valve and the poor plant steam quality.

"Lieberman, if you wreck that machine due to your incompetence, I'm holding you personally responsible. Don't ever come into this office again with your problems."

And I didn't.

Man-made trip malfunctions are far more common than mechanical trip failures. An operator wraps a piece of wire around the trip lever so it cannot move. This is called **wiring up a trip**. I've taught my troubleshooting seminar all over the world, in Australia, Africa, Asia, Europe, and South America. In no country was this concept not immediately understood by experienced plant operators. The practice is not just common; it's universal.

But why do operators disable fully functional overspeed trips in the first place? It has to do with malfunction of the speed control governor. The trip is mechanically simple, but the governor is more complex and is subject to failure.

If the governor is out of order, then the operators run the turbine with a constant steam supply. Meaning, they regulate the steam flow to the turbine in the field with the isolation gate valve on the turbine's motive steam supply. Meaning, the horsepower output from the turbine to the pump that it is driving is constant.

Let us now refer back to Figure 31-2. The discharge valve on the pump is controlling an upstream vessel level. This level is falling. Again, assume the steam supply to the turbine is constant, because the governor is not working. What then?

- The pump discharge valves begin closing.
- The pump moves from point B to point A on its performance curve.
- The work or load on both the pump and driver diminishes.
- At a constant driver horsepower, both the turbine and pump, which are coupled together, run faster.
- The turbine trips off due to the excessive speed, which is indeed its correct safety function. The operator resets the trip.

Since the upstream vessel level is naturally rising and falling, the turbine repeatedly trips off due to overspeed. The operator tires of the repeated pump shut-downs and wires up the trip. The lesson is that the trip is not a backup for the governor. If the governor is not working, sooner or later an operator will wire up the trip. Then the pump and turbine are completely unprotected from overspeed. This problem resulted in serious accidents at an Exxon Chemical plant and at the Coastal refinery, both in south Texas.

Accidentally Closed Port Valve

I'm going to work next week in Coffeyville, Kansas. The plant manager, Robert Haugen, likes me because I once fixed a coker wet gas compressor problem for him in Aruba. This compressor in Aruba was running slow. Wet gas was being flared as a result. The night sky over Aruba glowed romantically in the resulting giant yellow flare. It was a major tourist attraction for the West End of the Happy Island.

My calculations showed that the compressor was running too slow due to the lack of turbine horsepower. I calculated that the turbine efficiency was poor, likely because the turbine wheels were fouled with steam hardness silicate deposits. So I recommended to my client, Mr. Haugen, that the machine be shut down and the turbine case disassembled. This is a tremendous maintenance undertaking for a 5,000-horsepower turbine. Huge bolts secured by giant nuts have to be removed. Then the massive turbine case has to be lifted by a powerful crane.

But when the steam turbine rotor was exposed, it looked pretty clean, considering that it had been in service for 2 years. Rather disappointing. I surveyed the scene sadly. Then I happened to notice the three port valves exposed on the lower portion of the steam chest. There were no handles on the stubby, square valve stems. And to be partially honest, I had never focused on them before, as they had been covered by insulation.

Idly, I took my 6-inch pipe wrench and checked to see if they were open. The first two port valves could not be opened, as they were already, as they should have been, fully open. But with some considerable force, the third valve turned from closed to open (counter-clockwise opens). I looked around. No one was paying any attention to me.

"Okay guys," I called, "Let's wash those salt deposits down. Let's get that turbine wheel cleaned up. We need to get the machine reassembled and back in service by tomorrow morning."

"Excuse me, Mr. Norm," asked Johnny the Aruban shift supervisor, "What deposits? That rotor's real clean."

"Even minor fouling may appreciably degrade the efficiency of a multistage, dynamic machine, converting kinetic energy to polytropic head. Do you not even know that the rotating, volumetric, thermodynamic, isoentropic boundary layer can be disturbed?" I explained to Johnny.

"Okay, Mr. Norm. We'll close er up. You're the boss," conceded Johnny.

The next afternoon, the freshly restored turbine and compressor were rotating 400 rpm faster than their previous speed. The flaring of coker wet gas had stopped. And the soft, warm, yellow glow of the flare had almost vanished from the night sky.

"Good work, Norman," said Mr. Robert Haugen, at the evening operations meeting. "I'll bet that's $10,000 a day of gas that's not being flared. Good job getting the turbine rotor cleaned and back in operation so quickly. Norman, write up the calculation procedure used to predict turbine efficiency both before and after cleaning. I would like our entire technical staff to profit from this fine example of your expertise in. . ."

So maybe I forgot to mention at the meeting the port valve I opened on the turbine's steam chest in Aruba. But the hardness deposits on the 400 psig steam supply trip valve in the Amoco refinery in Texas City were not my fault, either. It kind of all evens out in the end.

Condensing Steam Turbines

Backpressure on the turbine exhaust steam will reduce the amount of work that may be extracted from the motive steam. This is especially important for condensing steam turbines. You can quantify the effect of increased exhaust steam pressure by referring to the Mollier diagram in your steam tables. Just follow the lines of constant entropy down the chart to the exhaust steam pressure line.

Malfunctions with a high turbine exhaust pressure for condensing steam turbines may be due to either:

- Problems with the two-stage ejector vacuum system. See Chapter 26, "Vacuum Systems and Steam Jets," and Chapter 28, "Excess Gas Overloads Vacuum System Ejectors."
- Problems with the surface condenser or interstage condenser are discussed in Chapter 27, "Vacuum Surface Condensers and Precondensers."

Both steam turbine and vacuum towers employ the same sort of converging-diverging steam jets, and both suffer from the same type of malfunctions related to steam properties.

Water in Turbine-Driven Pump Bearing Housing

Joe Petrocelli was the chief engineering officer in an aircraft carrier in the battle of Leyte Gulf, in the Philippines, during World War II. During the 1980 strike in Texas City, at the Amoco refinery, Joe taught me this useful technique to protect an air blower's inboard radial support bearing from water damage.

I had noticed that the lube oil level in the bearing's **oiler glass** was increasing quite a lot every day. My first thought was that I was creating lubrication oil in the bearing housing. However, I quickly realized that water must be accumulating in this bearing housing. The 500-horsepower blower was driven by a steam turbine using 400 psig motive steam and exhausting to the 100 psig steam header. The turbine's inboard bearing housing carbon seal was defective. Steam was blowing out along the shaft (the term *inboard* means the side of a piece of rotating equipment nearest to the coupling). The steam leak enveloped the entire inboard side of the air blower and forced its way into the air blower bearing housing. Hence the observed increase in the fluid level in the oiler glass.

Using a length of ¼-inch copper tubing, Joe blew a stream of plant air along the blower shaft. The air pushed the steam away from the blower's bearing housing. This immediately stopped the increase of the lube oil in the oiler glass. I used to adjust the copper tubing every day to make sure the air hit the shaft. We used it for a month until the carbon seal was fixed on the steam turbine's shaft.

Joe taught me many old naval maintenance tricks during the long 1980 strike in Texas City. But the main thing I learned was, in his own quiet way, Joe Petrocelli was a testament to a heroic American past that I was privileged to share.

Critical Speed Vibrations

K-805 was a 10,000-horsepower air blower in fluid catalytic cracking service at the Good Hope refinery in Norco, Louisiana. It was driven by a condensing steam turbine. Stamped on the nameplate were several numbers labeled "critical speed." I had been charged by the refinery owner to supervise the startup of the blower. After three days of effort, I had the blower in operation.

The definition of critical speed is that RPM (revolutions per minute) which corresponds to the **natural harmonic velocity** of a machine. If the equipment is run at such a velocity for a second or two, nothing much happens. But if the machine is operated at one of its critical speeds, the minor vibrations are reinforced. The analogy is lightly pushing a child on a swing. Careful timing of each push will send the child soaring.

To protect the turbine and air blower from the destructive vibrations resulting from operation at a critical speed, a vibration monitor should automatically activate the trip valve, which shuts off the turbine motive steam supply. But I failed to have the vibration trip connected that evening. I was in a rush to get to the pet shop. My son was out of parrot food. My plan was to have the trip connected first thing next morning.

But that evening there was a dip in the steam supply pressure. The turbine and blower slowed to a critical speed. The resulting vibrations ended the operational life of K-805 before it had actually begun. A sad but true story. Especially since the parrot escaped anyway.

Turbines are normally run well above their natural harmonic frequency, or their upper critical speed.

Turbine versus Motor Drivers

An operator has two equal pumps. One is driven by a motor; the other is driven by a steam turbine. Which one is better to run from the point of view of energy conservation? I have a few guidelines to recommend to analyze this complex question:

- If the turbine is a topping steam turbine, and the exhaust steam is vented into a steam header that is exhausting to the atmosphere, run the motor-driven pump.
- If the turbine is a topping steam turbine, and the exhaust steam is used as a source of process heat, run the turbine-driven pump.

- If the exhaust steam will be used in a condensing steam turbine, run the turbine-driven pump.
- If the refinery is flaring excess fuel gas and importing electric power, run the turbine-driven pump.
- If both pumps are oversized for the service, run the turbine-driven pump, but at a reduced speed.
- If the motor-driven pump operation requires that the spill-back be used, run the turbine-driven pump, but without using the spill-back.

Gas-Fired Turbines

My experience with natural gas-fired tubines was acquired in Laredo, Texas, in the mid-1980s, with three Solar Centaur turbines. I have documented this experience in my book, *Troubleshooting Natural Gas Processing*.

Gas-fired turbine rotors, like steam turbines, are subject to fouling. Typically, they have to be detergent washed periodically. As the rotor fouls, the turbine's flue gas exhaust temperature will increase. In Laredo, a turbine flue gas exhaust temperature above 1,100°F was deemed excessive. Higher exhaust temperatures could damage the turbine blades.

An airplane's jet engine is a gas turbine whose thrust is used to push a plane through the air, rather than to drive a refinery air blower or wet gas compressor. The most energy-efficient machine ever built is the "high bypass jet engine." It converts 40% of the energy of the fuel to work in the form of jet engine thrust. A typical process plant gas-fired turbine converts perhaps 30% of the fuel fired to work. But then, the hot exhaust flue gas could be used to generate medium- or low-pressure steam for use in a distillation tower reboiler.

CHAPTER 32

Centrifugal Compressors: Surge, Fouling, and Driver Limits

Bad things happen to those who are not careful.

—The idea my clients can't seem to grasp

I heard that Professor Peterson died last year. But mentally, he expired in 1962. Sorry if I seem bitter, but it seems unfair to devote 4 years of study, only to have to learn it myself after I entered the industry. As my wife Liz has explained, I should have been chatting up girls in 1962, rather than pondering why:

$$\text{Compressor work} = [(A) \div (B)] \times (D^B - 1) \qquad (32.1)$$

where A is a number proportional to the amount of gas compressed. Which is reasonable and just common sense.

D is a number proportional to the amount of the pressure increase of the gas, or the compression ratio. Which is also reasonable and common sense.

$B = (K - 1) \div (K)$. What is this? What is K? It's the ratio of the specific heats. What does the ratio of the specific heats of a gas have to do with compression work?

Ratio of Specific Heats of a Gas

Let's say I have a certain amount of gas (A) trapped in a cylinder below the piston shown in Figure 32-1. I'm going to heat this up by 100°F. This piston itself has no weight. I plan to measure the amount

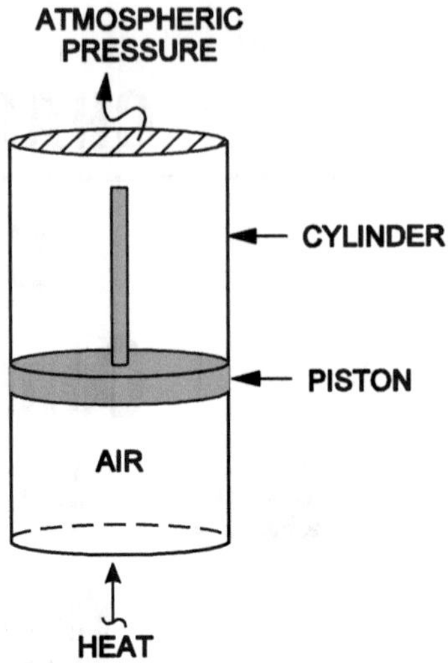

FIGURE 32-1 Apparatus to determine ratio (*K*) of specific heats (Cp/Cv) of air.

of heat that I need to increase the gas temperature by the 100°F. The question is, should I keep the piston in a fixed position (Case I), or allow the piston to move freely (Case II)? Maybe it won't make any difference—or maybe it will. Let's find out.

- Case I—I'm keeping the piston in a fixed position. The volume of my gas is constant. Of course, the pressure of the heated gas increases as the gas temperature increases from 100°F to 200°F. The energy required to heat the gas at a constant volume, I'll call Cv. It's 100 Btus.
- Case II—I'm allowing the piston to be pushed up. The pressure of my gas is constant. Of course, the volume of the heated gas increases as the temperature increases from 100°F to 200°F. The energy required to heat the gas at a constant pressure, I'll call Cp. It's 140 Btus.

Why is Cp 40% bigger than Cv? Because the piston, as it is pushed up by the gas in the cylinder, is doing work. It's slightly compressing the atmosphere of air surrounding our planet. Therefore:

$$\mathrm{Cp} - \mathrm{Cv} = \text{compression work}$$

The rest is simple:

- Divide Cp – Cv by Cp:

$$\frac{\mathrm{Cp}}{\mathrm{Cp}} - \frac{\mathrm{Cv}}{\mathrm{Cp}}$$

- Substitute $K = \mathrm{Cp}/\mathrm{Cv}$:

$$1 - \frac{1}{K}$$

- Multiply through by K/K

$$\frac{K-1}{K} \tag{32.2}$$

where Cp + Cv = K.

Can you now see that B in Equation (32.1) is a number directly proportional to the compression work?

When we compress any gas, some of the work performed by the compressor on the gas goes to heat. That's Cv.

When we compress any gas, some of the work performed by the compressor on the gas goes to heat plus the energy needed to compress the gas. That's Cp.

So, the difference between Cp and Cv is that portion of the compressor's work needed just to compress the gas. For example, compressing butane is easy. So its K value is about 1.26. On the other hand, compressing hydrogen is hard. So its K value is about 1.40. Does this mean it takes 10% to 20% less energy to compress 58 pounds (one mole) of butane than it takes to compress 2 pounds (one mole) of hydrogen?

Yes! It's an expression of the second law of thermodynamics, as it relates to gas compression. And if you read Chapter 31, "Steam Turbine Drivers," you will now understand what thermodynamics is all about. And you can use your knowledge of thermodynamics to attract girls—if it's the right girl.

The Dynamic Nature of Centrifugal Compressors

Reciprocating compressors are easy to understand. I discuss their operation in Chapter 33. Centrifugal compressors are complicated. They're dynamic machines, just like centrifugal pumps. Dynamic in the sense that they convert velocity into feet of head. Recip's directly increase the pressure of a gas by squeezing the molecules of gas closer together with a piston. Centrifugal compressors work as follows:

- Gas enters the center of the wheel at a low pressure (P_1).
- The wheel spins at a very high rate, perhaps 10,000 RPM (revolutions per minute).
- The spinning wheel imparts a very high component of velocity to the gas as it escapes from the outer edge of the wheel.
- The gas now flows into openings inside the compressor case (the case is the stationary part of the compressor surrounding the spinning wheels and shaft).
- The gas slows down. As the gas slows, the velocity imparted to the gas by the spinning impeller is converted, not into pressure, but into "polytrophic feet of head."

I will not explain "polytrophic." It's not relevant to my simplified explanation. But the idea that a centrifugal compressor works by producing feet of head and not pressure is the essence of understanding and troubleshooting these complex machines. Think about a centrifugal pump. They're also dynamic machines. A centrifugal pump produces the same feet of head, regardless of the density or SG of the liquid (see Chapter 30, "Centrifugal Pump and Driver Capacity Limits"). A centrifugal compressor produces the same feet of head, regardless of the density of the gas. To convert from feet of head to the differential pressure increase (delta P) of the gas being compressed, I have to multiply the density of the gas compressed (*PV*) by the feet of head developed by the spinning compressor wheels (*HP*):

$$\text{Delta P} = (PV) \times (HP)$$

Centrifugal Compressor Nomenclature

- Thrust bearing—Constrains the axial movement of the shaft.
- Radial bearing—Supports the shaft.
- Labyrinth seal—Prevents backflow of gas along the inside of the case or stator.
- Stator—The stationary elements where gas velocity is converted into feet of head.
- Rotor—The shaft and wheel assembly.
- Wheels—Like a pump impeller. Accelerates the gas. Three to six wheels per rotor are typical.
- Interstage cooler—Removes the heat of compression between stages.
- Stage—After partial compression, gas is cooled before it is further compressed in the next stage.
- Case—Covers the rotor. Consists of a top and bottom half bolted together.
- Inlet guide vanes—Directs the inlet gas into the first wheel.
- Surge—Gas flow instability caused by high discharge pressure, low flow, and low gas density.
- Stonewalling—Gas flow no longer increases as discharge pressure is reduced.
- Lube oil system—Circulates lubrication oil to bearings. Filters lube oil.
- Seal oil system—Circulates seal oil to the labyrinth seal. Filters seal oil.

Factors Affecting Vapor Density (*PV*)

- Molecular weight: Butane has a molecular weight of 58. Air has a molecular weight of 29. Does that mean that the same compressor, running at the same speed, will produce double the delta P with butane as it does with air? Yes and no. Yes, if it's a centrifugal compressor. Small changes in molecular weight have a dramatic effect on centrifugal machines, but a negligible effect on recips.
- Compressibility: (Usually referred to as *Z* in most thermodynamics textbooks.) The compressibility of air or hydrogen is pretty close to 1. The compressibility of butane is (depending on temperature and pressure) pretty close to 0.90. The smaller the compressibility, the denser the gas. Compared to the effect of molecular weight, it's not important. I typically ignore the effect of compressibility.
- Temperature: Expressed in °R (460 + °F), or °K (273 + °C). The cooler the gas, the greater its density.
- Pressure: Expressed in psia (psig + 14.7) or (BARG + 1.0). The higher the pressure, the greater the gas density.

To summarize:

$$\text{Gas density} = \frac{(\text{Molecular wt})}{(Z)} \times \frac{(\text{Pressure})}{(\text{Temperature})} \qquad (32.3)$$

None of these factors affect the feet of head (*HP*) developed by the centrifugal compressor. But they, especially the molecular weight, certainly do affect the delta P developed by the centrifugal compressor.

Factors Affecting Feet of Head (*HP*)

- The speed of the compressor—The greater the speed, the more feet of head produced by the compressor.
- The diameter of the wheels—The bigger the wheels, the more feet of head produced by the compressor.
- The number of wheels—The more wheels, the more feet of head produced by the compressor.

Let's say I wish to compress the wet gas stream shown in Figure 32-2 from P-1 to P-2. For practice, let's calculate the compression ratio:

- P-1 equals 1 bar gauge or 2 bar absolute.
- P-2 equals 5 bar gauge, or 6 bar absolute.
- Therefore, the compression ratio is $6 \div 2 = 3$.

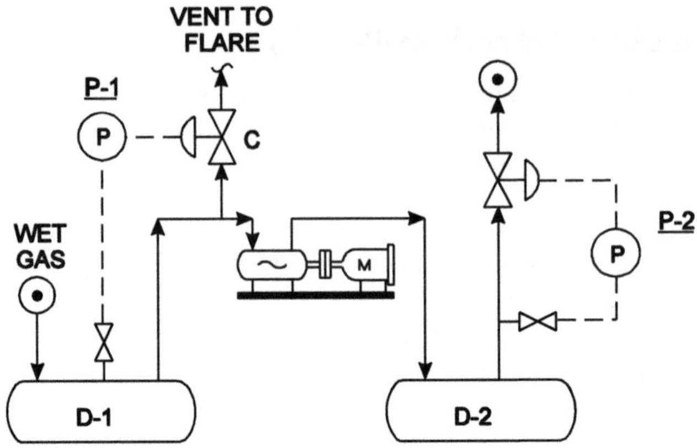

FIGURE 32-2 The minimum compression ratio of (P-2) ÷ (P-1) must be achieved to avoid surge.

This is a typical compression ratio for a "single stage" compressor. By a single stage, I mean the gas flows through about three wheels in series before flowing through an interstage cooler. These three wheels are inside a single compressor case. If I wanted to have a compression ratio of 9 (i.e., a discharge pressure of 18 bar absolute) I would have a second case or stage, also with three wheels. Why not put all six wheels in the same case? This could be done. But then I wouldn't be able to cool the gas after compressing it to 6 bar absolute. The hotter the gas, the bigger the volume and the less the density. Which would make it more difficult and expensive to compress the gas to the final 18 bar absolute discharge pressure.

Density versus Head Requirements

I've got to pump my gas pressure, from P-1 up to P-2. I have two ways of handling this requirement:

- More head (*HP*)
- Denser gas (*PV*)

Let's say I'm operating my compressor shown in Figure 32-2. Suddenly, some evil person injects 50% H_2 into my butane wet gas. The molecular weight drops from 58 to 30. This does not in any way affect *HP*, the feet of head developed by my compressor. But the compressor discharge pressure begins to drop radically because *PV* is reduced. Now what?

- One option would be to increase the number and diameter of the wheels. This will take 2 years to implement. Not a practical option.

- Or we could run the compressor faster. Likely another 30% to 50% speed would do the job. But Figure 32-2 shows a fixed-speed motor driver. Anyway, the compressor might have been running at 90% of its maximum rated speed. So this is also not practical, except for turbine drives and relatively smaller changes in the wet gas molecular weight.
- I could reduce the flow of wet gas. Centrifugal compressors, just like centrifugal pumps, operate on a performance curve, head versus flow (Figure 32-3). As the flow goes down, the feet of head produced by the compressor would increase. But if the flow of gas passing through the wheels gets too low, the flow of gas through the compressor will become unstable, and the compressor will **surge**. I'll describe surge in greater detail later. But even if the compressor does not surge, it may not have the operational flexibility to compress a lot less wet gas.
- I could lower the pressure in D-2 by reducing the backpressure control valve set point, perhaps from 5 bar gauge to 3 bar gauge. And this is exactly what I have done. In 1983, I was observing two operators commissioning a cracking unit at the Tenneco Oil refinery in Chalmette, Louisiana. During the startup, the wet gas unavoidably had a low molecular weight. To maintain the pressure in D-1 (Figure 32-2), the operators (who didn't like me because I came from New York City) were venting half the wet gas to the refinery flare. The entire Mississippi River was bathed in a soft yellow light.

"Why don't you guys lower the D-2 set-point pressure? We're only venting D-2 to fuel gas. No sense holding all that excess backpressure. Then you could stop the flaring, as the compressor runs out on its performance curve [Figure 32-3]," I suggested in a friendly tone.

FIGURE 32-3 If the flow through a compressor drops too low, then the compressor will surge.

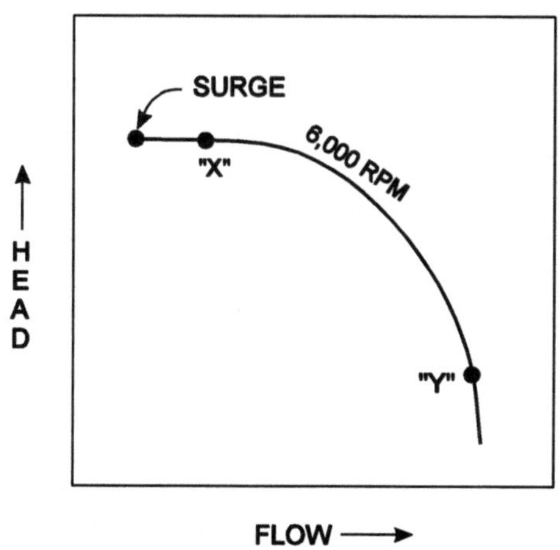

"Lieberman, why the (expletive deleted by publisher) don't you go back to New York City," my Cajun coworkers responded.

"Why not try it? If it doesn't work, I'll go away. You'll never see me again."

"Really?" both operators responded. "You promise you ain't never comin back?"

So the operators reduced the pressure in D-2. The compressor ran out on its curve. The pressure in D-1 started to fall. The spill valve vent to the flare on D-1 began to close. The yellow glare over the river dimmed.

You would have thought that these guys would have been grateful. You would have thought they would have said, "Golly, Mr. Lieberman. Could you explain this to us?"

But no. They just became even more hostile. "You damn engineers. They should ship you all back to Yankeeland. What you doing out here on graveyard shift anyway?"

What Is Surge?

If a centrifugal compressor does not develop enough delta P to push the wet gas from D-1 into D-2, then the wet gas flow slows down. Not to zero, but to a flow that is so small that it causes instability. That instability is called surge. Here's what happens:

- The gas flow stops flowing forward.
- As the pressure at the discharge is greater than the suction, the gas flows backwards through the wheels.
- The reverse gas flow causes internal damage to the compressor's internal components (see *Troubleshooting Process Operations*, 4th ed.).
- The reverse gas flow raises the suction pressure, which reduces the required delta P for the gas flow to move forward.
- The restored gas flow reduces the compressor suction pressure, which repeats the cycle or surge, typically 5 to 15 seconds per cycle. It's a really loud and quite terrifying sound.

Surge is caused by:

- A reduction in gas molecular weight.
- An increase in gas temperature.
- An increase in compressor discharge pressure.
- A reduction in suction pressure.
- Rotor or wheel fouling.
- Changes in speed. Sometimes surge is caused by slower speed; sometimes faster. I've seen it both ways. It depends on the design of each individual compressor and the range of operations.

Effect of Spill-Back on Surge

I first experienced compressor surge in August 1974. It was during a long strike at the Amoco refinery in Texas City. I was working the graveyard shift on No. 2 Alkylation Unit. It was 2:00 A.M. and I was standing next to our 6,000-horsepower steam turbine–driven centrifugal refrigeration compressor. Suddenly, it started to surge. With each roaring surge, there was a loud clanging bang. That was the sound of the **check valve** on the compressor discharge line slamming closed. I did not appreciate it at the time, but that check valve was moderating the destructive backflow of the gas. What had happened?

- The alky unit was operating at a reduced charge rate.
- Therefore, the flow of circulating refrigerant butane vapor was low.
- The suction pressure of the refrigerant compressor fell to zero psig. This is potentially dangerous. As I've explained to my clients, should a slight vacuum develop at the compressor suction, a small leak can draw air into the circulating refrigerant. The air will accumulate in the vapor space of the refrigerant receiver, on the discharge of the compressor (see Figure 32-4). When the mole% of air reaches about 85% (Caution: This value

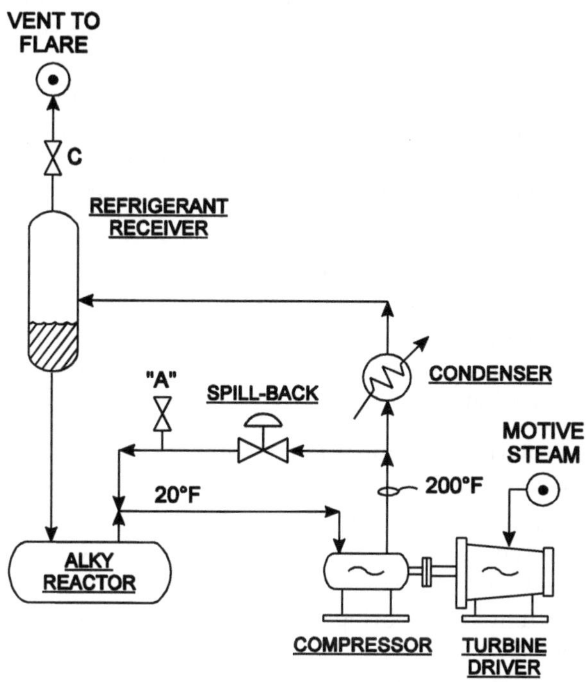

FIGURE 32-4 Opening the spill-back too far causes compressor to be trapped in surge.

is approximate, and varies with hydrocarbon composition), a spark originating from static electricity inside the vessel may cause the vessel to **detonate**.

- We knew this at Amoco Oil. The panel board operator began to open the compressor's discharge to suction spill-back control valve a lot to keep the suction pressure from falling below 0.5 psig. Referring to Figure 32-3, the compressor began producing less head as the spill-back increased the flow. This suggests, at least initially, that the compressor was moving away from surge, which is good.
- But the spill-back connection was on the compressor discharge line. And the discharge flow was hot. So, as the panel operator increased the spill-back flow, the compressor suction temperature became hotter and hotter.
- As the refrigerant vapor to the compressor suction also became hotter, the vapor density was reduced.
- The delta P produced by the compressor now became too low to overcome the backpressure from the refrigerant receiver shown in Figure 32-4.
- Thus, a progressively larger percentage of the compressor discharge flow recirculated to the suction, which lowered the vapor density even more, which lowered delta P more, which increased the percent of the compressor feed slipping back through the spill-back valve.
- The problem continued to feed upon itself. I call this a **positive feedback loop**. My book, *Troubleshooting Process Plant Control*, deals with this subject extensively. The end result of the declining vapor density was surge.

Suddenly, my boss, Frank Citek, rushed up. "Lieberman," he screamed above the roaring, clanging compressor, "Do something! Mr. Durland is watching."

I had been the supervisor of the alky unit for only 4 months when the strike started. I had no idea what to do. Now, 36 years on, I would know exactly what to do:

Step 1—Start closing the spill-back slowly.

Step 2—Start slowing the turbine down. But stay above its upper critical speed (which is stamped on the nameplate).

Step 3—Bypass the reactor feed-effluent cooler to generate more vapor flow from the reactor (see Figure 32-4).

Step 4—Temporarily bring in more feed to the reactor to generate more vapor.

Step 5—Vent the refrigerant receiver to the flare to lower the compressor discharge pressure (as per my previous story).

But in 1974, I didn't know what to do. So I ran up to Vincent Capadonno, one of the experienced shift supervisors, and yelled, "Cappy. Do something! Mr. Durland's here."

And Cappy blocked in the compressor discharge. It's true the clanging from the check valve stopped. But the roaring surge became louder and more frequent.

"Cappy. You've made it worse!"

"Lieberman, if you're so damn smart, why don't you run this (expletive deleted) unit?"

So I tripped off the steam to the compressor turbine and blocked in the olefin feed stream, and put the unit on isobutane circulation until 6:00 A.M. At which time Bobby Felts, a rather more insightful and experienced supervisor, came on shift.

"Bobby, I'm going to the house. Start up the unit. But whatever you do, don't open the spill-back more than 50% of the valve position."

This story had a happy ending. Bobby and I decided we could prevent the overheating of the compressor feed and loss in vapor density by spraying liquid butane into the spill-back vapor at point A, shown in Figure 32-4. The evaporating liquid would cool the superheated vapor.

I designed the system to use 1-inch pipe, and I showed my design to Bobby the following week.

"Okay, Norm. That looks good. Let's do it."

"Let's do what?" I asked Bobby.

"Let's get a couple hundred feet of 1-inch pipe, a threading machine, and a box of screwed fitting, and do it. Norm, you know how to thread pipe? No? Well, I'll teach you," said Bobby. "Then we can run at reduced feed rates without always worrying about putting that compressor into surge. I've seen a giant compressor come apart from surging. It's really quite dangerous, Norm. People could get killed."

The happy ending was that I learned that there are two types of people in this world:

- Type One—Cappy, Larry Durland, Frank Citek.
- Type Two—Bobby Felts.

I've calculated the ratio is about 11:1. Type One ÷ Type Two.

Suction Throttling versus Spill-Back

Once I was presenting a seminar to a group of operators, explaining the effects of gas molecular weight on the compressor's suction pressure. Here's what happened.

Let's make a few assumptions regarding Figure 32-5 to describe the next problem:

- The centrifugal compressor is driven by a fixed-speed motor.
- The motor is operating at its maximum power or amperage limit.

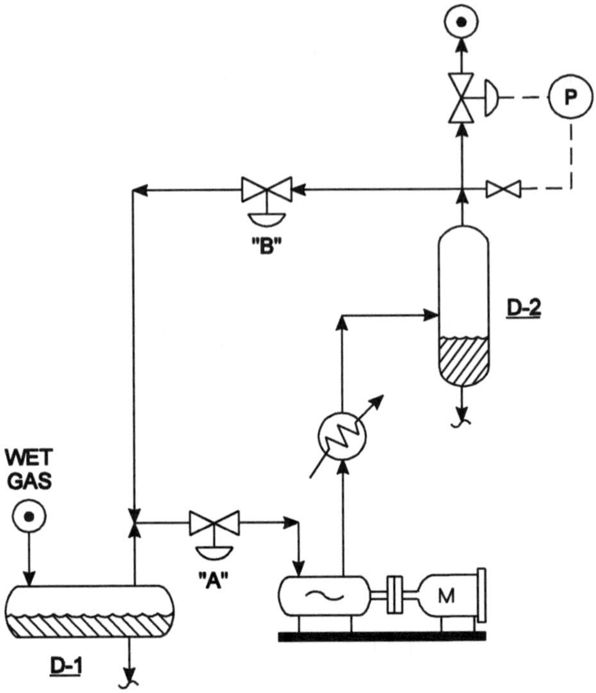

FIGURE 32-5 Depending on the operating point on the compressor curve, suction throttling can save driver energy.

- The discharge pressure is constant.
- The molecular weight of the gas is increasing.

The result of the increase in molecular weight is:

- The vapor density goes up (*PV*).
- The feet of head produced by the compressor (*HP*) remains the same.
- The compressor delta P increases.
- As the discharge pressure increases, the suction pressure must drop.
- The compression ratio (P-2/P-1) will increase.
- The increased compression ratio causes the amperage load on the motor driver to increase.
- The compressor motor now trips off due to high amps (i.e., power).

"Mr. Lieberman," Tommy, an FCU operator, objected, "I never told you to reduce the compressor's suction pressure. That's the reason the compressor's working harder."

"Tommy, I'd appreciate it if you'd stop interrupting the seminar. I'm talking."

"Okay, talk. But don't lower my suction pressure."

"Look, Tommy," I explained, "I didn't decide to lower the suction pressure. It's just a necessary consequence of the fact that the vapor density is going up, but the feet of head is constant. It's not my fault."

"Well, slow the sucker down."

"I can't, Tommy. It's an AC motor drive."

"You know, Mr. Lieberman, we got this same problem at our cat cracker," Tommy said as he fidgeted with his cell phone. "That compressor suction pressure goes down when the wet gas gets hotter."

"That's because the molecular weight of the wet gas goes up when the fractionator reflux drum (D-1) gets hotter. Tommy, put your cell phone away," I asked.

"Yeah. But, I'm expecting a call from my girlfriend. But what are we supposed to do to keep from over-amping the motor?"

"I thought you were married. Did you try suction throttling?" I referred to valve A on Figure 32-5.

"Yeah. We did. I read that in one of your books. Sometimes it works, but sometimes it don't. Anyway, it's okay. My girlfriend—she's also married. But just not to me." Tommy studied the giant flakes of snow drifting past the window. "How come sometimes throttling on the suction valve makes amps go down? And sometimes it makes amps go up? Don't seem to make no sense. Mind if I smoke? It's snowing out."

"Tommy! No smoking in class. Look," I said, "at Figure 32-3. It's a head-versus-flow curve for a centrifugal compressor."

"Kinda looks like an ole pump curve."

"Right. Same idea. Look, Tommy. If you can focus on Figure 32-5. Let's assume I start to close valve A. Which way will I move on my Figure 32-3 curve?"

"To the left. Towards surge. If you throttle on a flow, it's gotta go down. And the head will go up, cause you gonna have more pressure drop by throttling on that suction valve."

"Right, Tommy. But let's say I'm starting at point X on the compressor curve."

"That's the relatively flat portion of the curve, Mr. Lieberman," observed Tommy.

"Right. Where flow drops quickly, but head goes up slowly," I said.

Tommy removed the unlit cigarette from his lips and transferred it to behind his ear. "That's `cause of the shape of the curve. It's kinda flat around point X.

"Tommy, you know what power is:

$$\text{Power} = (\text{Head}) \times (\text{Flow}) \div (\text{Time})$$

"Sure, Mr. Lieberman. I know that. So power goes down `cause flow's dropping much faster than head's rising up."

"So Tommy, explain to the class what happens if I suction throttle, starting at point Y on Figure 32-3."

Tommy's face clouded as his cell phone displayed a new text message. "Women ain't nothin' but trouble. Anyway, Mr. Norm, if you start me out at point Y, and I suction throttle on valve A to hold backpressure, then I'm on the steep part of the curve. And then, looks like the head will go up faster than the flow will drop off, due to the shape of the curve. Then, motor power or amps will go up by suction throttling. I seen that also happen on our cat cracker centrifugal wet gas compressor."

"In general, Tommy, compressors are designed to operate on the flat portion of their performance curves. Just the same as centrifugal pumps. That's why it's best to suction throttle to maintain a constant pressure in the wet gas drum (D-1), shown in Figure 32-5."

"Mr. Lieberman, can we take a break now?"

"No! Any other questions?"

"Does this apply to reciprocating compressors?"

"No! Completely different. I'll cover that tomorrow (Chapter 33). What else?"

"Mr. Norm, it kinda looks like if I suction throttle too much, I'm going to put my compressor into surging. That happened on my shift one night."

"Right, Tommy. If you get the flow too low and the head too high, the compressor will slip into surge," I explained.

"Then what would I do at low wet gas flows, if my pressure in D-1 drops, and I'm close to the surge point, and the wet gas molecular weight increases?" Tommy asked.

"Open valve B, the discharge to suction spill-back. That will move you away from surge and increase the pressure in D-1," I answered.

"But Mr. Lieberman," Tommy objected, "opening the valve B spill-back is gonna increase power requirements. I'll be wasting motor amps. Ain't there a way to move away from surge without wasting electricity?"

"No, Tommy, it's like having a happy marriage and a girlfriend, too. You can't have it both ways. That's why the spill-back valve is called the anti-surge valve."

"Man! Ain't that the God's awful truth. Guess we need us both a suction throttle valve and a spill-back valve."

"Let's take a 10-minute break," I said, as Tommy rushed out into the raging blizzard for a smoke.

The problem that Tommy and I were discussing is what should be the response to a rising vapor density when trying to maintain a constant pressure in the upstream process. As Tommy first suggested, one should slow the compressor. A compressor driven by a variable-frequency, speed-adjustable motor or turbine would handle this problem easily.

If we restore this upstream pressure by opening the spill-back valve, then the load on the driver will increase and electric power consumption will rise. If we restore this pressure by throttling on the suction valve, then the compressor moves closer to surge and possible mechanical damage. It's a compromise. We try to optimize the position of the suction valve and the spill-back valve to minimize motor amps but without causing instability due to surging. To complicate matters, this all assumes that I'm operating on the relatively flat part of my head-versus-flow compressor curve.

Stonewalling

If one keeps reducing a centrifugal compressor discharge pressure, Figure 32-3 shows that flow barely increases, regardless of how much one reduces the need for delta P or head. That's called **stonewalling**. This sounds bad, but it's not. Stonewalling does not have any potential to damage the compressor. It has nothing to do with surge. As a young engineer, I thought both terms referred to a similar problem. But I was very wrong. Stonewalling would be encountered at point Y, on Figure 32-3.

Stonewalling limitations can be overcome by running the compressor faster, by increasing the density of the flowing vapor, or by reducing compressor fouling. A better name for stonewalling is choke flow. Reducing downstream pressure restrictions will not help with such choke flow limitations.

Rotor Fouling

Here's an interesting problem that I stumbled across in the gas fields in Laredo, Texas. The raw natural gas flowed into a variable-speed centrifugal compressor, driven by a gas-fired turbine. By "raw," I mean the natural gas was contaminated with:

- Salts
- Moisture
- Residual drilling mud
- A waxy grease

Let's assume that the driver horsepower output was constant. As the compressor rotor fouled—mostly with hard, gray salts—did the compressor go:

- Faster?
- Slower?
- Remain the same?

The correct and surprising answer is faster. As the compressor fouled, the compressor suction pressure increased and the gas flow rate decreased. The lower delta P and the reduced gas flow reduced the load on the turbine driver. Thus, the compressor could go faster. If the driver had been a fixed-speed AC motor, the rotor fouling would have caused the amp load on the motor to drop. I've described an interesting example of this phenomenon for an FCU combustion air blower in my book, *Process Engineering for a Small Planet*.

My experience with rotor fouling began in Texas City in 1974. I was supervising a process unit limited by the capacity of a large refrigeration compressor. I had tried several techniques to enhance refrigeration capacity:

- Optimizing the refrigerant composition, as described in my book, *Troubleshooting Process Operations*, 4th ed.
- Cleaning the discharge condensers.
- Raising the suction pressure.
- Improved cooling of the reactor feed.
- Maximizing the steam turbine driver speed.

These methods all had a positive but small effect on the refrigeration capacity. However, what I had failed to do was the most basic step in correcting process equipment malfunctions. That is, to determine if the centrifugal compressor was operating on its performance curve. To do this, we need:

- The suction and discharge pressures
- The composition of the refrigerant, which in this case was isobutane
- The suction temperature
- The compressor speed
- The vendor performance curve for the indicated speed
- The refrigerant flow rate

Typically in hydrocarbon processing and petroleum refining, if a centrifugal compressor is operating below its performance curve, the malfunction is fouling of the rotor's wheels. To remove a rotor for cleaning requires unbolting the compressor's case and lifting the top of the case. The rotor itself may then be lifted out. From start to final reassembly, I've had this done in three days. As my unit in Texas City was down several times during the year for a week, I could have conveniently cleaned the compressor. Except that I did not, for two reasons:

- The circulating refrigerant was clean, with no fouling mechanism possible.
- I never checked to determine if the machine was operating on its curve.

In 1977, I was demoted from operating supervisor to process engineer. I had proved to be a truly inept supervisor. Jay Golding replaced me. Based on a comparison of the compressor performance curve to current operations, Jay decided to inspect the rotor. He found that the wheels were full of grease.

At some point in the history of the compressor, a new rotor had been purchased. To protect it from atmospheric moisture and corrosion, it was shipped packed in grease, a common practice for high-quality machinery. When the new rotor was installed, it was not degreased. I later heard from Jay that he was able to circulate 10% more refrigerant as a consequence of removing the shipping grease from the compressor's rotor.

Salt Accumulation in Rotor Wheels

I have only carefully inspected a fouled compressor rotor that has been extracted from the compressor case on one occasion. The first wheel was pretty clean, the second wheel was moderately fouled with dirty, gray salt. The last wheel was quite badly fouled with salt. The reason for this fouling pattern has to do with moisture in the feed gas.

Compressed gas heats up as it is compressed. This is true regardless of the compressor efficiency. Even if the compressor was running at 100% efficiency, compressing gas from 10 to 60 psig would still heat the gas from 100°F to about 200°F. But since centrifugal compressors only operate at around 70% efficiency, the discharge temperature would be about 250°F. This increase in temperature always forces the gas above its dew-point temperature. It's true that a higher discharge pressure raises the dew-point temperature, but not enough to offset the heat of compression.

Process gas entering a compressor always contains entrained liquid, typically 0.1 to 1.0 wt%. The entrained liquid will contain dissolved solids. As the gas is heated by compression above its dew-point temperature, the entrained liquid evaporates. Solids are left accumulating on the rotor's wheels.

Or, in refinery processes, most process gases are contaminated with:

- Ammonia chloride
- Ammonia bisulfide

These salts, in vapor form, sublime out as solid salts at elevated pressures. Sometimes, when process equipment is opened on hydrodesulferizers, I've been able to smell the evolved NH_3. Higher temperatures, as a consequence of low compressor efficiency, may tend to prevent NH_3 salt sublimation.

If there is sufficient entrained moisture in the inlet to the compressor, then even the final rotor wheel would be kept wet and thus be protected from salt accumulation and fouling. The standard method that I have used in petroleum refineries is to spray a heavy naphtha mist into the suction of the compressor. The amount of naphtha required is typically about 1 wt% of the gas. However, do not use this value. It must be calculated taking into account the following factors:

- The molecular weight of the naphtha, which raises the dew point of the vapor.
- The latent heat of vaporization of the naphtha, which cools the vapor.

The objective is to add enough naphtha spray to bring the discharge flow from the compressor to its calculated dew-point temperature. On this basis, the final wheel of the rotor will remain wet.

Running without the naphtha spray for a few hours is not critical. So design the spray nozzle for on-stream removal for cleaning. The naphtha flow is best tripped off on auto, when the compressor shuts down.

Excessive Vibrations

As a rotor fouls, the thickness of the deposits increases. Efficiency and capacity suffer. Eventually and inevitably, the deposits will begin to break off in an uneven fashion. This is certain to unbalance the compressor's rotor wheels and cause vibrations. Hopefully, your compressor will then trip off due to the excessive vibrations. If it does not, very serious damage will result to the bearings and labyrinth seals. (These seals are inside the compressor case. They prevent gas from flowing backwards, as it is compressed. The labyrinth seals are part of the compressor's stationary component or stator.)

To prevent excessive rotor vibration, we have to either prevent excessive fouling to the rotor or shut down for cleaning in a timely fashion. That is, when the compressor has lost a certain amount of capacity due to fouling, the machine must be shut down, before it is too late.

By too late I mean the fouling deposits become too thick for removal without opening the compressor case and removing the rotor. There are a number of effective methods to clean the fouled rotor wheels without removing the rotor from its case. However, they all involve spinning the rotor while it is being cleaned. Cleaning the rotor will initially unbalance the wheels if the fouling deposits are too thick and break off during cleaning.

Preventing excessive vibrations on three 10,000-horsepower natural gas compressors was a big part of my job when I worked in the gas fields in south Texas. I have a detailed description of how to clean such rotors in my book, *Troubleshooting Natural Gas Processing*. Most importantly, I observed how a rotor is disassembled and then the individual wheels are rebalanced. Very much like having a tire rebalanced for your car, except small, very shallow holes are drilled near the edge of the wheels.

Work versus Flow in Dynamic Compressors

Let me pose a question based on the following assumptions:

- Suction pressure constant
- Discharge pressure constant
- Suction temperature constant
- Volume or number of moles of gas constant
- Compressor speed constant

Let's now assume the gas mole weight increases from 20 to 40. What would happen to the compressor's suction pressure?

- Increase
- Decrease
- Remain the same

The correct answer, as per my previous discussion with Tommy, is decrease.

In what direction should I then adjust the compressor speed to keep the suction pressure constant?

- Faster
- Slower
- Remain the same

The correct answer is slower. Meaning, I must respond to the doubling of the vapor density by slowing the compressor, to avoid dragging the suction pressure down.

Now the situation is:

- The compression ratio is kept constant by slowing the compressor.
- The volume of gas compressed is also constant.
- The pounds of gas compressed have doubled, because the molecular weight of the gas has increased from 20 to 40 (I'm assuming hydrocarbon vapors).

What now happens to the driver horsepower or motor amps needed to spin my compressor, assuming constant efficiency at the slower speed?

- Driver work increases.
- Driver work decreases.
- Driver work remains constant.

The correct—and somewhat counterintuitive answer—is decreased. Decreased by about 6%. Z, the compressibility, is reduced for heavier gas. The ratio of the specific heats is also reduced for heavier hydrocarbons. My rule of thumb is that each increase in carbon number reduces compression work by about 4%. Meaning, all other process conditions constant, it takes about 16% more electric power to compress 2 pounds of hydrogen than 58 pounds of isobutane. Strange! But that's just the nature of process operations.

CHAPTER 33

Reciprocating Compressors: Unloading and Efficiency

You people aren't just stupid, you're dangerous.

—Instrument tech Mondo Lira to my operators in Texas City, 1976

A reciprocating compressor is a direct volume reduction machine. Gas is squeezed into a smaller volume by a piston sliding through a cylinder. When the gas pressure is high enough, the piston pushes the gas out of the cylinder, into the discharge line. That's half the story. The rest of the story is the opposite. Gas is expanded inside the cylinder by pushing against the piston. When the gas pressure is low enough, new gas is drawn into the cylinder from the compressor's suction line.

A reciprocating compressor is called a positive displacement machine. Changes in gas velocity are of minor importance.

On the other hand, a centrifugal compressor, as described in Chapter 32, is a dynamic machine. For a dynamic machine, gas compression is achieved mainly through changes in gas velocity. Centrifugal compressors represent relatively new technology. Reciprocating compressors represent ancient technology. In general, for a similar service, a reciprocating compressor installation is much less expensive than a centrifugal compressor installation.

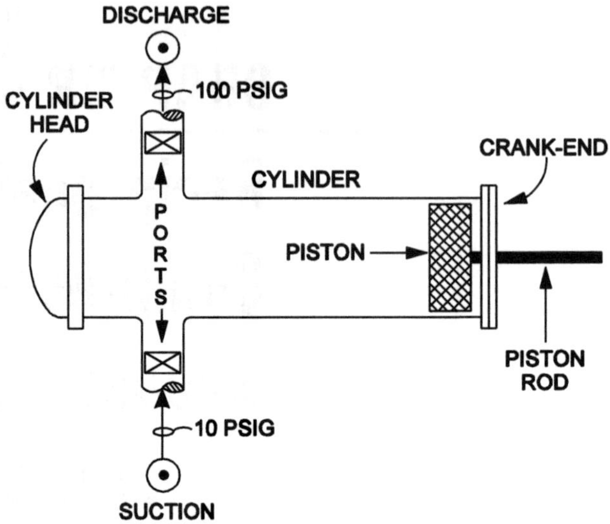

FIGURE 33-1 The ⊠ symbol is a spring-loaded check valve used in reciprocating compressors. Crank end valves omitted for simplicity.

The only dance I ever mastered was the foxtrot. It goes something like this:

- Step forward
- Step left
- Step back
- Step right

I found this terribly confusing, until my big sister Arlene explained that it's just like the four parts of the compression cycle. Referring to Figure 33-1:

- The piston begins its travel at the extreme right side of the cylinder.
- The piston moves to the left.
- The pressure of the gas trapped inside the cylinder, between the piston and the cylinder's head, increases.
- I say trapped because the check valve at the suction nozzle only permits the flow of gas in the indicated direction.
- Step A: The pressure in the gas in the cylinder increases from its initial 10 to 100 psig. This is called the compression portion of the cycle.
- With the gas pressure in the cylinder rising above the discharge line pressure of 100 psig, the discharge check valve is forced to open.

- Step B: Gas is now pushed into the discharge line, and the piston continues moving to the left. This is called the discharge portion of the cycle.
- The piston has to stop before it reaches the end of the cylinder. If the piston kept moving to the left, it would cover up most of the discharge port. Then the gas could not be pushed out into the discharge line.
- There's a certain volume of gas trapped at this point in the cycle between the piston head and the cylinder head. The pressure of this trapped gas is 100 psig.
- The piston now reverses its travel. It is pulled back to the right.
- Step C: The residual gas inside the cylinder now starts to expand. Its pressure falls from 100 psig down to 10 psig. This is called the expansion part of the cycle.
- Note that the check valve at the discharge nozzle also only permits the flow of gas in the indicated direction.
- Step D: Once the pressure in the cylinder falls to 10 psig, new gas is drawn into the cylinder through the suction nozzle. This is called the intake portion of the cycle.
- The piston has now returned to its starting position and is ready to repeat the cycle.

It's just like the foxtrot:

- Step A—Compression
- Step B—Discharge
- Step C—Expansion
- Step D—Intake

Unfortunately, my sister's efforts were wasted. The music at the dance was rock-n-roll. Gloria was only interested in doing the "Lindy." So she danced with Mark, and I became . . . that which I was destined to become.

The Carnot Cycle

Figure 33-2 repeats, in graphical format, that which I've just described with reference to Figure 33-1, which is a sketch of a reciprocating compressor cylinder. Let's start at point A on this chart.

Point A corresponds to the dotted line shown in Figure 33-3 (labeled Bottom dead center). The piston is now as far as possible from the cylinder head and hence as close as possible to the crank end. By the "crank end," I mean the compressor's driveshaft that's

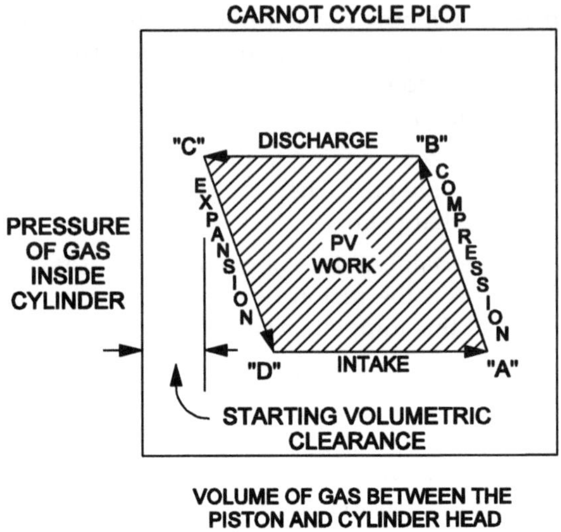

FIGURE 33-2 A Carnot cycle volume versus pressure diagram.

moving the piston rod. This position of the piston is called bottom dead center.

The piston now is pushed by the piston rod toward the cylinder head. The discharge valve is still closed, and it will remain closed until the pressure inside the cylinder increases a bit above the discharge line pressure. The purpose of the piston ring shown in Figure 33-3 is to minimize compressed gas leakage to the backside of the piston. Let me explain.

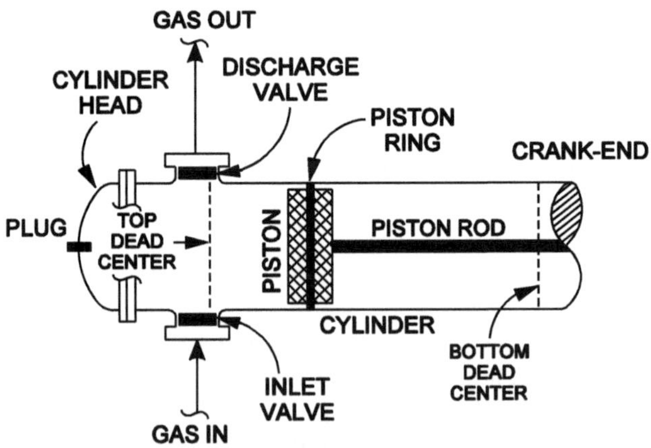

FIGURE 33-3 Components of a reciprocating compressor cylinder. The volume "top dead center" and the "cylinder head" is the "starting volumetric clearence."

All reciprocating compressors are "**double acting machines.**" That means that whatever is happening on one side of the piston, just the opposite is happening on the other side. Not shown in Figure 33-3 are a duplicate set of the valves, which ordinarily should be shown toward the end of the cylinder head, close to the cylinder's crank end. I've omitted showing these valves for simplicity. Thus, if the gas pressure is rising on the head end, the gas pressure is dropping on the crank end. Gas leakage across the piston would therefore reduce the overall compressor efficiency. The piston ring is intended to minimize such gas leakage.

When the piston reaches point B as shown in Figure 33-2, the gas pressure inside the cylinder is large enough to push the discharge valve open. This pressure must be a bit higher than the discharge line pressure, to force the discharge valve springs to permit the gas to escape into the discharge line. The work done by the piston on the gas between points A and B is called the compression part of the **Carnot cycle**.

The piston now continues its motion without change in velocity toward the cylinder's head. Gas is pushed out into the discharge line, until the piston reaches the second dashed line shown in Figure 33-3. That corresponds to point C, shown on my Carnot cycle curve. The work done by the piston on the gas between points B and C, is called the discharge part of the Carnot cycle.

The left-hand dashed line shown in Figure 33-3 is called top dead center. About 30% of the discharge port is covered by the piston head. If the whole port area was covered by the piston head, how could the gas escape from the cylinder into the discharge line? Of course, it couldn't.

The volume of gas remaining between the cylinder head and this dashed line must expand and drop to a pressure somewhat below the inlet gas pressure, before the gas from the inlet line can flow into the cylinder. As the piston head is pulled back by the piston rod from the dashed line, back toward the crank end, we move back from point C to point D on Figure 33-2. This is the expansion portion of the Carnot cycle. Until this residual gas left in the cylinder can expand and drop in pressure a little below the inlet line pressure, no new gas can be drawn into the cylinder.

The piston now continues its motion without change in velocity toward the crank end. Gas pressure pushes open the spring-loaded inlet valve. New gas is drawn or flows into the cylinder, until the piston returns to point A, or bottom dead center, on the Carnot cycle curve, shown in Figure 33-2. This is the intake portion of the Carnot cycle, which has now been completed.

The area inside the pressure-versus-volume curve is compression work. Let's ignore the following problems:

1. Frictional losses through both the intake and discharge valves.
2. The differential pressure between the cylinder and the discharge line required to force the discharge valve to open.

3. The differential pressure between the cylinder and the inlet line required to force the suction valve to open.
4. Leakage of gas back from the discharge line, back into the cylinder during the expansion and intake portions of the compression cycle. That is, a leaking discharge valve.
5. Leakage of gas back from the suction line, back into the cylinder during the compression and discharge portions of the cycle due to suction valve leakage.
6. Piston ring leakage.
7. Pulsation problems.

Work performed by the piston that is not used to compress the gas, due to these seven inefficiencies, appears as heat. Of course, even if we could entirely avoid such inefficiencies, the compressor discharge temperature would be hotter than the compressor suction pressure. That's called the heat of compression. The larger these seven inefficiencies become, the greater the temperature rise of compressed gas. We can use this concept in gauging relative efficiencies between individual machines (or even individual compression cylinders) that are operating in parallel. That is, compressor efficiency is inversely proportional to the gas temperature increase.

If these inefficiencies were ignored, then the reciprocating compressor would operate at 100% theoretical compression efficiency. The Carnot cycle curve assumes 100% theoretical efficiency. Therefore, it does not reflect actual reciprocating compressor operation. In particular, it ignores the big problem with recips. And that big problem is pulsation.

Pulsation Malfunctions

Pulsation is an acoustical phenomenon. It's a function of the geometry of the suction and discharge piping and the speed of the compressor. It is not a function of the gas properties. The amplitude of the pulsations but not their frequency can be influenced by the compressor valve design. I do not know how to design to avoid pulsation. Nor, I suspect, does anybody else. Pulsation is bad, in that it reduces the capacity and efficiency of a recip. It's not difficult to observe pulsation.

If you check Figure 33-3, I have shown a plug on the cylinder head. To observe the pressure pulsations inside the cylinder, the plug would be replaced with a **pressure transducer**. The pressure transducer converts a pressure signal into an electrical (i.e., milliamp) output signal. The electrical output is then connected to an oscilloscope where the pressure pulsations can be seen.

There are three ways of damping down the amplitude of such pulsations:

- One method is to install "pulsation bottles" on the suction and/or discharge. These are usually supplied with the compressor.
- My method (I only used it once at the Hebbronville, Texas, GHR Natural Gas Booster Station) is to install weaker springs in the compressor valves.
- The standard method is to use orifice plates to induce a substantial delta P in the suction and/or discharge piping. This method certainly suppresses vibrations, but it also parasitically wastes compressor capacity and horsepower.

Indicator Card (or Beta Scan)

The degree of pulsation varies with the portion of the Carnot cycle the recip is operating on. To determine the frequency and amplitude of pulsations, we need to plot the piston position versus the cylinder internal pressure. The piston position is a function of the crankshaft position. This can be measured by a magnet fixed to the crankshaft. The position of this magnet is measured electronically. Along with the electrical output from the pressure transducer measuring the cylinder pressure, both electronic signals are sent to an oscilloscope to produce the "indicator card," plot, as shown in Figure 33-4.

The straight lines on this chart are the Carnot cycle. They neglect all sources of inefficiency. The staggered lines are generated from the oscilloscope inputs. What is actually being plotted on this figure is the effect of pulsations on the gas pressure inside the cylinder,

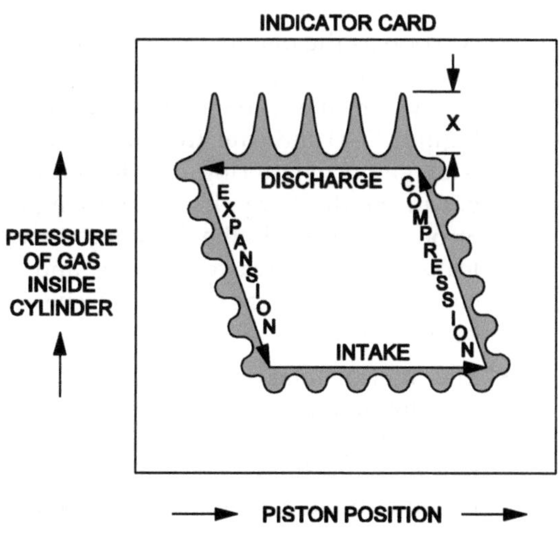

Figure 33-4 An indicator card plot used for troubleshooting. Dimension "X" is the pulsation amplitude.

between the piston and the cylinder head. The way to interpret this indicator card plot is as follows:

- Starting at the piston position shown as bottom dead center (the right-hand side dotted line in Figure 33-3), the piston begins to compress the gas. The peaks and valleys shown on the compression part of the cycle represent chatter of the intake or inlet valve. Pulsations are causing the higher-pressure gas in the cylinder to blow back intermittently into the suction line. Stronger springs in the inlet valve would diminish this leakage. This leakage wastes energy and capacity. You have just compressed the gas, and part of it now must be recompressed.
- The discharge valve is now forced open. The peaks and valleys shown in the discharge part of the cycle represent overcoming the spring tension of the discharge valve. The several peaks and valleys shown indicate that the discharge valve is opening and closing, or chattering. This requires the valve's spring tension to be overcome several times, which takes work and wastes more energy. This malfunction continues until the piston stops moving toward the cylinder head—that is, when the piston reaches top dead center, as indicated by the left-hand side dashed line in Figure 33-3. Weaker springs in the discharge valve would diminish this waste of energy.

NOTE *The only time I've personally applied the use of an indicator card was in Hebbronville. Then, I retrofitted the discharge valves of a 5,000-horsepower Enterprise-Delaval compressor with weaker springs. It did help based on the results of running a second indicator card.*

- As soon as the piston reverses its travel, the discharge valve closes as the pressure in the cylinder drops. The peaks and valleys in the expansion part of the Carnot cycle (see Figure 33-4) indicate that the discharge valve is blowing back gas or chattering due to pulsations. Higher-pressure gas in the discharge line is blowing back into the cylinder. Stronger springs in the discharge valve would diminish this leakage, which wastes energy and capacity. Having just compressed this gas into the discharge line, part of it has to be recompressed.
- The inlet or intake valve is now forced open by the higher pressure in the inlet gas line. The peaks and valleys shown in the intake part of the cycle represent overcoming the spring tension of the intake valve. The several peaks shown indicate that the intake valve is opening and closing, or chattering.

This requires the valve's spring tension to be overcome several times. This takes extra work and wastes more energy and continues until the piston reaches bottom dead center. The cycle is now completed.

Drawing Conclusions from Indicator Card

Referring to Figure 33-4, note that the number of pulsations during the discharge portion of the stroke is five. This frequency of pulsation cannot be altered for a motor-driven, fixed-speed recip. The amplitude of these pulsations is larger than for the other portions of the compression cycle. That's because the discharge valve springs are too tight. If I replace them with weaker springs, which I did in 1984 in Hebbronville, the amplitude will be reduced, but not the frequency. The amplitude of the pulsation peaks during the expansion portion of the cycle did increase. That's because the weaker springs used in the discharge valve increased blow-back or back-leakage from the discharge line back into the cylinder.

The distance between the Carnot cycle lines and the indicator card valleys shown in Figure 33-4 represent frictional losses through the open valves. These are called "valve velocity losses." These relatively small losses can be reduced by purchasing low delta P intake and suction valves. Usually this is a waste of money. Most of the inefficiency of a reciprocating compressor is associated with pulsation and leakage losses, not with the valve velocity losses.

The shaded area shown in Figure 33-4 is the difference between theoretical or Carnot cycle compression work and actual or real compression work. The shaded area appears as heat, or excess temperature of the discharge gas. If I divided the area inside the straight lines by the area inside the curved lines, I would be calculating the overall compressor efficiency.

Typical Reciprocating Compressor Efficiency

Recips usually have a bad efficiency when compared to their design efficiency, which is typically 95%. What is a realistic operating reciprocating compressor efficiency? It all depends on the compressor's physical condition.

Defective valves are the biggest problem. The moving parts—springs and valve plates—have a tendency to break due to liquids in the feed gas or vibrations. If a cylinder has four suction valves and four discharge valves, complete failure of any one of these eight valves will result in almost zero efficiency of compression and no net flow of compressed gas. Leaking valves create energy waste on the expansion and compression portions of the cycle. Excessively tight spring tension reduces leakage, but it creates energy waste on the intake and discharge portions of the cycle.

Leaking piston rings (see Figure 33-3) allow gas to slip back and forth between the head end and crank end, with a consequent loss in net gas being compressed.

Valve velocity losses are due to frictional losses as the compressed gasses pass through the valves. These losses, which are typically the only inefficiency considered by the equipment vendors, are relatively small.

And, beyond all of the above, amplifying the detrimental effects of the other losses are pulsation losses, sometimes called valve chatter.

Pulsation Damping Plates

Pulsation promotes vibrations. The larger the amplitude of the pulsations, the greater the destructive vibrations become. Reciprocating compressors are very high-maintenance items compared to centrifugal compressors. As an operating superintendent in Texas City in the mid-1970s, I had five large Clark isobutane refrigerant compressors on my sulfuric acid alkylation unit. I had a dedicated machinist, Big Mack, who worked on overhauling these machines, and nothing else. Big Mack, an experienced machinist, would finish up on one compressor and then go on to the next. Meaning, at best, I had 80% compressor availability from the five compressors, even with continuous maintenance.

One effective method to reduce required recip maintenance is to dampen out the amplitude of pulsations (and the corresponding vibrations), by the use of **pulsation damping plates**. These are just ordinary orifice plates, such as those used in a flow transmitter. They are inserted in the suction and discharge lines to induce a delta P that tends to suppress the pulsation's amplitude. A typical such delta P might be 15 psig, doubled for both suction and discharge. If your compressor is developing 300 psi of differential pressure, this will reduce the compressor efficiency by 10%:

$$(15 \text{ psi})\ (2) \div 300 = 10\%$$

It's always a battle between the process engineers, who are trying to save energy and maximize capacity, and the maintenance department guys, who are trying to reduce the recip's destructive vibrations and thus maintenance costs.

Breaking Piston Rods

Reciprocating compressors are double-acting machines. While the gas is expanding on the head end of the cylinder, the gas is being compressed on the crank end of the cylinder. There are two complete sets of valves on either end of the cylinder. The difference in pressure between the head end and the crank end of the cylinder creates a large pressure differential across the piston. You can see from Figure 33-4 that this differential is larger than the discharge pressure minus the suction pressure, due to valve losses and pulsations.

This difference in pressure across the piston, times the area of the piston face, equals the force on the piston rod. If this force exceeds the mechanical strength of the piston rod, then the rod will break. Which is really bad news, but not uncommon.

To protect against rod failure, recips have a high-discharge temperature trip (i.e., a self-actuated shut-down). The larger the compression ratio of any compressor, the higher the outlet temperature. And the compression ratio can be correlated with piston rod loading. So this makes good sense.

Reciprocating Compressor Nomenclature

- Starting volumetric clearance—Space between piston and cylinder head at top dead center.
- Top dead center—End of piston travel toward the cylinder head.
- Bottom dead center—End of piston travel toward the crank end of cylinder.
- Crank end—End of cylinder near the driver.
- Valves—Spring-loaded check valves at suction and discharge of the cylinder.
- Double-acting—Compression on both the forward and backward strokes of the piston.
- Piston ring—Reduces leakage across the piston.
- Piston rod—Pushes/pulls the piston.
- Carnot cycle—Ideal compression cycle.
- Indicator card—Measures the actual pressure versus volume inside cylinder.
- Adjustable head end unloader—Correct method to reduce capacity of the cylinder.
- Unloader bottle—Stepwise method to reduce compressor capacity.
- Pulsation amplitude—The height of the peaks on the indicator card.
- Pulsation frequency—The number of the peaks on the indicator card.
- Pulsation losses—Compression work wasted due to pulsations. Function of valve spring tension.
- Pulsation dampeners—Orifice plates inserted into the suction and discharge piping flanges.

But in the remote and wild natural gas fields in Laredo, Texas, this high-temperature trip caused me unending problems. My wellhead compressors were always shutting down for no apparent reason. They were tripping off on high discharge temperatures because the ambient temperatures would exceed 120°F, and not due to excessive compression ratios. I had continuous arguments with Halliburton Corporation, from whom we rented several hundred wellhead reciprocating compressors, about these trip settings. They wanted to set the trip temperatures really low to avoid piston rod failures, and I wanted to maximize gas production. Especially during hot weather, when air conditioner electric consumption soared and caused a surge in demand of natural gas supply to electric power generation utilities.

Gas Engine Drives

All of my 200-odd little wellhead compressors in Laredo were driven by gas engines. Sometimes I would lose power to a compressor because one of the engine cylinders would stop firing, either because of a broken ignition wire or a defective spark plug. But which was the nonfiring cylinder? Here's the trick:

- Disconnect the ignition wires, one at a time.
- Each disconnected ignition wire will cause the compressor to slow.
- Except for the cylinder that was not firing. Disconnecting its ignition wire will have zero effect.

Starting Volumetric Efficiency

If you really wish to understand reciprocating compressors, then you'll have to master the concept of volumetric efficiency. Recips have two unrelated measures of efficiency. The efficiency I've been explaining up to now is called the "adiabatic compressor efficiency." It's a measure of the amount of useful compression work performed on the gas, divided by the total amount of driver power output.

The volumetric efficiency is only a function of the amount of gas trapped between the cylinder head and the piston at top dead center (see Figure 33-3). Top dead center is the position of the piston when it is closest to the cylinder head. The volume inside the cylinder between the head and the piston at this point in the cycle is called the starting volumetric clearance. This is best defined by the Carnot cycle in Figure 33-2.

By leaving more gas trapped inside the cylinder at the end of the discharge portion of the cycle, the expansion part of the cycle increases, and the intake portion of the cycle decreases. This reduces the volume of gas compressed per cycle, but without wasting any compression energy. But why would we want to reduce the amount of gas compressed?

Controlling Compressor Gas Flow Rates

The best way to control gas flow through a reciprocating compressor is to simply slow it down. But with an ordinary fixed-speed AC motor (i.e., without frequency control), this is not possible.

The absolute worst way (but of course, the most common) is discharge to suction spill-back. Observing this in the field invariably makes me angry. (See my book, *Process Engineering for a Small Planet*, for a particularly frustrating story.)

Throttling on the discharge is almost as bad, but this option is rarely used. Suction throttling, as I've recommended for centrifugal compressors, is a better alternative. However, this option will also waste about half as much energy as throttling on the discharge or spill-back for an equivalent reduction in compressor throughput.

"Valve disablers" are usually used by compressor designers to reduce recip capacity. In effect, a valve disabler prevents a compressor valve from closing. The capacity of the cylinder associated with the disabled valve is now zero. The gas "sloshes" back and forth inside the cylinder, which gets quite warm. I've estimated (see my book, *Troubleshooting Natural Gas Processing*) that 20% of the compression work normally performed by this cylinder is still wasted by the valve disabler method of unloading a reciprocating gas compressor. The 20% was based on a direct field temperature measurement, which I doubt that anyone else has ever bothered to observe.

Head-End Adjustable Unloaders

What then is the correct way to reduce the capacity of a constant-speed reciprocating compressor? This adjustment is made by increasing the starting volumetric clearance that I explained earlier.

There are two ways to accomplish this. One method, which is essentially free, is to employ an "unloading bottle." This steel bottle, which we used at the Good Hope gas compression station in Laredo, was the size and shape of a liter bottle of Coke. I simply had a compressor shut-down to install the bottle. The plug shown on the cylinder head in Figure 33-3 was removed and the liter steel bottle screwed in. The bottles would reduce the capacity of the head-end compression stroke by about half. However, since all our recips are double-acting machines, the overall reduction in compression capacity was only about 25%.

The other way to unload a recip, without loss of compressor energy efficiency, is to use an adjustable head-end unloader. This device replaces the cylinder head entirely. Using a large wheel, a cavity in the new head can be enlarged, thus increasing the starting volumetric clearance and reducing the volumetric efficiency. Typically, this device also can only reduce capacity by 25%. However, the reduction in capacity can be achieved without a compressor shutdown and in small, adjustable increments. All recips I have seen can be retrofitted with head-end unloaders of this sort, with very little effort. But, they are very expensive and to be honest, none of my clients have ever taken my advice to retrofit their recips with such an energy-saving device. I've provided a detailed sketch for an adjustable head-end unloader, which I've used myself, in my book, *Troubleshooting Natural Gas Processing*.

And now for the truth. This past year, in both Durban, South Africa, and St. Croix (Virgin Islands), I noticed that the operators did have, but were not using the adjustable unloader, but instead they used the spill-back line. Even though, in both cases, the operators had a clear understanding of how to use the unloader valve and its energy-saving potential benefits.

"Norman," they explained, "Using the spill-back is easier for the operator. The adjustable unloader valve has become hard to turn over the years, due to lack of use." And that's all too often the reality in a process plant.

Effect of Molecular Weight on Compressor Performance

If we increase the molecular weight of the gas flowing to a centrifugal compressor, running at constant speed, by 10%,

- Delta P developed by the compressor goes up by about 10%.
- The motor amps will also increase by about 10%, or a bit less.

On the other hand, if we increase the molecular weight of a gas flowing to a reciprocating compressor, also running at a constant speed by 10%, then

- Delta P developed by the compressor will not change.
- The motor amps will decrease by about 1%, if we are working in the range of 30 to 40 molecular weight.

For my work, I take great care in determining the design range of molecular weight for centrifugal compressors. It's critical. For recips, I wouldn't even bother to get a gas sample. It's not important.

Incidentally, my clients sometimes observe that decreasing the molecular weight of a gas, say from 40 to 10, reduces flow by half from a reciprocating compressor. That's wrong. They forget to correct the orifice flow meter for the vapor density or the molecular weight (MW):

$$(\text{Indicated flow}) \text{ proportional to } (\text{MW})^{\frac{1}{2}}$$

Reciprocating versus Centrifugal Compressor Selection

Rotating equipment is historically preferred over reciprocating motion. For example, most of us have flown on a jet plane this year. But when was the last trip you made in a piston-driven airplane? Ninety-nine percent of our pumps are now centrifugal. But 150 years ago, there were only reciprocating pumps.

Centrifugal compressors are always preferred in process applications, but they are not always practical. If the gas density (molecular weight) is low, and the required delta P (compression ratio) is large, then the polytrophic feet of head required by the centrifugal compressor becomes excessive. Many wheels are required to produce the excessive amount of polytrophic head. The shaft becomes very long, and shaft alignment becomes a problem. Minor shaft misalignments will produce excessive vibrations. It's possible to reduce the number of wheels and hence the shaft length, with greater compressor speed. But that also demands a more perfect shaft alignment to avoid destructive vibrations.

Reciprocating compressors simply do not care about the molecular weight being too small. If more delta P is needed, one just adds another compression cylinder in series with the existing cylinders. Also, reciprocating compressors are cheaper to buy than centrifugal compressors. And, recips are extremely easy to install compared to centrifugal compressors.

Finally, at least in theory, a reciprocating compressor has a design compression efficiency of 90% to 95%. A centrifugal compressor has a design compression efficiency of only 65% to 80%. But that's all just in theory.

In reality, the reciprocating compressors that I have worked with are subject to a wide range of malfunctions, as discussed in this chapter. Unlike a centrifugal compressor (Chapter 32), they are not going to spin along happily for years, producing a constant feet of head and at a constant flow rate. Reciprocating compressors, rather like your car which has a reciprocating engine, require constant maintenance and operator attention. My ex-boss, Bill Duvall, always told us that if we specified a reciprocating compressor in our designs, it had to be spared to provide an acceptable degree of availability and reliability. Slow-speed reciprocating compressors are actually quite reliable, but the slower the speed, the larger and more expensive the compressor becomes.

Reciprocating versus Centrifugal Compressor Selection

CHAPTER 34

Summary: Typical Energy Savings and Efficiency Projects

If you got the money, I got the time.

—My company motto

I recently had a great idea to establish a joint venture with Bechtel Engineering in Canada to market engineering services. They would do all the marketing, and I would do all the engineering. So I wrote an outline of the projected services I had in mind, including a dozen examples.

Would you believe it, I never heard from Bechtel again. No! It's true. They apparently didn't like my examples. Or maybe they didn't like me. Probably both.

So I'm stuck with these examples. No one wants to buy them. I guess you can have them for free. To some extent, these examples are illustrations of the principles I have explained in detail in the preceding chapters. And in that sense, the examples are intended to summarize this text.

How to Harvest an Apple Orchard

I know that there is lots of fruit high up in my apple tree. But my ladder is broken, and I can't afford to buy another. No matter. There are lots of low-hanging apples that I can still reach. It's rather like petroleum refining. Sure, our profits sometimes have miscarried, but we can still make money and help protect the environment without capital investment if we will take a little extra trouble to harvest the available profits. So that you don't think you're wasting time reading this, I'll give you some free examples that I've implemented in several plants.

Steam Turbines

- Did you know that steam turbines are variable-speed drivers? I guess everybody knows that. Then why do you throttle on the discharge of your turbine-driven pumps to control the discharge flow or pressure? Perhaps you also drive your car with the gas pedal on the floor, and control the speed with the brake? Better to control the turbine's governor speed to develop the required pump discharge pressure and flow. How about the downstream FRC, or PRC, or LRC? Open their bypasses and forget them. Each 3% reduction in turbine speed will save 10% of motive horsepower and 10% of the turbine driver's steam.
- However, slowing down a steam turbine will also cause the governor to close. Which is good in that it saves steam. But it is bad in that the pressure of the motive steam is reduced as the steam enters the turbine's steam chest. That's what the hand valves (properly called horsepower or port valves) are for. Close a hand valve and the governor is forced back open. This raises the steam chest pressure and increases the amount of work that is extracted from each pound of steam.
- However, suppose you don't have any hand valves on your turbine case. Then what? Well, you could buy such valves, as all turbines have provision for two or three such valves. But that's like me buying a new ladder. It's not that you can't afford $10,000 for the new valves. It's that you have to get your "know-nothing, do-nothing" supervisor to sign off on the expenditure. Better do it my way. Open the turbine case. Unscrew one or two of the existing nozzles and screw in a steel plug. Same thing as closing a hand valve (see Chapter 31 for more details).

Centrifugal Pumps

- Did you know that amps on your motor driver vary with the diameter of the pump impeller raised to the third power:

$$\text{Work} \sim (\text{Diameter})^3$$

- How about optimizing impeller sizes? If you reduce an impeller from 10 to 9 inches, you'll save about 25% of the electricity needed to drive the pump, regardless of the size of the motor. Also, the motor will run cooler.

  How do you know which pumps can have their impellers downsized? Look at the control valve at the discharge of the pump. If it's 30% open, then trimming the impeller can save lots of electricity.

- But, you say, all of my control valves are 40% to 60% open. So I don't have this opportunity. You're wrong! Look on the opposite side of the tree for your low-hanging fruit. Your operators may have throttled back on the local isolation gate valve at the pump discharge, to keep the control valve in a reasonably open position for good process control. So they're wasting electricity, not across the control valves, but across the pinched gate valves.
- Then how does one decide how much to trim the pump impeller? You have to look at the family of pump curves issued by the pump's manufacturer. Deciding on the optimum reduced impeller size is a rather complex calculation (see Chapter 30 for more details).
- What do you think happens to an electric motor when it gets hot? The amps increase with the greater resistance in the coil's windings. So you would like to run motors cooler to save electric power. I have developed two high-tech solutions to this energy-saving opportunity:
    1. Relocate your plant to the Antarctic highlands.
    2. Or clean off the screen on the back of the motor. Actually, I only did this once, and it saved 3% of the amps on the motor driver.

Distillation

- The biggest energy consumer in most plants is the distillation tower reboiler. I'll bet you'll agree that most of the reboiler duty goes into generating reflux. How can we reduce reflux rates and save energy? Let's make this a multiple choice test.

A. Computerize your tower with a multivariable, cascade, feed-forward ultra-modern Laplace transformer?

B. On-stream gas chrome analyzers?

C. Optimizing tower pressure?

Choice C is correct. It doesn't cost anything and you can do it before lunch. Lower pressure enhances relative volatilities between the key components. But lower pressure also promotes entrainment and loss of tray efficiency. So how do I know what is the optimum tower pressure to get the required split, with a minimum reflux (and a minimum reboiler duty)? Just check the tower delta T (bottom minus top) temperature. That tower pressure that maximizes the temperature profile at a fixed reflux rate is the optimum tower pressure (see Chapter 3 for details).

Optimizing Opposing Forces

- Too little combustion air wastes energy, but too much air also wastes energy. There's an optimum balance between CO and O_2 in the furnace flue gas.
- Too little heat to a reboiler reduces thermosyphon circulation, but too much reboiler duty also reduces circulation. There's an optimum balance between creating a density difference driving force and reboiler frictional losses.
- Increasing reflux improves fractionation, but too much reflux makes fractionation worse. There's an optimum balance between internal reflux and entrainment.
- Larger-diameter pump suction piping reduces NPSH losses, but increases the starting NPSH requirement of the pump. There's an optimum balance between frictional losses and accelerating the fluid in the suction piping.
- Too little steam pressure to a jet hurts vacuum, but too much steam pressure also hurts vacuum. There's an optimum balance between steam velocity in the diffuser and overloading the discharge condenser.

Everything we do in process operations is a compromise, as we try to optimize opposing parameters. One has to see both sides of the problem. We're reaching out to touch the "best efficiency point." It's the basis of our natural world—the balance between the forces of creation and destruction. Understanding both aspects of these opposing forces is essential to optimizing process operations.

When I write about the "point of absolute combustion," or the "point of incipient flood," it's this optimum balance that I have in mind.

- Of course, anything that increases tray efficiency will reduce the reflux needed to achieve a given degree of fractionation. One such way to improve tray efficiency is to check that the tray decks are level. Out-of-level trays promote vapor–liquid channeling and poor contacting on the tray decks. Weir levelness is equally important for the same reasons. I use a carpenter's level laser light to check tray deck levels. The best type contains a pendulum and the light is self-leveling. We can sell you this instrument for $3,890, or you can buy one at Home Depot for $106.95.

FCU Air Blower Limitations

- Less coke on catalyst not only increases gasoline yields, but may also reduce coke make. So extra air to your cat cracker regenerator can have a synergistic effect. But how to get extra air at no cost? Check the suction pressure of your air blower.

Use a bottle of water and a piece of clear plastic tubing. Check the pressure right at the blower inlet. Let's say the water is drawn up by 8 inches in the tubing. If you can get rid of half this pressure loss, then the suction pressure will rise by 1% (i.e., there are 408 inches of H_2O in an atmosphere). This will give you around 1% more airflow.

- In one plant, we reduced the air intake pressure losses by removing the thick paint on the expanded metal suction screen. The screen had been painted repeatedly over the years. In another plant (with an air intake pressure of 11 inches of H_2O vacuum), we pulled off the 24-inch manway on the blower intake line. Fortunately, the pipe fitter had adequate mass to avoid getting sucked into the blower.

Vacuum Towers

Certainly the best way to make money in a refinery is to suck harder in the vacuum tower. This recovers more heavy gas oil from delayed coker fed and into cat cracker feed. But how to improve the performance of the overhead jet systems? I'll bet that at least one of these ideas will work on your vacuum tower:

- Try throttling back to the motive steam pressure to each jet. As the jet nozzles wear with time, the optimum motive steam pressure is always reduced.
- If the cooling water to the interstage or precondenser comes from a turbine-driven pump, increase its speed.
- Back-flush the water side of the above condensers. Blow some N_2 through the tubes with the water to create more turbulence.
- If running jets in parallel, try taking off one jet at a time. Many jets are subject to reverse flow.
- Lower the boot's liquid level in the vacuum tower bottoms. This reduces time-temperature cracking in the boot and reduces gas evolution.
- Use extra coil steam in the heater passes. This also will reduce gas evolution that has to be handled by the jets.
- Increase the stripping steam flow into the bottom of the upstream crude tower. This will reduce the saturated gas flow, which the vacuum jets have to handle and thus reduce the suction pressure to the jets.
- Blow out the barometric seal legs with steam. This reduces the steam condensate backup in the condenser.
- Block in the primary jet discharge-to-suction spill-back pressure control valve (see Chapter 26 for details).

Troubleshooting Experience of Author's Daughter

One day my 88-foot rock well kept losing pressure for no apparent reason. I called the company that originally installed my pump. They said the tank was losing pressure either because my pump was dying or my well was drying up.

"With the recent earthquake in Japan, the earth's tectonic plates might have shifted and that might have affected your well," they explained.

After running a few tests, the guys concluded the pump was worn out. I was relieved that I'd only have to spend $2,500 on a new pump instead of $20,000 drilling a new well. But one thing still didn't make sense: Why was the tank losing pressure if there was no water running in the house?

I walked around my house. And sure enough, I tripped over a mud puddle where a water pipe had sprung a leak. I dug up the pipe, cut off a section, and replaced it with a $2 PVC coupler.

"Now that's what my dad would call troubleshooting," I thought.

Lisa Lieberman is a freelance writer who lives in Three Rivers, CA.

Fired Heaters

Let's say I have a heater firing on TRC. Naturally, I wish to minimize the fuel consumption to meet my target heater outlet consumption. I do this by minimizing heat loss up the stack. This loss occurs in two ways:

- Partially oxidized hydrocarbons (CO, aldehydes, ketones, alcohols, etc.).
- Sensible heat in flue gas.

In order to minimize these combined losses, I'm going to adjust the combustion air rate by opening the stack damper and/or the burner's secondary air registers. But how much air is optimum? I'll give you another multiple choice test:

A. Purchase an O_2 analyzer. Target for x% O_2 in the stack. The x% target is a function of air–fuel mixing efficiency in the burner and tramp air leaks in the firebox. The value of x is certain to be between zero and 21%.

B. Purchase an expensive combustible analyzer. Target for x% CO in the stack. The x% target is a function of burner air–fuel mixing efficiency, which is unknown and unknowable.

C. Hire a combustion engineering consultant to apply Laplace transformations and paradoxical control to optimize operational variables.

D. Adjust the airflow to minimize the fuel consumption, as shown on the fuel gas meter.

Okay. I'm being stupid. But I'm not that stupid. I didn't buy that O_2 analyzer and stick it above the convective tubes, where it varies largely with draft and tramp air leaks. So the correct answer is D. But you still ought to try to minimize excess air. Not arbitrarily, but by improving the air–fuel mixing efficiency on your existing burners. I'll admit this is a task that does require some engineering know-how.

What then is the true function of the O_2 analyzer? The answer is that having established a target, it allows you to adhere to the target. Provided the burner air–fuel mixing efficiency is constant—which it usually is not. It's the same story with the combustible analyzer (see Chapter 22 for details).

- Opening the stack damper too far promotes tramp air leaks in the convective zone. The cold air quenches the hot flue gas and causes a reduction in heat transfer efficiency in the . . .

Don't think we're out of examples as to how to improve your refinery's profitability without any capital investment. I have lots more. However, my ink pen runneth dry, but not the supply of low-hanging apples.

Thanks for reading my books. Really, I've just stopped writing because I'm out of ink. Otherwise, I would keep on going.

What's the Problem?

"Dad, I finished typing your book. It doesn't make any sense."

"Irene, if you had attended engineering school like I wanted you to, maybe you would have understood it better."

"No, Dad! The whole book is dumb. Not the technical stuff. That's just boring. It's the stuff I understood that doesn't make any sense. It's all dumb."

"Well, Irene, maybe if you. . ."

"No, Dad, it's not me. It's you, and those other engineers, and the big oil refineries, and those giant chemical plants on River Road. That's what's dumb."

"How so?" I asked.

"They're blowing up and killing people. Like BP in Texas City. They're polluting the earth. Like BP in the Gulf near Shell Beach and Venice. They're wasting money, building stuff they don't need and burning up energy for no good purpose. I thought you engineers were supposed to be smart people."

"We are."

"No," she screamed, "You're not! You all probably cheated on your SATs and GREs." Irene thought for a while, "That's it," she decided, "You all cheated! That's how you became engineers."

"That's not the problem, Irene."

"Then what is the problem?

"It's like this: The process engineering profession has lost touch with reality. Engineers are no longer being trained to touch, see, and smell process equipment. The orientation is now toward complex control, computer technology, catalyst advances, exotic heat exchangers, and super-frac trays. The old ideas of discussing plant problems with hourly operators, running routine performance tests, conducting field pressure and temperature surveys, have slipped away from our profession. Now, it's the cell phones, and computers, and the texting, and the e-mails that have come between chemical engineers and process equipment. That's the problem, Irene."

"That's just what I mean, Dad. It's all dumb and doesn't make any sense. I wouldn't care. I know there are lots of stupid people around. But when you dumb engineers, who all cheated in school, spill 200,000,000 gallons of crude oil into the Gulf of Mexico because you don't cement up an oil well, and it blows out and kills all those turtles—well, then I get mad. You engineers need to be more careful. You shouldn't locate back-up diesel generators for nuclear plants in basements in Japan. You need to stop being so dumb." Irene studied the ceiling for a while. "Dad, maybe if your friends read this book, they'll see how dumb they really are. Then they can start to get smarter."

One lives in hope.

Glossary

There are a wide variety of terms that are used in the process industry that have an intent quite different from their meaning in the world at large. For example, the term *admiralty* has nothing to do with the British Royal Navy, but refers to a copper-based alloy used to make heat exchanger tubes.

When I speak with a young engineer, I'll often forget to define my terms. "Tommy, we've lost the level because the backing lights were out in T-303." With the aid of the following definitions, perhaps Tommy can better understand what I am trying to explain.

Adiabatic flame temperature The temperature inside a closed box where the only way heat can escape is with the flue gas.

Admiralty A copper-based alloy used to make tubes, usually for cooling water service. Better than carbon steel tubes.

Afterburn A fire in the convective section of a natural-draft or forced-draft heater. Also called secondary ignition.

Air to open Instrument air pressure is used to push open a control valve. On P&IDs, designated as AFC (air failure closes).

Approach temperature The delta T in heat exchangers between the inlet and outlet on the cold or hot end of the exchanger.

Autoignition temperature The temperature at which a substance will start burning, when dry, without a source of ignition.

Available NPSH The pressure at a pump suction, minus the vapor pressure of liquid at the pump suction, expressed in feet.

Back-flush Reversing the water flow direction in an exchanger to remove deposits stuck onto the tube sheets.

Backing lights Small lights in back of a gauge glass used to help observe the visible level in the glass.

Baffle trays Trays that have no perforations in the tray panel for vapor flow.

Balance line A small pipe used to equalize the pressure, usually for level control, between two adjacent vessels.

Ball-type check A check valve prevents fluid from flowing backwards through a line. Most check valves use flappers, but some use balls.

Barometric condenser Used to condense steam in vacuum service by direct contact with cooling water. Steam condensate is mixed with the cooling water.

Barometric leg Used to drain vacuum condenser. Both for surface condensers and barometric condensers.

Best efficiency point The operating point at which a pump, distillation tray, burner, compressor, turbine, or other device operates most efficiently. Often the design point.

Blinded In Europe, also called "spading." A flat metal plate is slipped between piping flanges to preclude leakage through an isolation valve.

Blowdown The flow of boiler feed water or cooling tower water drained to the sewer to control the accumulation of hardness deposits.

Blow-through Refers to steam blowing out of the condensate drain line in a steam heater or reboiler. Greatly reduces heat transfer rates.

Bubble cap An archaic vapor–liquid contacting device used in distillation trays. Normally provides better tray efficiency than valve, sieve, or grid-type tray decks.

Carbon seals Used to preclude air from being drawn into the bearing housing of centrifugal pumps.

Carnot cycle Pressure-versus-volume plot for a reciprocating compressor.

Channel head cover The front part of a shell-and-tube heat exchanger. It's removed to clean the tube side of a heat exchanger.

Check samples A nonroutine sample sent to the lab to follow the trend of a particular parameter.

Check valve Also called a "nonreturn valve." Prevents fluids from flowing backwards through a pipe.

Cherry picker Used to gain access to an elevated process valve or instrument.

Coded vessels Process vessels are licensed by governmental authorities as to allowable pressures and temperatures at which they may be operated.

Condensing steam turbine Motive steam exhausts to a condenser operating at subatmospheric pressure.

Constant entropy The ability of steam to do work is not diminished as the steam expands to a lower pressure.

Continuous phase A small volume of the dispersed phase is contacted with the larger volume continuous phase.

Convective section In a fired heater, the bank of tubes above the firebox. Convective tubes should not be exposed to radiant heat transfer (i.e., fire).

Critical discharge pressure Refers to vacuum steam ejectors. Above this discharge pressure, the jet will not develop its design compression ratio.

Critical flow At some vapor flow, a steam ejector will start to surge as it becomes overloaded and loses its sonic boost.

Critical speed For a centrifugal compressor, the RPM at which the compressor will vibrate in a self-destructive manner. This speed is noted on the compressor's nameplate.

Cross-flow velocity In a shell-and-tube heat exchanger, the shell-side fluid velocity between the tubes at the edge of the tube support baffle.

Cycle of concentration The percentage increase in the salts or solids between the blowdown and the supply water.

Dead man's valve A valve that closes by a spring as soon as the operator releases the valve. An important safety device.

Delta P The pressure difference between two points, usually measured across the same piece of equipment or process line.

Desalter A vessel with an electric grid used to separate brine and hydrocarbons. The actual extraction of salt into the brine is accomplished and water injection and an upstream mix valve.

De-superheating station Water is injected into superheated steam to reduce the temperature close to saturation.

Detonate A detonation generates a pressure wave front 10 times as powerful as an explosion. A butane vapor cloud will likely detonate with great destructive force.

Distillation test Measures the percent evaporated from a sample versus the temperature in the boiling flask.

Double acting machines Reciprocating compressor pistons compress gas on both the forward and backward strokes.

Double pipe Heat exchangers built from a small pipe inserted in a larger pipe. Results in true countercurrent flow. Also called a "pipe-in-pipe" heat exchanger.

Downcomer bracing brackets Used to stiffen the bottom edge of wide downcomers. Not required if the downcomer width is less than 4 feet.

Downstream Flow goes from upstream equipment to downstream equipment, except when there is a flow reversal.

Dry basis A lab term meaning that the composition of a sample is reported after all water has been removed.

Dumping Distillation trays are subject to tray deck orifices weeping or leaking. If trays are not level, vapor–liquid channeling results.

Duplex Two parallel strainers or filters. When one plugs, the other is put into service. Duplex filters should never have a bypass.

Endothermic reaction Heat is absorbed by the reaction.

Eutectic mixture A mixture of two compounds or elements so that the combined melting point temperature is lower than that of either component.

Explosion doors Pressure relief doors used to protect distillation tray decks from damage due to a surge in vapor flow.

Film resistance A stagnant layer of fluid close to a tube that inhibits heat or mass transfer. High velocity and turbulence will minimize film resistance.

Flash specification The temperature at which a hydrocarbon will ignite when exposed to a flame. Typically 122°F for jet fuel, 150°F for diesel oil.

Floating head The back end of a shell-and-tube heat exchanger bundle. Accommodates differential rates of thermal expansion between the tube and shell sides.

Flooded condenser pressure control The preferred method of tower pressure control when the overhead product is fully condensable.

Flooding In distillation, when tray efficiency is degraded, whether due to downcomer backup or entrainment, as vapor and liquid rates increase.

Flow path length The distance between the bottom edge of the downcomer from the tray above to the outlet weir. An optimum flow path length equals the tray spacing.

Fluid cracking unit (FCU) The most important unit in a refinery. Converts gas oil to alky feed, gasoline, and diesel oil. Same as *fluid catalytic unit*.

Fractionation Research Incorporated (FRI) Organization that evaluates tray innovations to determine their capacity and efficiency. Ultimate authority on distillation hardware.

Full limit amperage load (FLA point) The trip point set for electric motors.

Galvanic corrosion Placing two dissimilar metals into physical contact will create an electrical current that will corrode the less noble metal.

Gas test meter A portable meter that is used to measure the hydrocarbon content of vessels prior to issuing an entry permit to a vessel.

Governor The mechanical device that controls the speed of a turbine.

Governor speed control valve Controls the flow of motive steam into the turbine steam chest.

Gravity distributor A liquid distributor used in packed towers.

Hand jack A wheel that is used to manually move a control valve, regardless of the air supply pressure to the control valve's diaphragm.

Head loss Pressure drop expressed in feet, rather than psi.

Heat transfer coefficient Heat transfer efficiency expressed in Btus per hour per unit of heat transfer surface area, per unit of the log mean temperature difference.

High-level alarm An alarm that sounds to alert operators that a level has reached its maximum permissible height.

High-temperature creep At some temperature, a furnace tube becomes plastic and starts to bulge due to internal pressure.

Hogging jet A diverging jet used to evacuate air from a vacuum system on startup. Not intended for continuous use.

Hole watch The safety person assigned to stand by a vessel manway while the engineer or operator inspects the vessel's internals.

Horizontal pass partition baffle Located inside the channel head or inside the floating head of a shell-and-tube heat exchanger.

Horsepower valve Properly called a port valve. Used to open or shut ports in the steam chest of a steam turbine. Operated manually from outside the turbine.

Hot tap The practice of cutting a hole and then welding on a new valve on the side of a process line, but without shutting down the flow. Should be done with great care and after a special review of the circumstances.

Hydrogen blister A delamination-type failure to vessel walls caused by accumulation of molecular hydrogen from corrosion inside the metal plates of the vessel walls.

Hydrotested The practice of pressure-testing vessels or valves with high-pressure water. Typically 50% above M.A.W.P. (maximum allowable working pressure).

Impingement plate Placed in front of a feed nozzle to dissipate excessive momentum of fluid at the nozzle inlet inside a vessel.

Industrial fuel oil Also called Bunker C or Number 6 oil. The common low-value, black fuel oil produced from crude oil. Typically, refineries try to minimize its rate of production.

Internal reflux Cooled top reflux typically generates a higher reflux rate after leaving the top tray of a distillation tower.

Isoenthalpic expansion Expansion of steam without increasing its kinetic energy.

Isoentropic expansion Expansion of steam so that its kinetic energy is maximized, but its enthalpy is minimized. Isoentropic expansions are good. Isoenthalpic expansions are wasteful.

Joule-Thompson expansion coefficient Expanding a gas so that its energy and velocity are unchanged. Most gases cool upon such an expansion, except for H_2 and CO_2.

Level-trol A 3- or 4-inch empty pipe connected to a vessel used to attach gauge glasses and pressure transducers for level indication purposes.

Lift gas Physically lifts a liquid or solid to a higher elevation by means of reduction in their mixed phase density. Example: an air lift pump.

Liquid hold-up The hold time in a vessel. Typically 3 to 5 minutes. Required to obtain steady liquid level control.

Mesh Number of wires per inch. Example: a 16 mesh filter has 16 wires per inch of filter circumference.

Mollier diagram Describes the relationship between pressure, temperature, enthalpy, and entropy of a pure component such as water or steam.

Natural harmonic velocity The speed of a machine where the vibrations reinforce themselves. Also called the critical speed or self-destruct RPM.

Nitrogen barrier seals A modern mechanical seal used for a centrifugal pump that requires no seal flush oil. Very expensive and built to very close tolerances, but eliminates wasteful use of an external seal flush oil. A dry seal.

Nucleate boiling limitation The rate-limiting situation when boiling a relatively pure component. Increasing heat transfer surface temperature results in a reduction in heat exchange duty.

Oiler glass Used on the bearing housing of a centrifugal pump that employs the wet sump system of bearing lubrication.

Online spalling Technique of removing coke accumulations from inside heater tubes by slowly heating, then rapidly cooling the tubes with steam.

Optimum drainability point Providing adequate drainage, but without blowing the condensate seal and without promoting condensate backup in a shell-and-tube heat exchanger.

Pass partition baffle Located in the channel head and floating head of a heat exchanger. Determines the number of tube-side passes.

Pig A small cylinder with an abrasive exterior, forced through a process line with water pressure, used to clean a fouled line, section of piping, or furnace tube.

Pitch Tar; vacuum tower bottoms; asphalt; vacuum resid.

Point of absolute combustion The amount of combustion air that maximizes heat absorption into the process. Is not the same as complete combustion.

Point of incipient flood The vapor rate through a distillation tray that maximizes fractionation efficiency by minimizing tray deck leakage and entrainment. The optimum tray efficiency.

Positive feedback loop A response in a control system that reinforces the divergence from the desired set point, rather than restoring the system to the set point.

Post-weld heat treated Partly reheating a weld to relieve thermal distortion in the metal lattice structure caused by the weld. Process vessels often must be so treated after new nozzles are installed.

Pressure transducer A mechanical device that converts a pressure indication into an electrical output, usually in milliamps.

Pulsation damping plates Used in discharge and suction of reciprocating compressors to dampen out pressure pulsations in the flowing gas.

Pumparound Used in distillation towers to remove heat from intermediate points in the tower.

Pump displacement A measure of centrifugal pump vibration, which may damage the seal.

Push valves A modern grid-type tray pushes the liquid toward the outlet weir to even out the liquid level on the tray deck.

Radiant section The portion of a fired heater that contains the tubes exposed to direct radiant heat transfer from the flames and from the heated refractory walls.

Ram's horn An archaic device used to sense a liquid level at a point by the change in the wall temperature. The most reliable temperature indication available.

Required NPSH The net positive suction head needed to keep a centrifugal pump from cavitating and damaging its mechanical seal.

Restriction orifice A disk with a small hole used to control a flow.

Restrictive steam sparger A steam distributor with small holes sized to develop a substantial pressure drop.

Reynolds number A dimensionless value used to predict whether a flow is laminar or turbulent.

Sacrificial anode Used to protect a metal component (such as carbon steel) from galvanic corrosion.

Screwed connection Small pipe is often screwed into a threaded connection on a vessel. But then the connection should be back-welded.

Secondary ignition Afterburn. A fire that takes place in the convective section of a fired heater.

Shot coke Formed in refinery delayed cokers. Looks like black, shiny B-Bs. Some shot coke balls are the size of softballs or basketballs.

Shroud In an air cooler, the curved metal cylinder containing the fan. The tube bundle sits atop the shroud in forced-draft air coolers.

Slug flow in risers Vapor and liquid will separate into two separate phases in vertical lines if velocities become too low to maintain emulsified or mixed-phase flow.

Sonaray A device that is used to check metal thickness on pipes and vessels during normal operations.

Space velocity A ratio of catalyst volume to the volumetric flow of reactor feed.

Spalling Rapidly changing the temperature of a furnace tube or a heat exchanger tube can cause the deposits in the tube to break off and thus be removed from the tube.

Speed control governor In gas or steam turbines, the mechanical device that regulates the turbine or compressor speed by varying the flow of fuel gas or motive steam.

Split range control Two different control valves are controlling the same parameter, but over slightly different ranges of the parameter.

Spool piece A section of process piping with a flange at either end. An alternate, but more costly method to assemble piping as compared to welded piping sections.

Stab-in Refers to a reboiler tube bundle inserted into the bottom of a vessel itself, as opposed to the more standard external shell-and-tube reboilers.

Steam-air decoking Removing coke deposits from the interior of furnace tubes by a combination of steam and air. Steam is used continuously and air is used intermittently.

Steam jacketed pipe Process fluid flows through an interior pipe, and steam flows in the annular space between an outer pipe and the interior pipe. Widely used in refinery and gas field sulfur plants.

Steam-out Refers to a connection in the bottom of a vessel used to purge out air on startup, and hydrocarbons on shut-down, with low-pressure steam.

Steam sparger A grid array of piping with orifice holes used to distribute stripping steam in the bottom of a distillation tower.

Steam tables The chart that defines the properties of steam regarding pressure, temperature, density, enthalpy, and entropy. Steam velocity is assumed to be zero.

Step-down A distillation tray with descending levels on the tray decks. Used mainly with old-style bubble-cap trays.

Stilling well An empty pipe to which level-sensing instruments are appended.

Stokes' law Correlates droplet size entrainment with vapor and liquid density and the viscosity of the vapor phase.

Stonewalling A characteristic of a centrifugal compressor, where a reduction in the discharge pressure no longer increases the gas flow. Also called choke flow. Does not damage the compressor and is not dangerous.

Stress relieved Also called post-weld heat treating. After welding, a vessel or pipe is partially reheated to relieve the stresses to the metal lattice structure introduced by the rapid cooling of the melted metal on either side of the weld.

Subcooled Cooling a liquid below its saturated liquid or bubble point temperature.

Subcooling baffles Baffles used in a condenser to intentionally increase liquid backup so as to subcool the liquid with the purpose of providing more NPSH to a centrifugal pump.

Sublime out The tendency of gases of a salt to change directly from a vapor to a solid salt. In refineries, NH_4Cl is a common example.

Suction pressure control Varying the discharge flow from a centrifugal pump to maintain an adequate pressure at the pump suction with the objective of preventing cavitation of the pump.

Surface condenser Used in vacuum systems. Has both a liquid outlet draining to a seal drum and a vapor outlet flowing to a vacuum steam ejector.

Surge The erratic and dangerous unstable flow of gas through a centrifugal compressor, usually due to low gas flow rates, low gas density, or high discharge pressure.

Tapped out When the liquid level in a vessel rises above the top-level tap connection.

Temperature cross In a heat exchanger, when the hot side outlet temperature is lower than the cold side outlet temperature. This is an indication of good heat transfer efficiency.

Thermal cracking The degradation of hydrocarbons due to high temperature. Typically above 740 to 760°F.

Thermosyphon driving force Circulation through a reboiler due to vaporization that creates a density difference between the inlet line and the outlet riser.

Throttling calorimeter A portable lab device used to measure the moisture content or quality of steam.

Total dissolved solids (TDS) A measure of the solids or silicate content of boiler feed water or the blowdown from a steam boiler.

Total trap-out chimney tray A tower tray that does not fractionate, but draws off all the liquid from a tower at an intermediate elevation.

Tray deck manway A bolted section of a tray panel that is unbolted and removed when the tower is inspected.

Tray ring The ring that supports the tray panels. It is an integral part of the vessel itself, and not part of the tray.

Tri-cocks An old way of checking levels by opening a drain on the side of a vessel and observing what type of fluid drains out.

Tripped position Operating a piece of rotating equipment with the overspeed trip disengaged. A dangerous malfunction.

Tube support baffles The baffles in a shell-and-tube heat exchanger that support the tubes and encourage greater turbulence in shell side flow.

Turnaround The planned periodic shut-down that takes place on all process units for maintenance.

Turndown The ability of process equipment to operate efficiently at less than design capacity.

Vapor lock When liquids partially flash to vapors in process piping, the fluid flow is largely stopped.

Virgin Uncracked hydrocarbons. Typically free of olefins and aromatics, but not hydrotreated products.

Vortex shedding The turbulence introduced in a fluid as it flows across a horizontal tube at a 90° angle.

Wash oil grid A packed section in a distillation tower used to promote de-entrainment.

Weir length The length of the outlet overflow weir, located at the end of a distillation tower tray panel.

Wire up a trip The practice used on steam turbines of disabling the overspeed trip because the speed control governor is not working correctly. A very dangerous, but not uncommon, practice.

Zero order reaction A reaction rate that proceeds independently of the reaction component concentrations. Only a function of time and temperature. For example, thermal cracking of hydrocarbons is a zero order reaction.

Index

Note: Page numbers in *italic* reference figures.

C

G

The Norm Lieberman Video Library of Process Operations

Intended for experienced operating personnel or process engineers.
Each video is between 35 and 55 minutes in length.

Distillation

- Distillation Fundamentals
- Jet Flood & Dry Tray Pressure Drop
- Incipient Flood—Basic Concept
- Identifying Flooding Due to Foaming
- Tray Dumping and Weeping
- Tower Pressure Drop Evaluation
- Plugging and Fouling of Trays
- Onstream Cleaning of Trays
- Packed Tower Fundamentals
- Structured Packing
- Optimizing Pumparound
- Coker, FCU & Crude Fractionators
- Inspecting Tower Internals
- Steam Strippers

Heat Exchangers & Reboilers

- Heat Transfer Fundamentals
- Once-Through Thermosyphon Reboilers
- Circulating Thermosyphon Reboilers
- Kettle Reboilers
- Forced Circulation Reboilers
- Foam Formation in Reboilers
- Steam & Condensate Flow
- Water Hammer—Causes & Cures
- Shell & Tube Exchanger Fouling

Fired Heaters

- Fired Heater Fundamentals
- Draft Balancing Heaters
- Optimizing Excess Air in Heaters
- Preventing Heater Tube Coking
- Air Preheaters

Condensers

- Fundamentals of Condensation
- Back-Flushing & Acid Cleaning Condensers
- Pressure Control of Columns
- Subcooling Robs Condenser Capacity
- Surface Condensers

Vacuum Towers

- Improving Vacuum Tower Operation
- Packing and Grids in Vacuum Towers
- Steam Jets
- Vacuum Tower Overhead Systems

Pumps

- Fundamentals of Centrifugal Pumps
- Optimizing Impeller Size
- Cavitation & NPSH
- Causes of Seal & Bearing Failures

Compression

- Fundamentals of Gas Compression
- Surge in Centrifugal Compressors
- Variable Speed Centrifugal Compressors
- Steam Turbines
- Reciprocating Compressor Efficiency
- Compressor Motor Over-Amping

Treating

- Amine Regeneration & Scrubbing
- MEA Degradation
- Sulfur Plant Start-Up
- Sulfur Plant Pressure Drop
- Jet Fuel Treating

Refinery Processes

- Troubleshooting Sulfuric Acid Alky
- Crude Tower Overhead Corrosion
- HF Alky Pressure Problems
- Troubleshooting Delayed Cokers—Part I
- Troubleshooting Delayed Cokers—Part II

Safety

- Field Testing Alarms & Trips
- Failure of an Amine C3-C4 Scrubber
- Boiling Water Causes Coker Fatality
- Sulfur Plant Safety Hazards
- Sources of Autoignition

Printed by Libri Plureos GmbH in Hamburg,
Germany

9 780071 770200